T0332124

Intelligent Computational Paradigms in Earthquake Engineering

Nikos D. Lagaros
University of Thessaly, Greece

Yiannis Tsompanakis
Technical University of Crete, Greece

IDEA GROUP PUBLISHING

Hershey • London • Melbourne • Singapore

Acquisition Editor:	Kristin Klinger
Senior Managing Editor:	Jennifer Neidig
Managing Editor:	Sara Reed
Assistant Managing Editor:	Sharon Berger
Development Editor:	Kristin Roth
Copy Editor:	Julie LeBlanc
Typesetters:	Sharon Berger and Jennifer Neidig
Cover Design:	Lisa Tosheff
Printed at:	Integrated Book Technology

Published in the United States of America by
Idea Group Publishing (an imprint of Idea Group Inc.)
701 E. Chocolate Avenue
Hershey PA 17033
Tel: 717-533-8845
Fax: 717-533-8661
E-mail: cust@idea-group.com
Web site: http://www.idea-group.com

and in the United Kingdom by
Idea Group Publishing (an imprint of Idea Group Inc.)
3 Henrietta Street
Covent Garden
London WC2E 8LU
Tel: 44 20 7240 0856
Fax: 44 20 7379 0609
Web site: http://www.eurospanonline.com

Library of Congress Cataloging-in-Publication Data

Intelligent computational paradigms in earthquake engineering / Nikos D. Lagaros and Yiannis Tsompanakis, editors.
    p. cm.
  Summary: "This book contains contributions that cover a wide spectrum of very important real-world engineering problems, and explores the implementation of neural networks for the representation of structural responses in earthquake engineering. It assesses the efficiency of seismic design procedures and describes the latest findings in intelligent optimal control systems and their applications in structural engineering"--Provided by publisher.
  Includes bibliographical references and index.
  ISBN 1-59904-099-9 (hardcover) -- ISBN 1-59904-100-6 (softcover) -- ISBN 1-59904-101-4 (ebook)
  1. Earthquake engineering--Mathematical models. 2. Neural networks (Computer science) I. Lagaros, Nikos D., 1970- II. Tsompanakis, Yiannis. 1969-
  TA654.6.I5635 2006
  624.1'762--dc22
                              2006032165

British Cataloguing in Publication Data
A Cataloguing in Publication record for this book is available from the British Library.

Intelligent Computational Paradigms in Earthquake Engineering

Table of Contents

Section I:
Structural Optimization Applications

Chapter I

Michalis Fragiadakis, National Technical University of Athens, Greece
Nikos D. Lagaros, University of Thessaly, Greece
Yiannis Tsompanakis, Technical University of Crete, Greece
Manolis Papadrakakis, National Technical University of Athens, Greece

Chapter II

Ricardo O. Foschi, University of British Columbia, Canada

Chapter III

Arzhang Alimoradi, John A. Martin & Associates, Inc., USA
Shahram Pezeshk, University of Memphis, USA
Christopher M. Foley, Marquette University, USA

Foreword

Earthquake engineers are often criticized by others for being too constrained by old traditions. We are told that we have a limited vocabulary consisting of beams, columns, walls, and few other words; rather narrow means of analyses; and a restricted vision of possibilities for evaluation and retrofit of structures. Such criticism, of course, is much more descriptive of the ignorance of those who assert it than it is reflective of the current status, vision, and complexity of earthquake engineering research and practice. Over the past several decades, earthquake engineers have advanced the field of applied physics by leaps and bounds. Earthquake engineers developed the finite element analysis currently used by all branches of science and engineering; pioneered the response spectrum concept and visualized and refined the techniques for analysis of nonlinear systems.

If something is true, however, about the above-mentioned criticism, it is the fact that earthquake engineers have not been very effective in relaying their technical achievements to those outside their field. In other words, we have been preaching to the converted. As a result, we have been at least partially culpable for the myths that demean our field and limit our recruitment of the brightest, most technically savvy young minds who are about to enter various fields of science and technology.

This book represents a critical turning point because it demonstrates how the most current, advanced, and revolutionary computational techniques can be put to effective use in earthquake engineering not only to satisfy our intellectual aspirations, but to save precious lives and limbs. As such, it is not only useful to the practitioners and researchers of earthquake engineering but it is also invaluable to demonstrate the enormous capacity of earthquake engineering to those who may become its future leaders. If one is interested in computational challenges with an eye on solving life-threatening problems, one does not need to ponder far away from earthquake engineering.

As co-editors for this book, professors Nikos D. Lagaros and Yiannis Tsompanakis should be congratulated for achieving the monumental task of engaging experts from all over the world to produce this book with great success. The book is filled with methodologies and techniques that can motivate young researchers to employ them to solve their unique problems and can be useful to practitioners to respond to their everyday challenges. No other book contains such a comprehensive coverage of techniques such as application of neural networks, fuzzy logic theory, and evolutionary tools such as genetic algorithms, and other modern optimization methods, all engaged to solve various earthquake engineering problems.

This book is bound to find a prominent place on the bookshelf of every serious student, researcher and practitioner of earthquake engineering. And for that I salute the editors and their distinguished contributors for a job well done.

Farzad Naeim, PhD, SE, Esq.
Editor, Earthquake Spectra

Vice President and General Counsel
John A. Martin & Associates, Inc.
Los Angeles, CA, USA

Farzad Naeim *is the editor-in-chief of Earthquake Spectra, the professional journal of the Earthquake Engineering Research Institute (EERI). He is also the vice president and a member of the board of directors of EERI. Naeim is also the vice president and general counsel for John A. Martin & Associates, Inc. (JAMA) in Los Angeles, one of the largest structural consulting firms in the U.S. He joined the firm as a seismic design analyst in 1982 after obtaining a PhD in structural engineering from the University of Southern California. In 2002, Naeim obtained his JD with highest honors, and he has been admitted to practice law in California. In addition, he is also licensed patent attorney. Naeim serves as an advisor to several national and state organizations and major universities. He is the editor of* The Seismic Design Handbook, *now in its second edition, and the co-author of* Design of Seismic Isolated Structures. *He has published more than 120 papers on various aspects of earthquake engineering and has developed more than 45 different software systems for earthquake engineering design and education. Three of his software systems, Earthquakes—Be Prepared, Northridge Earthquake Information System, and CSMIP-3DV have been funded and distributed by public agencies in the U.S.*

Preface

The enormous advances in computational hardware and software resources over the last 15 years resulted in the development, of new, nonconventional data-processing and simulation methods. Among these methods soft computing (SC) or artificial intelligence (AI) has to be mentioned as one of the most eminent approaches to the so-called intelligent methods of information processing that present a great potential for engineering applications. Most SC methods are inspired by natural paradigms and therefore differ significantly from conventional mathematical approaches. Artificial neural networks (ANNs), expert and fuzzy systems, as well as evolutionary methods are the most popular soft-computing techniques. Artificial-intelligence methods are used either in order to reduce the computational cost, or when the complexity and/or the size of the problem forbids the use of conventional techniques. Especially, ANNs have been widely used in many fields of science and technology as well as in an increasing number of problems in structural engineering in general.

From among general problems that can be analyzed by means of AI techniques *simulation*, *inverse simulation* and *identification* problems are the most popular paradigms. Simulation is linked with direct methods of numerical analysis, that is, for known inputs and characteristics of the system under investigation the unknown outputs (responses of the system) are searched. On the other hand, inverse simulation (e.g, the identification of an unknown load of a given structural response) takes place if inputs represent known responses of the system and the "excitations" that caused this behavior of the certain system are searched as outputs. In addition, identification is also associated with the inverse analysis of systems, including structures and materials. In this case excitations and responses are known and characteristics of the system are searched. In many cases the application of AI techniques is focused on the simulation, that is, structural reliability analysis and/or optimization problems. SC methods have been proven also very effective at solving inverse engineering problems. Another very promising field of SC applications in computational mechanics is flaw or damage detection, which in essence can be considered as an inverse problem. For example, material or parameter identification problems, which can be formulated as output-error optimization problems, can be solved very efficiently with AI techniques such as ANNs.

This volume is a "multicollective" book of 16 chapters containing various applications of AI methods in earthquake engineering. Contributions cover a wide spectrum of very important real-world engineering problems. The chapters can be classified into three categories of soft-computing applications in earthquake engineering, namely optimization, assessment, and identification. A short description of the proposed chapters (sequence is according to the numbering in the table of contents) is presented in the following paragraphs.

In the first chapter by the editors of this book and their co-authors, M. Papadrakakis and M. Fragiadakis, the main aim is to assess the efficiency of seismic design procedures, as imposed by the current prescriptive codes, by integrating them in the framework of structural optimization. The comparison is based on the European seismic design code (EC8), where procedures based on both linear and nonlinear response history analysis are adopted. For the solution of the optimization problem a highly efficient evolutionary algorithm, the evolution strategies (ES), is adopted. The presented results demonstrate the potential of structural optimization when applied to the design of structures under earthquake loading. The main conclusion of the presented evaluation of optimum seismic design procedures is that improved control on the structural response and weight is achieved when nonlinear analysis is performed, compared to optimum designs obtained when linear analysis procedures based on EC8 requirements were used.

In the sequence, R. Foschi explores the implementation of neural networks for the representation of structural responses in earthquake engineering, and their subsequent use in reliability evaluation and optimization for performance-based design. An efficient treatment of random variables is presented in order to quantify in an effective manner the probability that a given structure may not perform as intended via a set of performance criteria or limit states in a performance-based design framework. The necessary reliability evaluations, and the optimization involved in performance-based design, have been efficiently performed via simulation. Neural networks have been used in this chapter as a means of simulation in order to represent structural responses in terms of input variables and design parameters.

The objective of the contribution by A. Alimoradi et al. is to provide an overview of structural design procedures for optimal seismic performance. The chapter introduces some of the latest advancements in the field of earthquake engineering by briefly discussing the concept of performance-based design. Since optimal seismic design for multiple performance-levels is computationally expensive an efficient naturally inspired computational agent is proposed for design automation of structural systems considering multiple performance states and various earthquake scenarios. A very descriptive example of genetic algorithms in probabilistic performance-based seismic design of steel moment frame buildings is described.

J. E. Hurtado emphasizes in reliability-based optimization (RBO), considering both the two basic factors of the problem: reliability and optimization. The solution of a RBO problem in earthquake engineering faces the following difficulties: (a) limit-state functions are normally given in implicit form, (b) the failure probabilities are very low and their estimation by some standard reliability procedures may be inaccurate in many cases, (c) there is a need of solving an entire reliability problem in each step of the optimization procedure. In order to overcome the computational difficulties associated with the minimization of cost functions with probabilistic constraints that involve the computation of very small probabilities an efficient methodology is presented. The proposed RBO model is based on the combination of a computational learning method (support vector machines) and a rather novel AI technique (particle swarm optimization). The former is selected because of its information encoding properties, as well as for its elitist procedures that complement those of the AI optimization method. The later has been chosen due to its advantages over classical genetic algorithms. The practical application of the procedure is demonstrated with earthquake engineering examples.

E. Salajegheh and A. Heidari develop a new structural optimum design concept for earthquake induced loading by a modified genetic algorithm (MGA) in which some features of the simulated annealing (SA) technique have been used in order to control more efficiently various parameters of the standard genetic algorithm (GA). In order to reduce the excessive computational cost during dynamic analysis of the structure for every design configuration tested by MGA a fast wavelet transform has been applied, where each seismic record is decomposed into a low-frequency and a high-frequency part. Subsequently, using a specially tailored wavelet neural network, the dynamic responses of the structures could be approximated. The fast wavelet transforms have managed to reduce the computational cost and to enhance the efficiency of the optimization process of the overall optimisation process

S. F. Ali and A. Ramaswamy describe the latest findings in intelligent optimal control systems and their applications in structural engineering. They provide a short introduction on this topic by starting with the shortcomings of conventional vibration control techniques and explain the need for intelligent control systems. In addition the basic tools required for intelligent control such as evolutionary algorithms, fuzzy rule base, and so forth, are outlined. In order to provide a better insight on the subject hybrid intelligent control techniques have been applied into benchmark examples on vibration control of building and bridge structures under seismic excitation.

The chapter by M. L. Carreño et al. is a contribution to the understanding of how soft computing applications, such as artificial neural networks and fuzzy sets, can be used in structural assessment and urgent processes of engineering decision-making, like the building occupancy after a seismic disaster. The authors focus on development and the implementation of a hybrid neuro-fuzzy system, developed under the perspective of assisting nonexpert professionals of building construction, to evaluate the damage and safety of buildings after strong earthquakes, facilitating decision-making during the emergency response phase on their habitability and reparability. The inputs to the system are fuzzy sets, taking into account that the damage levels of the structural components are linguistic variables, defined by means of qualifications of the damage levels. The presented hybrid AI tool is suitable in practice because building damage evaluation deals with subjective and incomplete information which requires the use of linguistic qualifications that are appropriately handled by fuzzy sets. In addition, the artificial neural network has been used to calibrate the system using the judgment of specialists since its training was performed by using a database of real evaluations made by expert engineers. The proposed neuro-fuzzy expert system enabled the development of a user-friendly computer program that has been used as an official tool for disaster risk management in Colombia.

M. R. Hernandez-Garcia and M. Sanchez-Silva cope with the application of learning machines—for example, artificial neural networks (ANNs) and support vector machines (SVM)—as highly efficient pattern recognition tools, for structural damage detection. This chapter presents an overview of structural health monitoring techniques and statistical learning theory. Furthermore, it describes a methodology for damage detection based on extracting the dynamic features of structural systems from measured vibration time histories by using independent component analysis. Based on this type of analysis, statistical learning theory (SLT) is used to determine and classify damage. The main novelty of this work is that ANNs and SVM were trained using only information from the healthy system and were accordingly tested trying to classify damaged and undamaged feature vectors correctly.

Engineering decisions related to pre- or postearthquake evaluation of structural systems are most frequently based on fuzzy information that is vague, imprecise, qualitative, linguistic or incomplete. M. Mezzina et al. have developed a fuzzy expert system that has been successfully applied for the structural assessment of reinforced concrete constructions. Particular attention has been paid to seismic risk mitigation by elaborating on two important aspects, namely the appraisal of the actual conditions of the structure (material deterioration, preexisting damages, etc.) and the evaluation of the structural "vulnerability," that is, the propensity to suffer damage because of the intrinsic geometric and structural arrangement, boundary conditions, specific structural details. Attention is focused at first on the damage investigation protocol, which is organized through a multilevel, hierarchical scheme that includes visual inspections, surveys, experimental testing in situ and in laboratory, and so forth. In the sequence, the development of a genetic-fuzzy expert system tool for uncertainty management and decision-making is performed. The proposed methodology can handle the procedure of the assessment very effectively, accounting for uncertainty and errors, and is able to tune the parameters involved, on the basis of experts' knowledge, and in so doing to "train" the system.

H. Furuta and K. Koyama present an AI-based evaluation of bridge structures using life-cycle cost (LCC) principles and considering seismic risk. LCC has become nowadays an important tool for the design and/or the evaluation of structures and infrastructure. The cost of a structure during its functioning (LCC) consists of its initial cost, maintenance cost, and renewal cost. However, when considering LCC in a region with increased natural hazards such as typhoons and earthquakes, it is necessary to account for the effects of such natural threats. This can be achieved by using the probability of damage occurrence. The solution of the formulated optimization problem is performed via Genetic Algorithms. The proposed LCC method can be applied for the optimal maintenance planning and can be extended to the LCC analysis of an entire road network.

Many sources of uncertainty (material, geometry, loads, etc.) are inherent and unavoidable in structural design. Probabilistic analysis of structures leads to safety measures that a design engineer has to take into account due to the aforementioned uncertainties. Therefore, earthquake-resistant design of structures using probabilistic analysis techniques is an emerging field in earthquake engineering. On the other hand, probabilistic analysis problems, especially when seismic loading is considered, are highly computationally intensive tasks. The objective of the contribution by N. D. Lagaros et al. is to investigate the efficiency of SC methods, that is, ANNs metamodels, when incorporated into the solution of computationally intensive stochastic structural problems. Two metamodel-based applications are considered, in which the efficiency of a trained ANN is demonstrated. In the first application the probability of exceedance of a limit-state as specified by Eurocode 8 and calculated by means of Monte Carlo simulation (MCS), is obtained. In the second application, fragility analysis of moment resisting steel frame is performed where limit-state fragilities are determined by means of nonlinear time-history analysis.

D. Assimaki emphasizes in the use of AI methodologies in a geotechnical earthquake engineering application. In particular, inverse analysis of weak and strong motion downhole array data using genetic algorithms is presented. A seismic waveform inversion algorithm is proposed for the estimation of elastic soil properties using low amplitude, downhole array recordings. Based on a global optimization scheme in the wavelet domain, complemented by a local least-square fit operator in the frequency domain, the hybrid scheme can efficiently identify the optimal solution vicinity in the stochastic search space, whereas the best fit model detection is substantially accelerated through the local deterministic inversion. The inversion algorithm has provided robust estimates of the linear and equivalent linear impedance profiles, while the attenuation structures are strongly affected by scattering effects in the near-surficial heterogeneous layers. Results show that the hybrid optimization scheme can efficiently provide estimates of the shear wave velocity, attenuation and density profiles of near-surficial soil formations using downhole array recordings, reflected in the illustrated agreement between the synthetic response of the global optimal profiles and the corresponding ground surface observations.

C. G. Koh and M. J. Perry explain the set up of a suitable genetic algorithm (GA) that has been applied in structural identification and damage detection structural problems, where damage detection is considered as a natural extension of identification. The authors stress the fact that an elaborate consideration of a specially tailored GA strategy that involves a search space reduction method (SSRM) using a modified genetic algorithm based on migration and artificial selection (MGAMAS) is a very efficient choice for the identification structural properties in multiple degree-of-freedom systems. The proposed GA approach works on multiple populations or "species" and balances the search with both broad and local search capabilities. The challenge of identifying mass, stiffness and damping properties from incomplete, noisy measurements and data is the focus of this chapter.

S. Chakraverty develops a neural network based procedure for the identification of structural parameters in multistory buildings by solving the forward vibration problem instead of the inverse vibration problem. This chapter includes the definition of neural architectures and system identification of multistory structure. An efficient identification algorithm for the multistory structure subject to initial condition and ground displacement is presented. Response identification subject to real earthquake data has also been discussed. Several example problems are incorporated to show the efficiency and robustness of the proposed algorithm.

In the contribution by L. Ziemiański et al., the main aim is to describe advanced SC methods that can be applied in order to deal more efficiently with parameter identification problems. More specifically they address the application of neurocomputing to parametric identification using dynamic structural responses. The analysed problems cover a wide range of applications: (a) implementation of dynamic response to parameter identification of structural elements with defects that are modelled as local changes of stiffness or material loss; (b) updating of FEM models of beams, including the identification of material parameters and parameters describing possible defect; (c) identification of circular void or supplementary mass in vibrating plates; (d) identification of a damage in frame structures using both eigenfequencies and elements of eigenvectors as input data. In the examples that involve experimental measurements, the application of a random noise is proposed in order to increase the not sufficient number of data.

The basics of ANNs theory are briefly presented in the first part of the last chapter by K. Kuźniar and Z. Waszczyszyn in order to form a base for other chapters of this book in which ANNs have been used. In the sequence of their contribution the authors introduce the use of ANNs for the identification of dynamic properties of actual buildings, the simulation of building responses to paraseismic excitations as well as for the analysis of response spectra. Mining tremors were the sources of these kinds of vibrations. On the basis of the experimental data obtained from the measurements of kinematic excitations and dynamic building responses of actual structures the training and testing patterns were formulated. It has been stated that the application of neural networks enables us to predict the results with accuracy quite satisfactory for engineering practice. The results presented in this chapter lead to a conclusion that the neurocomputing provides new prospects of efficient analysis of structural dynamics problems under seismic or paraseismic excitations.

Nikos D. Lagaros
Yiannis Tsompanakis
September 2006

Acknowledgments

The editors of this book would like to express their deep gratitude to all the contributors for their time and efforts for the preparation and the reviewing of the chapters. We are also most appreciative to Professor Farzad Naeim for preparing the foreword of the book. In addition, we would like to thank the series editor of this book, Professor Lakhmi Jain, for his kind invitation to edit this volume. Finally, the editors are most appreciative to Kristin Roth and the personnel of Idea Group Inc. for their support during the preparation of this book.

Nikos D. Lagaros
Yiannis Tsompanakis
September 2006

Section I

Structural Optimization Applications

Chapter I

Improved Seismic Design Procedures and Evolutionary Tools[1]

Michalis Fragiadakis, National Technical University of Athens, Greece

Nikos D. Lagaros, University of Thessaly, Greece

Yiannis Tsompanakis, Technical University of Crete, Greece

Manolis Papadrakakis, National Technical University of Athens, Greece

Abstract

Four alternative analytical procedures are recommended by the design codes for the structural analysis of buildings under earthquake loading. The objective of this chapter is to assess these procedures by integrating them in the framework of structural optimization. The evaluation is based on the European seismic design code, where procedures based on both linear and nonlinear response history analysis are adopted. In order to realistically simulate seismic actions, suites of both natural and artificial ground-motion records are used. For the solution of the optimization problem an evolutionary algorithm is adopted. The results obtained demonstrate the advantages of using more elaborate seismic design procedures, based on a detailed simulation of the structural behaviour and the applied seismic loading, as opposed to the commonly used simplified design methodologies. Designs with less material cost combined with better seismic performance are obtained when nonlinear response history analysis is performed.

Introduction

During the last three decades structural optimization has been the subject of intensive research and several different approaches have been advocated for the optimal design of structures in terms of optimization methods or problem formulation. Most of the attention of the engineering community has been directed towards the optimum design of structures under static loading conditions with the assumption of linear elastic structural behaviour. For a large number of real-life structural problems, assuming linear response and ignoring the dynamic characteristics of the seismic actions during the design phase may lead to structural configurations that are highly vulnerable to future earthquakes. Furthermore, seismic design codes suggest that under severe earthquake events the structures should be designed controlled inelasticity due to the large intensity inertia loads imposed and economy of engineering design.

The objective of this work is to examine the influence of three analytical procedures, namely the response spectrum modal analysis, the linear and the nonlinear analysis (Lagaros, Fragiadakis, Papadrakakis, & Tsompanakis, 2006). These design procedures are suggested by the European seismic code Eurocode 8 (EC8; 1994), while the provisions of the FEMA-356 (2000) guidelines are also used complementarily for the nonlinear analysis cases. The structural performance is investigated, under the objective framework provided by structural optimization. Several studies have appeared in the literature where seismic design procedures based on nonlinear response (e.g., Bazzuro, Cornell, Shome, & Carballo, 1998; Han & Wen, 1997) are presented and compared. However, this task can be accomplished in a complete and reliable manner in the framework of structural optimization, where the designs obtained with different procedures can be directly evaluated by comparing the seismic performance of the optimum solution achieved.

During the last 15 years there has been a growing interest in optimization algorithms that rely on analogies to natural processes such as evolutionary algorithms (EA). For complex and realistic structural optimization problems, EA methods appear to be very robust and reliable compared to most mathematical programming optimizers. EAs do not require the calculation of gradients of the constraints, as opposed to the "old fashion" mathematical programming algorithms, and thus structural design-code checks can be incorporated within an optimization environment as constraints in a straightforward manner.

Progress on Seismic Design Using Structural Optimisation Procedures

Structures built according to the provisions of contemporary seismic design codes are designed to respond nonlinearly under strong earthquakes. However, even nowadays, very frequently in engineering practice this behaviour is taken into account during the design phase only implicitly. This practice can be mainly attributed to the fact that nonlinear response history analysis results in increased computational complexity and requires excessive computational resources. In general, a comparatively limited number of studies have been performed for the solution of structural optimization problems under dynamic loading conditions considering nonlinear behaviour. One of the earliest studies on the subject is the work of Polak, Pister, and Ray (1976) who minimized the cost of multistory frame structures using nonlinear response history analysis of simplified structural models. Bhatti and Pister (1981) and Balling, Ciampi, Pister, and Polak (1981) introduced optimization procedures based on nonlinear response history analysis. Several limit-states were considered, while different performance objectives are adopted for each limit-state. Pezeshk (1998) presented an integrated nonlinear analysis and optimal minimum weight design methodology, for a simple 3D frame under an equivalent pseudo-static loading scheme.

Following recent developments in structural design procedures (Bozorgnia & Bertero, 2004), a number of researchers have stressed the need to integrate structural optimization and performance-based earthquake engineering (PBEE) (Charney, 2000). Ganzerli, Pantelides, and Reaverley (2000) implemented a performance-based optimization procedure of RC frames using convex optimum design models and a standard mathematical optimizer. Esteva, Díaz-Lopez, García-Perez, Sierra, and Ismael (2002) presented a performance and reliability-based optimization under life-cycle considerations using special type damage functions in conjunction with pushover analysis. Gong (2003) combined pushover analysis and a dual optimization algorithm for the optimal design of steel building frames for various performance levels. Chan and Zou (2004) proposed a two-stage, optimality criteria-based, optimization procedure for 2D concrete frame structures.

The advancements achieved by evolutionary algorithms over the past years made possible the solution of real-scale structural optimization problems incorporating design code-based constraints. Two of the earliest studies, where EA were employed for the optimum seismic design of structures, are those of Kocer and Arora (1999, 2002) for the optimal design of H-frame transition poles and latticed towers conducting nonlinear response history analysis. They proposed the use of genetic algorithms (GA) and simulated annealing (SA) for the solution of discrete variable problems although the computational time required was excessive. Beck, Chan, Irfanoglu, and Papadimitriou (1999) proposed a multicriteria GA-based structural optimization approach under uncertainties in both structural properties and seismic hazard.

Cheng, Li, and Ger (2000) used a multiobjective GA-based formulation incorporating game and fuzzy set theories for the optimum design of seismically excited 2D frames. Game theory was used in order to achieve a compromise solution that satisfied all competing objectives in the multiobjective optimization problem, while with the implementation of fuzzy-set theory the constrained optimization problem was transformed into an unconstrained one which subsequently was solved with a Pareto GA-based optimizer. Liu, Burns, and Wen (2003) proposed a GA-based multiobjective structural optimization procedure for steel frames using pushover analysis considering minimum weight, life-cycle cost and design complexity criteria as the objective functions of the optimization problem. The same authors recently extended their methodology in the framework of PBEE (Liu et al., 2005; Liu, Wen, & Burns, 2004).

As opposed to the aforementioned studies on the optimum seismic design of structures and the different approaches proposed in the past, this work is focused on using structural optimization as a tool for comparing alternative design procedures. The formulation of the optimization problem differs according to the design procedure examined, while several issues regarding nonlinear response history analysis and structural optimization in the framework of earthquake engineering practice are discussed. European seismic design regulations for steel moment resisting frames are used as the testbed of the optimum design procedures, while the provisions of U.S. guidelines FEMA-356 (2000) are also taken into consideration supplementary to EC8 (Eurocode 8, 1993). In total three alternative seismic design procedures are evaluated in the framework of structural optimization: response spectrum modal analysis, linear and nonlinear response history analysis. The dynamic analysis based procedures are considered using both natural and artificial ground motion records. The various designs are compared in terms of their total weight.

Seismic Design Procedures

Analytical Procedures

According to FEMA-356 (2000), four alternative analytical procedures, based on linear and nonlinear structural response, are available for the structural analysis of buildings under earthquake loading. The linear procedures can be either the linear static procedure (LSP) or the linear dynamic procedure (LDP). For the linear dynamic procedure there are two alternatives, the response spectrum method, which in Eurocode 8 (EC8; 1994) is referred as multi-modal response spectrum (MmRS) analysis, and the linear response history method. In a similar fashion the nonlinear methods are distinguished to the nonlinear static procedure (NSP), also known as pushover analysis, and the nonlinear dynamic procedure (NDP), also referred as nonlinear response history analysis. Procedures that are not based on dynamic response history analysis, when applied in the framework of a seismic design code, usually resort to a regional response spectrum. Design procedures based on nonlinear analysis are less preferred due to their computational cost and the requirements for highly trained engineers to put into practice more elaborate analysis methods. However, when a linear analysis method is employed simplifying assumptions of the structural response are made, which may result to conservative and therefore to more expensive designs, or to designs with reduced safety since phenomena that have not been accounted during the design phase may influence the capacity of the structure.

Three different analysis procedures are considered in this study and implemented in the framework of structural optimization. The first corresponds to the EC8 design procedure where the MmRS analysis method is employed. The second and third correspond to linear elastic and nonlinear response history analysis, respectively. Most current design practice is based on the first procedure, while the second takes directly into consideration the dynamic characteristics of earthquake loading. The third procedure is considered as the "exact" analysis method requiring excessive computational demands.

Design-code procedures adopt a number of simplifying assumptions, the most critical of which is the use of a behavior factor in order to take into account implicitly the ductile response. This can be done either at a member level (FEMA-356, 2000) or at the structure level (EC8, 1994). The use of a behaviour factor may lead to a gross overestimation of the design loads, since it does not take explicitly into account the specific characteristics of the structure at hand. In an effort to ensure that the structure will have a desirable nonlinear response under major earthquakes, a number of checks are suggested in order to ensure ductile response and adequate rotation capacity for the members and the joints of the structure at hand. Furthermore, the design codes indicate checks based on maximum allowable stresses or forces, while a number or factors are adopted to account for various phenomena, for example, reduced compressive strength due to potential loss of stability. On the other hand, design methods based on nonlinear structural behaviour do not require the use of behaviour factors and allow more direct (high-level) design criteria to be implemented (Foley, 2002).

Therefore, nonlinear analysis methods, such as pushover analysis, and nonlinear response history analysis, are more suitable compared to linear methods for the realistic evaluation of structural response against strong ground motions. Although, the latter method is capable of capturing the true behaviour of the structure, the obtained designs are sensitive to the characteristics of the ground motion history and details of structural modeling. On the other hand, pushover analysis does not depend on the ground motion history and does not require excessive computational time compared to response history analysis. However, despite the latest improvements of the method (e.g., Antoniou & Pinho, 2004; Chintanapakdee & Chopra, 2003), its simplifications make it appropriate only for structures with specific characteristics. Nonlinear static procedures, in general, cannot capture the dynamic structural behaviour with the same accuracy, as the dynamic response history analysis, especially when structural response is sensitive to higher modes.

Earthquake Loading

Load Combinations

In the present study, each candidate design is checked against the provisions of Eurocode 3 (EC3; 1993) and Eurocode 8 (1994) during the optimization process. Two design load combinations are considered:

$$S_d = 1.35 \sum_j G_{kj} \,"+"\, 1.50 \sum_i Q_{ki} \tag{1}$$

$$S_d = \sum_j G_{kj} \,"+"\, E_d \,"+"\, \sum_i \psi_{2i} Q_{ki} \tag{2}$$

where "+" implies "to be combined with", the summation symbol "Σ" implies "the combined effect of," G_{kj} denotes the characteristic value "k" of the permanent action j, E_d is the design value of the seismic action, and Q_{ki} refers to the characteristic value "k" of the variable action *i*, while ψ_{2i} is the combination coefficient for quasi permanent value of the variable action *i*, here taken equal to 0.30. Furthermore, for 3D problems earthquake actions must be considered in MmRS analysis method using the following combinations:

$$\begin{aligned} E_d &= E_{dx} + 0.30 E_{dy} \\ E_d &= 0.30 E_{dx} + E_{dy} \end{aligned} \tag{3}$$

where E_{dx} and E_{dy} represent earthquake loading at two perpendicular to each other directions. When natural or artificial records are used in response history analyses (linear or nonlinear) they must consist of two perpendicular components from the same earthquake. All design code checks are implemented into the optimization problem as constraints.

Seismic Loading

The design procedures adopted in the present investigation differ in the way they take into account earthquake loading. The first procedure (MmRS) is based on the EC8 response spectrum, while the two response history procedures, related to linear and nonlinear structural response, use suites of natural or artificial earthquake records. The selection of the proper seismic loading for design purposes is not an easy task due to the inherent uncertainties in seismic loading. For this reason a more rigorous treatment of seismic loading is to assume that the structure is subjected to a set of preferably natural or artificial records that match the characteristics of the site where the structure is to be located. In addition, "automatic" record selection methods have been proposed in the literature (Naeim, Alimoradi, & Pezeshk, 2004).

The dynamic analysis procedures in the current U.S. seismic design provisions and guidelines specify that a series of nonlinear response history analyses are conducted with pairs of horizontal ground motion components selected from not less than three seismic events. If three pairs of records are used, then the maximum value of the response parameter of interest (e.g., peak lateral displacement) is used for the evaluation of the current design, while if seven or more pairs of records are used, then the median value of the response parameter may be used. However, these are minimum requirements and in practice it is advised to use a larger number of records (e.g., 10 or more), since the seismic response is very sensitive to the choice of records. Moreover, the use of the median response is preferable compared to the mean, since the median is not sensitive to outliers.

Selection of Natural Records

In order to determine the damaging potential of each record a classification is performed according to the ratio of the peak ground acceleration (PGA) to peak ground velocity (PGV), the so-called a/v ratio (Zhu, Heidebrecht, & Tso, 1988). This ratio reflects many source characteristics, travel path and site conditions, as well as structural response parameters. With regard to structural response, high a/v ratios will be more critical for stiffer, short period structures, while more flexible, long period structures will be strongly shaken by earthquakes with low a/v ratios. These ranges approximately can be determined as:

$$0.8\,\text{g/ms}^{-1} \le a/v \le 1.2\,\text{g/ms}^{-1} \quad (\text{Normal})$$
$$1.2\,\text{g/ms}^{-1} < a/v \qquad\qquad (\text{High}) \tag{4}$$
$$a/v < 0.8\,\text{g/ms}^{-1} \qquad\qquad (\text{Low})$$

Table 1. Ten natural records

Earthquake name (Date)	Site \ Soil Conditions	M_s	PGA (g)	PGV (m/sec)	a/v (g/msec^{-1})
Victoria Mexico (06.09.80)	Cerro Prieto \ Alluvium	6.40	0.62	0.31	1.96
Kobe (16.01.95)	Kobe \ Rock	6.95	0.82	0.81	1.01
Duzce (12.11.99)	Bolu \ CAB: D, USGS: C	7.30	0.82	0.62	1.32
San Fernando (09.02.1971)	Pacoima dam \ Rock	6.61	1.22	1.12	1.08
Gazli (17.05.1976)	Karakyr, CWB: A	7.30	0.72	0.71	1.00
Friuli (06.05.1976)	Bercis \ CWB: B	6.50	0.03	0.013	2.18
Northridge (17.01.94)	Jensen filter Plant \ CWB: D, USGS: C	6.70	0.59	0.99	0.59
Athens (07. 09.99)	Sepolia \ Not classified	5.60	0.24	0.179	1.34
Cape Mendocino (25.04.92)	Petrolia \ CWB: D, USGS: C	7.10	0.66	0.89	0.73
Loma Prieta (18.10.89)	Hollister Diff Array \ CWB: D	7.10	0.28	0.35	0.78

where: Normal (a/v) refers to ground motions time histories with significant energy content for a broad range of frequencies. High (a/v) refers to ground motions with many large-amplitude, high-frequency oscillations. Low (a/v) refers to ground motions in which the significant response is contained in a few long duration acceleration pulses.

In the present study the ground motion records listed in Table 1 were used, where PGA and PGV values refer to the stronger, in terms of PGA, of the two horizontal components of each record. These records cover a sufficient range of a/v ratios. Figure 1 shows the 2.5%-damped EC8 elastic response spectrum and the corresponding spectra of the records indicated in Table 1. Scaling the ground motion records is necessary in order to make all records compatible with the design spectrum of EC8. Among the many ways of scaling ground motion records, a widely accepted choice is the 5%-damped first mode spectral acceleration. Shome, Cornell, Bazzuro, and Carballo (1998) found that when using this scaling approach the bias of damage estimation is not statistically significant. However, in cases where the participation of higher modes has significant impact on the structural response, a scale factor that covers a wide range of modes should be used. In this investigation each record is scaled by a factor that minimizes the error with respect to the EC8 spectrum at a number of N_T structural periods (typically from 0.1 to 4 seconds). Thus, the scaling factor λ is computed as follows (Al-Ali & Krawinkler, 1998):

$$error = min\left(\sum_{i=1}^{N_T} \left(1 - \frac{\lambda \cdot S_a(T_i)^{record}}{S_a(T_i)^{EC8}} \right) \right) \qquad (5)$$

Figure 1. EC8 2% damped spectrum and scaled spectra of Table 1 records

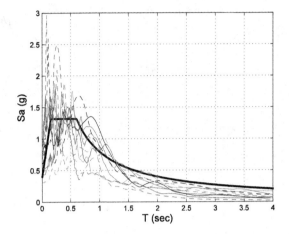

where $S_a(T_i)^{EC8}$, $S_a(T_i)^{record}$ are the spectral accelerations for the ith period value from the EC8 and the record response spectrum, respectively.

Artificial Accelerograms

Apart from enriching the catalogues of earthquake records of a region, a second reason for using artificial records is that they usually present a larger number of cycles than natural records and therefore they result to more severe earthquake loading for a greater range of structural periods. Although one may argue that artificial records are unrealistic in terms of input energy demand, and thus they do not simulate an earthquake, they can still be considered as an envelope ground motion and hence as a more conservative loading scenario. Five uncorrelated artificial accelerograms, produced from the smooth EC8 spectrum are used in the present study. The methodology adopted for the creation of artificial accelerograms from a given smooth spectrum was proposed by Gasparini and Vanmarke (1976). A typical artificial accelerogram and its corresponding 2.5%-damped spectrum are shown in Figure 2.

Finite Element Modeling

In order to perform dynamic analysis considering nonlinear behaviour there is a need for a detailed and accurate simulation of the structure in the areas where plastic deformations are expected to occur. Given that the plastic hinge approach has limitations in terms of accuracy, especially under dynamic loading; the fiber approach was adopted in this study. A fiber beam element based on the natural mode method has been employed (Argyris, Tenek, & Mattssonn, 1998). This finite element formulation is based on the natural mode method where the description of the displacement field along the beam is performed with quantities having a clear physical meaning.

Figure 2. (a) A typical artificial accelerogram, (b) 2.5% damped response spectra of EC8 and of the artificial accelerogram of Figure 2a

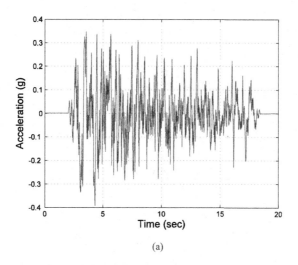

(a)

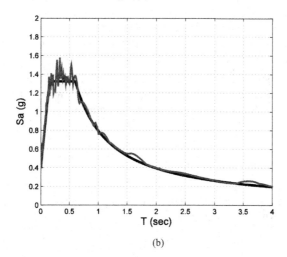

(b)

Each structural element is discretized into a number of sections, and each section is further divided into a number of fibers (see Figure 3), which are restrained to beam kinematics. The sections are located either at the centre of the element or at its Gaussian integration points. The main advantage of the fiber approach is that each fiber has a simple uniaxial material model allowing an easy and efficient implementation of the nonlinear behaviour. This approach is considered suitable for nonlinear beam-column elements under dynamic loading

Figure 3. Modeling of nonlinear behaviour: The fiber approach

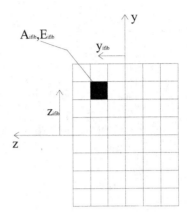

and provides reliable solution compared to other formulations for nonlinear analysis of frame structures. However, it results to higher computational demands both in terms of memory storage and CPU time. An adaptive discretization is used where a detailed simulation is restricted to the regions of the joints. Thus, the vicinity (usually 2.5% to 10% of element's length) of beam and column joints is discretized with a denser mesh of beam elements, while the remaining part of the member is discretized with linear elastic elements.

A simple bilinear stress-strain relationship with kinematic hardening is adopted. Studies have shown that this law is adequate and gives accurate results for many practical applications. The stress-strain relationship that allows for strain hardening is:

$$\sigma = E \cdot \varepsilon, \qquad \text{for } \varepsilon \leq \varepsilon_y$$
$$\sigma = E \cdot \varepsilon_y + E_{st} \cdot (\varepsilon - \varepsilon_y), \text{ for } \varepsilon > \varepsilon_y \tag{6}$$

where σ_y and ε_y are the yield strain and the corresponding yield stress, E is the elastic Young's modulus and $\dot{E}_{st}$ is the strain-hardening modulus.

Formulation of the
Structural Optimization Problem

In sizing optimization problems usually the aim is to minimize the weight of the structure under certain behavioural constraints on stress and displacements. The design variables are most frequently chosen to be dimensions of the cross-sectional areas of the members of the structure. In addition, engineering practice demands that the members should be considered in groups having the same design variables. This linking of elements results in a trade-off between the use of more material and the need of symmetry and uniformity of structures according to practical considerations. Furthermore, due to fabrication limitations the design variables are not continuous but discrete.

A discrete structural optimization problem can be formulated in the following form:

$$
\begin{aligned}
\min \quad & F(s) \\
\text{subject to} \quad & g_j(s) \le 0 \quad j=1,...,m \\
& s_i \in R^d, \quad i=1,...,n
\end{aligned}
\tag{7}
$$

where $F(s)$ and $g(s)$ denote the objective and constraints functions respectively. R^d is a given set of discrete values. The design variables s_i ($i = 1, ..., n$) can take values only from this set. For the steel frame considered, the objective function is the structural weight, the design variables are the cross-sections of the members and the constraints are imposed by the codes according to the analysis type that is used, as described in subsequent sections.

Solving the Optimization Problem with Evolution Strategies

The sensitivity analysis phase, which is an important part of all mathematical programming optimization methods, is the most time-consuming part of the optimization process when a gradient-based optimizer is used (Papadrakakis et al., 1996). On the other hand, the application of random search optimization methods, such as evolutionary algorithms, does not need gradient information and therefore avoid performing the computationally expensive sensitivity analysis step (Papadrakakis, Lagaros, Thierauf, & Cai, 1998; Papadrakakis, Tsompanakis, Hinton, & Sienz, 1999). Furthermore, it is widely recognized that evolution-based optimization techniques are in general more robust and present a better global behaviour than mathematical programming methods. Moreover, the feasible design space in structural optimization problems under dynamic response constraints is often disconnected or disjoint which causes difficulties for many conventional optimization algorithms. EA, however, may suffer from a slow rate of convergence towards the global optimum.

In the current study, evolution strategies (ES) have been employed for the solution of the optimization problem. ES imitate biological evolution in nature and have three characteristics that make them differ from conventional optimization algorithms: (a) in place of the usual deterministic operators, they use randomized operators: mutation, selection and recombination; (b) instead of a single design point, they work simultaneously with a population of design points in the space of variables; (c) they can handle continuous, discrete and mixed optimization problems. The second characteristic allows for a natural implementation of ES on parallel computing environments. The ES algorithm for structural optimization applications under seismic loading can be stated as follows:

1.   *Selection step:* selection of s_i (i = 1, 2, ..., μ) parent vectors of the design variables.

2.   *Structural analysis step*

3.   *Constraints check:* all parent vectors become feasible.

4.   *Offspring generation:* generate s_j, (j = 1, 2, ..., λ) offspring vectors of the design variables.

5.   *Structural analysis step*

6.   *Constraints check:* if satisfied continue, else change s_j and go to *step 4*.

7.   *Selection step:* selection of the next generation parents.

8.   *Convergence check:* If satisfied stop, else go to *step 4*.

There are two different types of selection schemes: ($\mu+\lambda$)-scheme: The best μ individuals are selected from a temporary population of ($\mu+\lambda$) individuals to form the parents of the next generation, and (μ,λ)-scheme: The μ individuals produce λ offsprings ($\mu\leq\lambda$) and the selection process defines a new population of μ individuals from the set of λ offsprings only. In the second type, the life of each individual is limited to one generation. The optimization procedure terminates when the mean value of the objective values from all parent vectors in the last $2\times n\times\mu/\lambda$ generations has not been improved by more 0.01%.

Optimum Seismic Design Procedures and Constraints

Regardless of the analysis procedure adopted, there are a number of behavioural constraints that has to be taken into consideration in order to ensure that during the optimization procedure the structure fulfils design code requirements. FEMA-356 (2000), for example, makes a clear distinction between deformation-based and force-based actions and specifies separate performance criteria for each type of action. For structural optimization problems under earthquake loading the constraints adopted are derived from seismic performance criteria specified by structural codes or guidelines. These criteria are placed into categories according to the analysis procedure, linear or nonlinear, used for the design. The criteria are further classified according to the performance level they refer to. In the present study, a single performance level is adopted, namely "life safety," due to the increased computational cost to perform structural optimization for a wider range of performance levels.

In both linear and nonlinear procedures, before proceeding to the structural analysis step of the optimization procedure, the strength ratio of column to beam is calculated and a check whether the sections chosen are of class 1, as EC3 suggests, is carried out. The later check is necessary in order to ensure that the members have the capacity to develop their full plastic moment and rotational ductility, while the check on the ratio of the column strength over the strength of the beam is necessary in order to have designs consistent with the 'strong column-weak beam' philosophy. In addition, the numerical model does not account for shear failure, while rigid moment connections are assumed for the joints. Subsequently, the structural analysis step is performed with linear or nonlinear structural response followed by the specified design checks.

Constraints for Linear Analysis Procedures

Ultimate limit state checks of EC8 are associated with specified forms of structural failure that may violate the life safety performance level. When a linear procedure is adopted, deformation controlled actions, or in other words, actions that the members can resist by deforming nonlinearly, are reduced by a behaviour factor. This behaviour factor depends on the structural system and also on the performance level considered. For moment resisting frames capable to develop their full plastic moment and rotational ductility considering the Life Safety performance level, EC8 suggests the use of a behaviour factor equal to 6. For beams, it is required that the full plastic moment resistance and rotation capacity is not decreased significantly by compression and shear forces:

$$\frac{V_{G.Sd} + V_{M.Sd}}{V_{Pl.Rd}} \leq 0.5 \tag{8}$$

where $M_{pl.Rd}$ is the plastic shear capacity of the section, $V_{G.Sd}$ is the shear force due to non-seismic actions and $V_{M.Sd}$ is the shear force due to the application of resisting moments with opposite signs at the extremities of the beam. Moreover the applied moment should be less than $M_{pl.Rd}$ and the axial load less than the 15% of $N_{pl.Rd}$.

According to EC3, for columns subjected to bending with the presence of axial load the following formula should be satisfied:

$$\frac{N_{sd}}{\chi_{min} N_{pl.Rd}} + \frac{\kappa_y M_{sd.y}}{M_{pl.Rd.y}} + \frac{\kappa_z M_{sd.z}}{M_{pl.Rd.z}} \leq 1 \tag{9}$$

where the χ_{min} factor is taken equal to 0.7 and while the parameters κ_y, κ_z are set equal to 1 (EC3) for the loading scheme considered. Moreover the shear capacity should be at least double than the applied shear force. Plastic capacities for each section are determined from the expressions:

$$M_{pl,Rd} = \frac{W_{pl} \cdot f_y}{\gamma_{M0}} \tag{10}$$

$$N_{pl,Rd} = \frac{A \cdot f_y}{\gamma_{M1}} \tag{11}$$

$$V_{pl,Rd} = \frac{1.04 \cdot h \cdot t_w \cdot f_y}{\sqrt{3} \cdot \gamma_{M0}} \tag{12}$$

in which W_{pl} is section's plastic modulus and parameters γ_{M0} and γ_{M1} are taken both equal to 1.10. The stability coefficient θ is evaluated for each story using the formula:

$$\theta = \frac{P_{tot} \cdot d_r}{V_{tot} \cdot h} \tag{13}$$

where P_{tot} is the total gravity load at the story considered, d_r is the interstory drift, V_{tot} is the total seismic shear and h is the story height. P-Δ effects need not be considered when the stability coefficient is less than 0.1.

Constraints for Nonlinear Analysis Procedures

The use of nonlinear analysis procedures allows more direct performance criteria to be adopted. Performance criteria that refer to the local member level, such as plastic hinge rotations or member chord rotations can be used. Alternatively, story level criteria, such as maximum interstory drift values, can also be adopted. Suggested values for plastic hinge rotations and maximum interstory drift for steel moment resisting frames are given by FEMA-356 guidelines. Since nonlinear analysis is performed the P-Δ effects can be taken into account explicitly. In the present study, a maximum interstory drift value of 2% is considered for the Life Safety performance level.

Furthermore, another restriction is that the applied axial force on columns should not exceed 50% of the member capacity given by equation (11) in order to allow ductile structural behaviour. If the nonlinear dynamic analysis fails to converge it is considered that the structure has lost its stability during the ground motion, and thus the design is rejected.

Numerical Results

The six-story space frame, shown in Figure 4, has been considered for the purpose of the current study. The space frame consists of 63 members, simulated with 383 beam elements (six adaptively distributed beam-column elements per member and approximately 100 fibers in each member) and about 2100 d.o.f. The modulus of elasticity is 200 GPa and the yield stress is $f_y = 250$ MPa. The structural members are divided into five groups, each of them having one design variable (i.e., element's cross-section), thus corresponding to 5 design variables. All sections are W-shaped given by standard AISC tables.

The structure is loaded with G = 3kPa permanent action and a live load of Q = 5 kPa in the vertical direction. Earthquake loading is taken into account using the EC8 response spectrum (Figure 1) for a PGA value of 0.40g. Both LDP and NDP design procedures are considered with either a suite of 10 natural records, scaled to the EC8 response spectrum according to equation (5), or five artificial accelerograms generated from the same spectrum. For all linear procedures, including the MmRS procedure, the design spectrum is reduced by a behaviour factor q = 6.0 as suggested by EC8 for steel moment resisting frame structures.

All optimization procedures were performed in this study with the ES(5+5) scheme, while the same initial population was employed for all cases examined. The objective function of the optimization problem is the structural weight, while the constraints for each type of

Figure 4. Six-story space frame

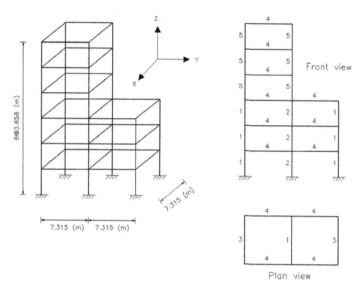

analysis were chosen according to previously described philosophy. In total five cases were examined; the MmRS procedure using the EC8 spectrum, and the LDP and the NDP using either 10 natural records (cases LDP-NAT and NDP-NAT) or five artificial records (cases LDP-AR and NDP-AR). All runs were performed using a Pentium-IV personal computer with a 2.8MHz CPU.

The main objective of this work is to study the influence on the final design of the analysis procedures previously discussed by comparing the optimum designs obtained with regard to their structural performance and material volume. In Table 2 the optimum designs obtained and the corresponding structural weights are shown together with the CPU times required by each procedure. It can be seen that the NDP procedures give the most economic designs, in terms of the material weight, while the MmRS procedure lead to the most expensive solution. Furthermore, comparing the dynamic procedures with natural and artificial records it can be seen that natural records lead to less material weight. The LDP procedures compared to the MmRS procedure achieved a volume reduction of 12% and 4% for the natural records (NAT) and the artificial records (AR) case, respectively, while the corresponding reduction of the NDP procedures is 60% and 57%, respectively. Furthermore, it can be seen that the number of ES optimization steps and the number of FE analyses required are reasonable for such complicated, nonconvex, problems.

In Figure 5 the seismic performance of the five optimum designs is compared in terms of their profiles of maximum interstory drifts. The drifts shown are the maximum values for each story obtained from all time steps. More specifically, Figure 5 shows the median of the maximum interstory drift values for each of the five optimum designs along the height of the frame when nonlinear response history analysis is performed using the records of Table 1.

Table 2. Optimum designs obtained for the five alternative design procedures

Procedure	Volume (m³)	Optimum Design	Gens.	Analyses	time (hrs)
MmRS	7.43	W14×132, W14×342, W14×61, W14×61, W14×132	66	164	2.7
LDP-NAT	6.56	W14×120, W14×270, W14×22, W14×74, W14×109	61	130	8.7
LDP-AR	7.14	W14×120, W14×283, W14×61, W14×74, W14×109	46	133	5.8
NDP-NAT	2.93	W14×43, W14×99, W14×22, W14×34, W14×48	50	142	23.7
NDP-AR	3.19	W14×43, W14×109, W14×22, W14×34, W14×61	49	144	15.6

Figure 5. Seismic performance of the five optimum designs: Profiles of median maximum recorded interstory drifts of each story

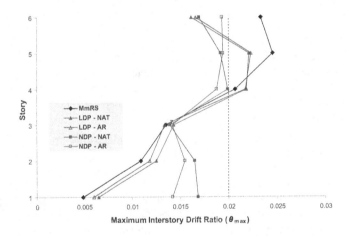

Examining Table 2 and Figure 5 it is clear that even though the MmRS procedure leads to the most conservative design in terms of volume achieved (7.43m³), it results to the highest interstory drift values compared to the remaining four optimum designs. Similarly the optimum designs of the LDP procedure exhibit interstory drift values slightly larger than the 2% drift limit imposed on the NDP designs but in terms of material weight the LDP designs are considerably heavier. This is attributed to the nature of the constraints imposed on the linear procedures, as opposed to the direct design criteria adopted in the nonlinear analysis procedures as it is described earlier. The designs obtained with artificial records both for the

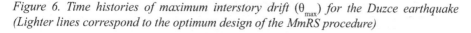

Figure 6. Time histories of maximum interstory drift (θ_{max}) for the Duzce earthquake (Lighter lines correspond to the optimum design of the MmRS procedure)

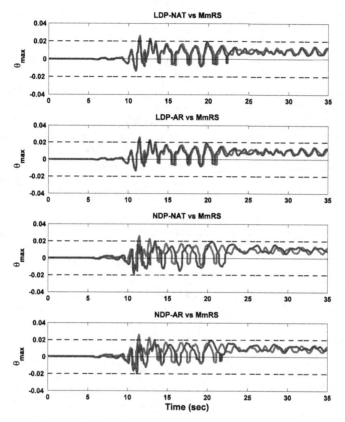

LDP and the NDP procedures resulted in slightly heavier designs with approximately the same maximum interstory drift values. This can be partially explained by the large number of significant cycles that artificial records contain and thus they pose a more conservative loading scheme.

Similar conclusions arise from Figure 6 where the optimum designs of the dynamic procedures are compared against the optimum design of the MmRS procedure, when subjected to the Duzce earthquake. It can be seen that the optimization procedure is dominated by two or three cycles, where the maximum interstory drift values occur. The response of the optimum design of the MmRS procedure results to drift values comparable to those of the optimum designs of the linear dynamic procedures where for at least two cycles the 2%-drift threshold is exceeded. However, the nonlinear procedures do not exceed this threshold and in general exhibit maximum interstory drift values close to those of the much heavier optimum design of the MmRS procedure.

Conclusion

The present study demonstrates the potential of structural optimization techniques when applied to the design of structures under earthquake loading. A number of procedures based on seismic design codes are implemented in the framework of structural optimization and the obtained optimal designs are compared both in terms of material volume (cost) and seismic performance. In optimized structures under linear analysis procedures the influence of nonlinear deformations can have a significant effect if the structure is subjected to earthquake loading, especially to their stability, since heavier linearly elastic structures do not, in general, have a better postlimit response. In order to circumvent this problem the use of alternative design procedures based on more elaborate analysis procedures is suggested. The numerical results demonstrate that more elaborate analysis procedures take into account explicitly the characteristics of the seismic response and thus lead to designs with better seismic performance. Furthermore, by avoiding conservative designs checks and assumptions made by simplifying procedures, designs with less material weight can be obtained.

The main conclusions of the presented evaluation of optimum seismic design procedures are the following: (a) improved control on the structural response and weight is achieved when nonlinear analysis is performed, compared to optimum designs obtained when linear analysis procedures based on EC8 requirements were adopted; (b) a reasonable number ES optimization steps, and thus with affordable computational cost, designs with substantially less material cost can be obtained; (c) the use of artificial instead of natural ground motion records leads to slightly heavier structures since they constitute a more 'demanding' loading scheme. It was also demonstrated that the Evolution Strategies algorithm can be considered not only a robust but also an efficient tool for design optimization of real world space frames under earthquake loading. Further research would include fully integrating the PBEE framework (i.e., use multiple performance levels), which of course would increase dramatically the computational cost.

References

Al-Ali, A. A. K., & Krawinkler, H. (1998). *Effects of vertical irregularities on seismic behaviour of building structures* (Rep. No. 130, John A. Blume Earthquake Engineering Center). Stanford, CA: Stanford University.

Antoniou, S., & Pinho, R. (2004). Development and verification of a displacement-based adaptive pushover procedure. *Journal of Earthquake Engineering, 8*, 643-661.

Argyris, J., Tenek, L., & Mattssonn, A. (1998). BEC: A 2-node fast converging shear-deformable isotropic and composite beam element based on 6 rigid-body and 6 straining modes. *Computer Methods in Applied Mechanics in Engineering, 152*, 281-336.

Balling, R. J., Ciampi, V., Pister, K. S., & Polak, E. (1981). *Optimal design of seismic-resistant planar steel frames* (Tech. Rep. No. UCB/EERC-81/20). Berkelely, CA: University of California, Berkeley, Earthquake Engineering Research Center.

Bazzuro, P., Cornell, C. A., Shome, N., & Carballo, J. E. (1998). Three proposals for characterizing MDOF nonlinear response. *ASCE Journal of Structural Engineering. 124*(11), 1281-1289.

Beck, J. L., Chan, E., Irfanoglu, A., & Papadimitriou, C. (1999) Multi-criteria optimal structural design under uncertainty. *Earthquake Engineering and Structural Dynamics. 28*, 741-761.

Bhatti, M. A., & Pister, K. S. (1981). A dual criteria approach for optimal design of earthquake resistant structural systems. *Earthquake Engineering and Structural Dynamics. 9*, 557-572.

Bozorgnia, Y., & Bertero, V. V. (Eds.). (2004). *Earthquake engineering: From engineering seismology to performance-based engineering.* CRC Press.

Chan, C-M., & Zou, X-K. (2004). Elastic and nonlinear drift performance optimization for reinforced concrete buildings under earthquake loads. *Earthquake Engineering and Structural Dynamics. 33*, 929-950.

Charney, F. A. (2000). Needs in the Development of a comprehensive performance based optimization process. In *Proceedings of the ASCE Structures 2000 Conference*, Philadelphia.

Cheng, F. Y., Li, D., & Ger, J. (2000). *Multiobjective optimization of seismic structures.* Proceedings of the ASCE Structures 2000 Conference, Philadelphia.

Chintanapakdee, C., & Chopra, A. K. (2003). Evaluation of modal pushover analysis using generic frames. *Earthquake Engineering and Structural Dynamics, 32*, 417-442.

Esteva, L., Diaz-Lopez, O., Garcia-Perez, J., Sierra, G., & Ismael, E. (2002). Life-cycle optimization in the establishment of performance-acceptance parameters for seismic design. *Structural Safety, 24*, 187-204.

Eurocode 3. (1993). *Design of steel structures. Part1.1: General rules for buildings.* CEN-ENV, European Committee for Standardization, Brussels.

Eurocode 8. (1994). *Design provisions for earthquake resistant structures.* CEN, ENV, European Committee for Standardization, Brussels.

FEMA-356. (2000). *Prestandard and commentary for the seismic rehabilitation of buildings.* Washington, DC: Federal Emergency Management Agency/SAC Joint Venture.

Foley, C. M. (2002). Optimized performance-based design for buildings. In S.A. Burns (Ed.), *Recent advances in optimal structural design* (pp. 169-240). ASCE.

Ganzerli, S., Pantelides, C. P., & Reaveley, L. D. (2000). Performance-based design using structural optimization. *Earthquake Engineering and Structural Dynamics, 29*, 1677-1690.

Gasparini, D. A., & Vanmarke, E. H. (1976). *Simulated earthquake motions compatible with prescribed response spectra* (Tech. Rep. No. R76-4). MIT, Department of Civil Engineering.

Gong, Y. (2003). *Performance-based design of steel building frameworks under seismic loading* (Unpublished doctoral thesis). Department of Civil Engineering, University of Waterloo, Canada.

Han, S. W., & Wen, Y. K. (1997). Method for reliability-based seismic design: Equivalent nonlinear systems, II: Calibration of code parameters. *ASCE Journal of Structural Engineering, 123*(3), 256-270.

Kocer, F. Y., & Arora, J. S. (1999). Optimal design of H-frame transmission poles for earthquake loading. *ASCE Journal of Structural Engineering, 125*(11), 1299-1308.

Kocer, F. Y., & Arora, J. S. (2002). Optimal design of latticed towers subjected to earthquake loading. *ASCE Journal of Structural Engineering, 128*(2), 197-204.

Lagaros, N. D., Fragiadakis, M., Papadrakakis, M., & Tsompanakis, Y. (2006). Structural optimization: A tool for evaluating seismic design procedures. *Engineering Structures, 28*(12), 1623-1633.

Liu, M., Burns, S. A., & Wen, Y. K. (2003). Optimal seismic design of steel frame buildings based on life-cycle cost considerations. *Earthquake Engineering and Structural Dynamics, 32*, 1313-1332.

Liu, M., Burns, S. A., & Wen, Y. K. (2005). Multiobjective optimization for performance-based seismic design of steel moment frame structures. *Earthquake Engineering and Structural Dynamics, 34*, 289-306.

Liu, M., Wen, Y. K., & Burns, S. A. (2004). Life-cycle cost oriented seismic design optimization of steel moment frame structures with risk-taking preference. *Engineering Structures, 26*, 1407-1421.

Naeim, F., Alimoradi, A., & Pezeshk, S. (2004). Selection and scaling of ground motion time histories for structural design using genetic algorithms. *Earthquake Spectra, 20*(2), 413-426.

Papadrakakis, M., Lagaros, N. D., Thierauf, G., & Cai, J. (1998). Advanced solution methods in structural optimization based on evolution strategies. *Engineering Computations Journal, 15*, 12-34.

Papadrakakis, M., Tsompanakis, Y., Hinton, E., & Sienz, J. (1996). Advanced solution methods in topology optimization and shape sensitivity analysis. *Journal of Engineering Computations, 13*(5), 57-90.

Papadrakakis, M., Tsompanakis, Y., & Lagaros, N. D. (1999). Structural shape optimization using evolution strategies. *Engineering Optimization, 31*, 515-540.

Pezeshk, S. (1998). Design of framed structures: An integrated nonlinear analysis and optimal minimum weight design. *International Journal for Numerical Methods in Engineering, 41*, 459-471.

Polak, E., Pister, K. S., & Ray, D. (1976). Optimal design of framed structures subjected to earthquakes. *Engineering Optimization, 2*, 65-71.

Shome, N., Cornell, C. A., Bazzuro, P., & Carballo, J. E. (1998). Earthquakes, records, and nonlinear responses. *Earthquake Spectra, 14*, 469-500.

Zhu, T. J., Heidebrecht, A. C., & Tso, W. K. (1988). Effect of peak ground acceleration to velocity ratio on the ductility demand of nonlinear systems. *Earthquake Engineering and Structural Dynamics, 16*, 63-79.

Endnote

---

[1]   Work published in the paper by: Lagaros, N. D., Fragiadakis, M., Papadrakakis, M., & Tsompanakis, Y. (2006). Structural optimization: A tool for evaluating seismic design procedures. *Engineering Structures*.

Chapter II

Applying Neural Networks for Performance-Based Design in Earthquake Engineering

Ricardo O. Foschi, University of British Columbia, Canada

Abstract

This chapter discusses the application of neural networks for the representation of structural responses in earthquake engineering, and their subsequent use in reliability evaluation and optimization for performance-based design. An approach is proposed by means of which the intervening random variables (including the design variables) are separated into two sets: a basic one and, another, grouping all the variables related to the ground motion. Structural responses are deterministically obtained for different combinations of all variables, and neural networks (with the basic set as input) are trained to represent, for example, either the mean or the standard deviation of the responses over the grouped set. Reliability evaluations, and the optimization involved in performance-based design, can then be efficiently performed via simulation. Examples are used to illustrate the approach, and the corresponding advantages are discussed.

Background

Reliability theory has been in continuing development since the 1950s. The objective in its application to structural engineering is to take into account the unavoidable uncertainty in the variables influencing the structural responses, and to quantify the associated probability that a given structure may not perform as intended in a set of performance criteria or limit states. Reliability levels may be expressed either by the corresponding probability of failure or nonperformance, or by an associated reliability index. Much research has gone into the development of effective methods and software for the calculation of the failure probability and, over the years, their effectiveness has been greatly improved by continuing developments in computer power (Schüeller, 2006). These methods, which allow limit state definitions with many random variables, not necessarily normally-distributed and with a correlation structure, range from computer simulations (Monte Carlo or importance sampling) to approximate, but very efficient procedures like FORM or SORM (respectively, first or second order reliability methods) (Ditlevsen & Madsen, 1996; Melchers, 1987).

Conversely, a structure can be designed to meet minimum target reliability levels for specified multiple performance criteria. This is the objective of performance-based design, an approach being proposed as the standard for design methodology, including seismic applications (Bertero & Bertero, 2002; Fajfar & Krawinkler, 1997; Ghaboussi & Lin, 2000; Krawinkler, 1999; Wen, 2001). The performance-based design process may also involve the minimization of an associated function; for example, the structural weight or the expected total cost of the system (initial plus the cost of system failure). Performance-based design is thus an optimization problem and, in the search for optimum design parameters, structural reliability analysis must be applied to evaluate the nonperformance probability associated with each of the limit states, given the current value of the parameters. This is a quantitative definition of performance-based design, but a more qualitative implementation is sometimes used on the basis of subjective or expert opinions regarding nonperformance probabilities. Here we are only interested in the formulation of the quantitative approach.

Introduction

The calculation of the probability of nonperformance in a given limit state requires the formulation of an associated performance or limit state function $G(x)$, in the form:

$$G(x) = C(x_c, d_c) - D(x_d, d_d) \tag{1}$$

containing the vectors of random variables x_c and d_c, associated with the capacity C, and x_d and d_d associated with the demand D. The variable vectors d_c and d_d are the design parameters, or the objective of the performance-based design.

When the functions C or D can be written explicitly, reliability calculations or performance-based design can be readily implemented, as shown in a following example. In general, however, these functions can not be given an explicit form, and only discrete values may be available for specific combinations of the intervening variables. For example, for a given variable combination, a finite element or a nonlinear dynamic, time-stepping analysis may be required to calculate the demand D, or a similar type of analysis may be needed for the capacity C. As these types of calculations are normally computationally intensive, their direct implementation in a reliability analysis or performance-based design optimization could be computationally demanding. However, a database of discrete values for C or D, obtained a priori in a deterministic manner, can be used to develop an approximating mathematical representation for those functions. In general, these types of mathematical approximations are called response surfaces (Breitung & Faravelli, 1996; Bucher & Bourgund, 1990; Faravelli, 1989; Schüeller, Bucher, & Bourgund, 1989). Bucher and Bourgund proposed, for example, representing the a priori database by means of a global quadratic surface, with or without interaction terms. Special forms can also be adopted for the response surfaces, depending on the characteristics of the problem at hand (Möller & Foschi, 2003). In practice, however, the form of an adequate approximating surface, capable of representing important details of the actual functions, is sometimes difficult to define. This is a particularly important point for dynamic problems in earthquake engineering, where the structural responses of interest may show peaks and valleys as a result of resonances.

As an alternative to fitting a global response surface, a local interpolation technique of the database has been proposed (Foschi, Li, & Zhang, 2002), but the robustness of this approach depends on the density and the design of the database. A response surface representation by neural networks (Patterson, 1996) has been found to be much more robust, as shown by Zhang and Foschi (2004), permitting a generally good representation of the actual responses.

Neural networks are commonly used to provide a mathematical expression to an unknown underlying relationship. In our case, however, neural networks provide a robust interpolating tool, since the underlying relationship is known and can be calculated by a further execution of, for example, the finite element or dynamic analysis involved.

Having adjusted a response surface or trained a neural network to the database, the probability of nonperformance can be obtained by FORM, SORM or by simulation techniques. The application of FORM procedures, following a representation of the structural responses by neural networks, has been recently studied (Deng, Gu, Li, & Qi Yue, 2005). However, response surfaces, particularly based on neural networks, permit the use of simulation as a very efficient tool in reliability estimation, with the advantage that simulation procedures do not have the accuracy limitations of FORM regarding the nonlinearity of the actual surface.

Performance-Based Design

Performance-based design is an optimization problem that is formulated as follows: to find the optimal design parameter vector d (that is, d_c and/or d_d) by minimizing the objective function Ψ:

$$\psi = F(d) \rightarrow \text{Minimum} \tag{2}$$

subject to reliability constraints $\beta_j(d) \geq \beta_j^T (j = 1, ND)$

and geometric constraints $d_i^l \leq d_i \leq d_i^u$;

or, in a more restrictive problem,

$$\Psi = \sum_{j=1}^{ND} (\beta_j^T - \beta_j(d))^2 \rightarrow \text{Minimum} \tag{3}$$

subject to geometric constraints $d_i^l \leq d_i \leq d_i^u$.

In these formulations:

   $F(d)$ is a function expressing the total expected cost or weight of the structural system;
   ND is the number of performance criteria or limit states;
      β_j^T is the target reliability index for the performance criterion j;
      $\beta_j(d)$ is the calculated reliability index for performance criterion j, given
         the design parameter vector d;
      d_i^l is the lower bound for the design parameter d_i;
      d_i^u is the upper bound for the design parameter d_i.

The reliability indices $\beta_j(d)$ are related to the probabilities of failure or nonperformance P_j, in the j limit state, according to:

$$P_j = \Phi(-\beta_j(d)) \tag{4}$$

in which Φ is the cumulative distribution for a standard normal variable.

The implementation of performance-based design requires then the following essentials:

1.   Reliable deterministic models for the calculation of the functions C and/or D ;
2.   A definition of minimum target reliability for each performance criteria;
3.   Probabilistic representation of the uncertainties in the intervening variables;
4.   Software implementing algorithms for the calculation of the reliability indices $\beta_j(d)$, given a vector of design parameters d;
5.   A calculation model for the objective function $F(d)$;
6.   An algorithm for the minimization of the objective function.

Although the formulation of the problem is mathematically simple, the solution may require a substantial amount of engineering judgment. The validity of the results obtained is dependent on the validity of the calculation models for the capacity C and/or demand D. It is obvious that one must use the best models available. The solution of problems in earthquake engineering would require the availability of dynamic analysis models that take into account the nonlinearities resulting from structural degradation during the shaking, the interaction between axial loads and bending deformations (P-Delta effects), an accurate representation of the hysteretic behaviour of connections (accounting for slack development, subsequent pinching and loss of stiffness and strength) and possibly the effects of interactions between the structure and the soil foundation. In particular, since hysteretic behaviour is not a material property but a structural response to the demand history, hysteretic models must be able to represent and adapt to any type of demand history. Shortcomings of the models used must be approximately overcome with the introduction of additional random variables accounting for model errors, the statistics of which may need to be subjectively estimated.

The probabilistic descriptions of the intervening variables are also subject to uncertainty and error due to, for example, limited data. In earthquake engineering, this problem is most severe in relation to the description of the ground motion. This is influenced by earthquake magnitude, epicentral distance, earthquake depth and soil characteristics between the site and the epicenter. While there are approaches to quantify the statistics of peak ground acceleration in terms of those variables, it is more difficult to estimate the range of accelerogram functions or records which may occur at a site and that are needed for the dynamic analysis. A set of historical earthquakes may be available at a given location, each representing a sample of the combined effect of the different ground motion variables. Normally, however, the number of significant historical earthquakes at a site might not be large, perhaps with a small representation of strong earthquakes. Nevertheless, one should use the largest set of historical records available at the location of the structure, perhaps augmented with those from a site with similar seismic characteristics. A set of artificially generated earthquakes can also be used to complement the set of historical accelerograms. The reliability estimates, and the results from the performance-based design optimization, will then be conditional on the strong motion representation used.

Finally, the performance functions $G(x)$ for seismic applications must be formulated in terms of structural damage (partial or collapse). Since the output from the dynamic analysis provides information on displacements or strains, member actions or stresses, a relationship is then required between those calculated quantities and the resulting damage. This may be represented by an index indicator of damage, itself a function of the calculated quantities. The uncertainty in the relationship must be quantified through experimental observations from, for example, shake table testing. The damage relationship can then also be expressed in mathematical form through a neural network.

The optimization requires the calculation of the achieved reliability indices β_j, which, in turn, and according to equation (4), requires the estimation of the failure probability P_j. In this regard, optimization algorithms that avoid the calculation of gradients are most efficient and less prone to numerical difficulties.

Figure 1. Timber beam (normal service and fire conditions)

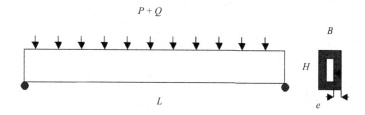

An Example of Performance-Based Design for Explicit Limit State Functions

It is useful to illustrate performance-based design using first a case of multiple performance criteria with explicitly given performance functions. Consider the problem shown in Figure 1, a timber beam with cross-sectional dimensions B and H and a span L, carrying loads P and Q. Both are uniformly distributed but have different statistical properties: P is a dead load, with low variability, Q is a live load with higher variability. The bending strength of the beam is f and the modulus of elasticity is E.

Two performance requirements are specified under normal service conditions: (a) the beam must carry the loads without bending failure and (b) the maximum deflection should not exceed 1/180 of the span. A third performance requirement specifies that, in case of a fire, the beam should not fail in bending before 45 minutes of exposure. In performance-based design, the reliability levels with which those three performance criteria are to be met must also be specified.

For the fire situation, it is necessary to know the rate at which wood burns, decreasing the dimensions of the beam by an amount e, as shown in Figure 1. Of course, the relationship between e and time might be complicated, depending on the environmental conditions. For this example, however, it is assumed that the burning rate is a constant α, so that after time T the amount burnt is $e = \alpha T$.

Using linear elastic beam theory, the three performance functions are written as follows:

1. Bending failure under normal service conditions:

$$G_1 = f - \frac{(Q+P)L^2}{8} \frac{1}{(BH^3/12)} \left(\frac{H}{2}\right) \tag{5}$$

2. Deflection under normal service conditions (ignoring, for this example, the shear contribution):

$$G_2 = \frac{L}{180} - \frac{5(Q+P)L^4}{384}\frac{1}{E(BH^3/12)}$$

(6)

3.   Bending failure under fire conditions:

$$G_3 = f - \frac{(Q+P)L^2}{8}\frac{1}{((B-2e)(H-2e)^3/12)}(\frac{H}{2}-e)$$

(7)

The corresponding specified, minimum target reliability levels are:

* $\beta_1 = 4.0$ for bending failure under normal service;
* $\beta_2 = 2.0$ for maximum deflection under normal service;
* $\beta_3 = 3.0$ for bending failure under fire after $T = 45$ minutes of exposure.

The problem is to find optimum dimensions H and B for the cross-section, achieving the specified reliabilities or having minimum deviations from them. Following equation (3), the solution requires the minimization of the objective function Ψ:

$$\Psi = \sum_{j=1}^{NP}(\beta_j^T - \beta_j(B,H))^2$$

(8)

in which NP = 3 is the number of performance requirements or reliability constraints, β_j^T are the corresponding target reliabilities and $\beta_j(B,H)$ are the achieved reliabilities for the dimensions B and H. The results presented here were obtained with a performance-based design package, IRELAN, developed at the University of British Columbia. The optimization algorithm implemented in IRELAN is a gradient-free search within the bounds for B and H. This algorithm starts with the calculation of Ψ for a number of random combinations of B and H, chosen within their bounds, locating the combination corresponding to the lowest value of Ψ. Starting from this combination, used now as an initial anchor, a random search is executed within a specified radius from the anchor point. Whenever a lower value of Ψ is found, the anchor is moved to the corresponding combination of design variables and the process is repeated. The sequence is stopped when all values of Ψ within the search radius are larger than the corresponding value at the anchor. The entire process is repeated starting from a different random combination as the initial anchor. The reliability indices $\beta_j(B,H)$, required during each calculation of Ψ, were obtained by equation (4), after the corresponding probability P_j was obtained by simulation.

The problem has five intervening random variables and these are identified in Table 1. It is assumed here that the statistics for the loads P and Q correspond to those that may be present at the time of the fire.

Table 1. Statistics and distribution for variables, timber beam under fire conditions

Random Variable	Mean	Std. Deviation	Distribution
f, bending strength (kN/m²)	40,000	8,000	Lognormal
P (kN/m)	2.0	0.2	Normal
Q (kN/m)	8.0	2.0	Extreme Type I
α (burn rate, mm/minute)	0.5	0.125	Lognormal
E, modulus of elasticity (kN/m2)	$10,000 \times 10^3$	$1,500 \times 10^3$	Lognormal

Table 2. Optimum dimensions, $\overline{H}$ and $\overline{B}$, timber beam under fire conditions

$\overline{H}$ (m)	$\overline{B}$ (m)	β_1 Achieved	β_2 Achieved	β_3 Achieved	Optimization Strategy
0.959	0.171	4.33	1.88	2.92	Unconstrained
0.971	0.172	4.41	2.03	3.00	Penalty on G_2 and G_3
Target β		4.00	2.00	3.00	

The dimensions B and H are also considered random variables (Normal), with a very small coefficient of variation. The design parameters are then the mean values $\overline{B}$ and $\overline{H}$. The results are shown in Table 2 for a span $L = 15$m.

As expected, the results from the optimization show that it is not generally possible to find values of $\overline{H}$ and $\overline{B}$ that will exactly achieve the prescribed target reliabilities. Optimum design parameters are obtained which minimize the differences between the values achieved and the target values. If the minimization is unconstrained, then the targets are not reached for the limit states G_2 and G_3, while the target for the limit state G_1, of bending failure under normal service, is exceeded. Penalty functions can then be introduced in the original objective Ψ for the limit states G_2 and G_3, in order to bring the corresponding reliability indices more in line with the minimum targets. As shown in Table 2, the new design parameters $\overline{B}$ and $\overline{H}$ can now meet the minimum conditions for all three limit states albeit with an increased reliability for bending failure under normal service conditions, which is not then a controlling design situation.

Representing a Function with a Neural Network

If the capacity C and/or the demand D in the performance function $G(x)$ cannot be written explicitly, as in the previous example, then a mathematical representation is required for an available set of discrete values. This can take the form of response surfaces or neural networks. These types of representations may only need to be used for some component of C and/or D. In the case of earthquake engineering, a nonlinear dynamic analysis for an input ground motion will yield all the outputs of relevance to the formulation of the different performance functions: for example, the maximum overall drift displacements, maximum interstory drifts, maximum moments and maximum shear forces. The dynamic analysis may be restricted to the structure itself or, in a more comprehensive study, it may include the interaction between the structure and the soil foundation. The input contains the variables associated with the ground motion and also the set of structural variables (including the design parameters).

The first step in the representation is to choose an adequate set of variable combinations to develop the discrete database. This implies an experimental design (Koehler & Owen, 1996; Sacks, Welch, Mitchell, & Wynn, 1989) and several techniques are available: a random selection may be used within given bounds for each variable, or a deterministic grid approach may be utilized. A combination of the two would use a random selection within a grid of cells for each variable. Regardless of the selection policy, the variable bounds for database development must be chosen sufficiently wide, to accommodate the range of values implied by the statistics of each variable. In this manner, during a subsequent simulation for reliability estimation, the response surface or neural network will always be used to interpolate an output and not to extrapolate.

The relevant outputs are obtained for each variable combination in a database, and this could imply the execution of possibly computational demanding software. Thus, the number of combinations must be chosen with care. In fact, building the database is the most time-demanding part of the process, since, having accomplished this task, subsequent reliability estimation or performance-based design optimization are made very efficient through the implementation of a response surface or a neural network.

The influence of the number and location of combinations used is illustrated in the following example. Consider a sinusoidal function:

$$f(x) = sin(\pi x) \qquad \text{with} \qquad 0 \leq x \leq 1 \tag{9}$$

and suppose that one has a set of N values for the function $f(x)$ but does not know the nature of this relationship. A global response surface approach, based on quadratic functions, will not provide a good representation of the true relationship. Neural networks, on the other hand, can produce almost a perfect match as N increases and, even for a low N, a relatively acceptable match. Here the input contains only one variable (x), and there is only one output (f). A network with one hidden layer containing a minimum of one neuron, requires a minimum of four data. Suppose that $N = 10$ data are available, corresponding to 10 values of x chosen equally spaced between 0 and 1. A neural network can be trained by back-propagation, using 80% of the data, with the remaining used for validation. The resulting optimal network had one hidden layer with two neurons. A comparison between the predictions from such a network and the actual $f(x)$, over all 10 data, is shown in Figure 2. A perfect agreement or training would require all points shown in Figure 2 to lie on the 45° line. Although, in this sense, the lack of agreement does not appear to be substantial, Figure 3 shows the comparison between the actual shape of $f(x)$ and that predicted by using the neural network for x between 0 and 1. It is to be noted the larger errors near the ends, where the network is extrapolating beyond the lowest or highest values of x used for training.

Figure 2. Neural network agreement, N = 10

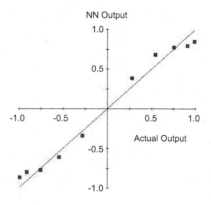

Figure 3. Neural network prediction, N = 10

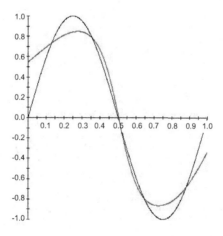

The neural network effectiveness improves rapidly with N. Figures 4 and 5 show the results for $N = 50$, results which were not substantially different from those for $N = 20$. The resulting network had one hidden layer with 11 neurons.

The choice of a sinusoidal for this example relates to the form of a typical structural response to dynamic excitation, with ups and downs resulting from resonances. Although there is no general rule about the number of variable combinations to be chosen in such a case, it is important to allow a larger number of samples for those variables deemed to be more important. Results can be obtained for a given database, which can then be augmented with outputs from additional combinations and the results checked for acceptable convergence.

Figure 4. Neural network agreement, N = 50

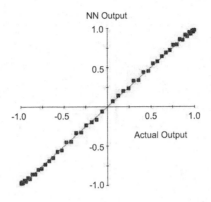

Figure 5. Neural network prediction, N = 50

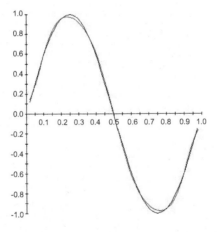

The neural network representation of functions appearing in either the capacity C or the demand D can also involve a strategy that is useful for a certain kind of problems, of which those in earthquake engineering are an example. In this case, the variables associated with the ground motion are many (peak intensity, duration of strong phase, random phase angles for each frequency, etc.). The general strategy involves the identification of a group or subset of input variables, a subset over which one can calculate statistics of the represented function. In order to illustrate this approach, consider the demand $D(x_d, d_d)$ and let the variables x_d be separated into two sets: x_{du} and x_{dg}. For fixed values of x_{du} and d_d, the demand D can be calculated for different combinations of the variables grouped under x_{dg}. This allows the calculation of response statistics over the subset x_{dg}, for example, the mean response

$\overline{D}(x_{du}, d_d)$ and the corresponding standard deviation $\sigma(x_{du}, d_d)$. Each of these statistics, in turn, can be represented by a corresponding neural network of a lesser input dimension. In the case of earthquake engineering, the ground motion variables would be grouped and their effect calculated for fixed x_{du} and d_d, essentially calculating the effect of each earthquake *record*. The mean response $\overline{D}(x_{du}, d_d)$ and the corresponding standard deviation $\sigma(x_{du}, d_d)$ would then be obtained over the set of records considered. Finally, the demand $D(x_d, d_d)$ may be written using the two neural networks and the assumption of, for example, a lognormal distribution (if appropriate) for the response over the grouped set:

$$D = \frac{\overline{D}}{\sqrt{1+(\sigma/\overline{D})^2}} \exp \; (R_N \sqrt{\ln \; [1.0+(\sigma/\overline{D})^2]}) \tag{9}$$

in which R_N is a standard normal variable. Should another type of distribution be deemed more appropriate, for example, an extreme type I, equation (9) would be replaced by:

$$D = B + \frac{1}{A}[-\ln(-\ln R_u)] \tag{10}$$

in which R_u is a random variable uniformly distributed between 0 and 1. The distribution constants A and B are obtained from the corresponding neural network representations $A(x_{du}, d_d)$ and $B(x_{du}, d_d)$. The same approach can be used for the capacity C, if necessary.

The choice of grouped variables in x_{dg} depends on the situation at hand. As just discussed, in an earthquake engineering problem the grouped variables in the demand D could be those associated with the characteristics of the ground motion. Thus, neural networks can be constructed both for the mean and for the standard deviation of the demand over a set of earthquake records likely to occur at a site. The peak acceleration of the earthquake is a random variable which could either be left within the basic set x_{du} or could also be made part of the grouped set x_{dg}, with its influence then reflected in the statistics represented by the two networks.

For another example on the application of the grouped strategy, in this case to vibrations of floors due to occupancy, see Foschi (2005). In this case, the grouped variables included the different coordinate locations of occupants on the floor, and the stiffness properties of the multiple supporting beams.

Having expressed the demand D and/or the capacity C in equation (1) by representations as per equation (9) or equation (10), using the trained neural networks for the statistics over the grouped variables, the probability of nonperformance (or "failure probability") can be efficiently obtained by direct Monte Carlo simulation, or by importance sampling simulation around a "design or anchor point." The location of this point can be approximately obtained by using the neural networks to develop values for D and/or C, which would permit the adjusting of a conventional quadratic response surface. FORM can then be applied to this surface to approximately locate the anchor point for the final Importance Sampling simulation.

Performance-Based Design in Earthquake Engineering Using Response Databases and Neural Networks

To illustrate the application of neural networks to earthquake engineering, following the grouping approach previously discussed, consider the reinforced concrete frame structure shown in Figure 6. It is a two-bay, 20-story frame, with a story height of 4 m and two spans of 8 m each. The beams have a constant cross section with dimensions 350 mm×700 mm. The column cross-sectional dimensions vary along the height of the building: from stories 1 to 7, $B_1 \times H_1$; from stories 8 to 14, $B_2 \times H_2$; and from stories 15 to 20, $B_3 \times H_3$.

The frame is required to meet six criteria with the following performance functions:

$$G_1 = 0.8m - D_{20} \tag{11}$$

$$G_2 = 0.5g - A_{20} \tag{12}$$

$$G_3 = 0.010 - \theta_{15} \tag{13}$$

$$G_4 = 0.010 - \theta_5 \tag{14}$$

$$G_5 = M_0 - M \tag{15}$$

Figure 6. Frame under earthquake excitation

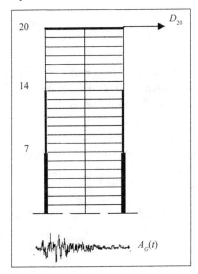

$$G_6 = V_0 - V \tag{16}$$

in which D_{20} is the roof displacement (in m), A_{20} is the roof acceleration (in g), θ_{15} the interstory distortion at floor 15, θ_5 the interstory distortion at floor 5, M the bending moment at the base of the columns and V the corresponding shear force. M_0 and V_0 are, respectively, the moment and shear capacities of the column cross section at the base. All responses correspond to the maximum values achieved during the earthquake. These six performance criteria are considered here only as an example, in reality more criteria can be incorporated (for example, maximum bending moments or shear forces at different levels). The corresponding target minimum reliability levels are:

$$\beta_1^T = 3.000; \ \beta_2^T = 2.500; \ \beta_3^T = 2.500;$$
$$\beta_4^T = 2.500; \ \beta_5^T = 2.500; \ \beta_6^T = 2.500 \tag{17}$$

The design parameters are the mean values corresponding to the three sets of column dimensions: $\bar{B}_1, \bar{H}_1, \bar{B}_2, \bar{H}_2, \bar{B}_3, \bar{H}_3$. These dimensions are considered as random variables, each allowing a small coefficient of variation 0.01.

For this illustration, ground motions $A_G(t)$ were generated using Clough and Penzien (1975) power spectral density, and were then modified to account for the duration of the strong motion. A total of 20 earthquake records were thus generated for the site. Thus, the complete set of 15 intervening random variables in this problem included:

1. the ground frequency ω_g in the power spectral density;
2. the earthquake strong motion duration, T_d;
3. the peak ground acceleration Ag;
4. the distributed load applied to the beams, q;
5. the steel yield strength, f_y ;
6. the concrete compression strength for columns from story 1 to 7, f_{c1};
7. the concrete compression strength for columns from story 8 to 14, f_{c2};
8. the concrete compression strength for columns from story 15 to 20, f_{c3};
9. the concrete compression strength for beams, f_b;

and, in addition, the column dimensions:

10. width of columns from story 1 to 7, B_1;
11. depth of columns from story 1 to 7, H_1;
12. width of columns from story 8 to 14, B_2;
13. depth of columns from story 8 to 14, H_2;

Table 3. Variable bounds for database development

Variable	Lower Bound	Upper Bound
Ag (g)	0.10	1.00
ω_g (rad/sec^2)	π	12π
T_d (sec)	1	60
q (KN/m)	15	60
fy (MPa)	400.0	450.0
fc_1 (MPa)	35	45
fc_2 (MPa)	25	35
fc_3 (MPa)	15	25
f_b (MPa)	15	25
B_1 (mm)	700	1000
H_1 (mm)	900	1200
B_2 (mm)	500	700
H_2 (mm)	700	900
B_3 (mm)	400	500
H_3 (mm)	500	700

14. width of columns from story 15 to 20, B_3;

15. depth of columns from story 15 to 20, H_3.

Response databases were constructed with a nonlinear dynamic analysis (CANNY) for 300 combinations of the variables, using with each the 20 earthquakes records. CANNY incorporates nonlinear reinforced concrete material properties and P-Delta effects. Table 3 shows the lower and upper bounds used for each variable for database development.

The mean and the standard deviation of the responses, over the 20 earthquake records, were then obtained for each of the 300 variable combinations. Two neural networks, with one hidden layer, were trained using the 300 data to represent, respectively, the mean and the standard deviation of the responses over the records. The training used the back-propagation algorithm, using 80% of the data for training and the remaining for validation. The distributions of responses over the set of records were well represented by lognormal distributions, so that the demands in the performance functions in equations (11) to (16) were written following equation (9).

The calculation of an optimum vector of design parameters was done for the variable statistics shown in Table 4. In addition, the capacity M_0 was assumed Normally distributed with a mean 54000 KNm and a standard deviation 5400 KNm. Similarly, the shear capacity V_0 was assumed Normally distributed with a mean 1440 KN and a standard deviation 144KN.

The optimization objective was to find design parameters, here the column dimensions, such that the achieved reliabilities in the six performance criteria were as close as possible to the target reliabilities prescribed. The optimization algorithm used was, again, as discussed previously for the first example in this chapter. The results of this optimization are shown in Table 5, while Table 6 shows the comparison between the target and achieved reliability levels in each of the performance criteria. In this example the reliability indices β required in the optimization were calculated, each time, by importance sampling simulation using the trained neural networks. Other approaches to this problem would differ either in the optimization algorithm used or in the calculation of the reliability indices β in the objective function. As discussed, problems in earthquake engineering benefit from the use of gradient-free optimization procedures. The simple search procedure used here compared favourably with other gradient-free methods like genetic algorithms. The calculation of the reliability indices is made very efficient and robust by simulation using the neural networks, avoiding possible problems of accuracy and convergence associated with other methods like FORM.

Table 4. Variable statistics

Variable	Distribution	Mean	Standard Deviation
A_g (gal)	Lognormal	290.0	191.0
ω_g (rad/sec^2)	Normal	5π	π
T_s (sec)	Normal	30.0	5.0
q (KN/m)	Normal	45.0	4.5
f_y (MPa)	Lognormal	400.0	10.0
f_{c1} (MPa)	Lognormal	40.0	1.5
f_{c2} (MPa)	Lognormal	30.0	1.5
f_{c3} (MPa)	Lognormal	20.0	1.5
f_b (MPa)	Lognormal	20.0	1.5
B_1 (mm)	Normal	900	$0.01\bar{B}_1$
H_1 (mm)	Normal	1100	$0.01\bar{H}_1$
B_2 (mm)	Normal	600	$0.01\bar{B}_2$
H_2 (mm)	Normal	800	$0.01\bar{H}_2$
B_3 (mm)	Normal	450	$0.01\bar{B}_3$
H_3 (mm)	Normal	600	$0.01\bar{H}_3$

Table 5. Results with optimum values for the column dimensions

Design parameter	Optimum
$\bar{B}_1$(mm)	830
$\bar{H}_1$(mm)	1129
$\bar{B}_2$(mm)	536
$\bar{H}_2$(mm)	747
$\bar{B}_3$(mm)	497
$\bar{H}_3$(mm)	608

Table 6. Comparison of target and achieved reliability levels

Performance Criterion	Target Reliability, β	Achieved Reliability, β
1	3.0	2.952
2	2.5	2.525
3	2.5	2.505
4	2.5	2.543
5	2.5	2.923
6	2.5	2.574

Final Discussion and Conclusion

Neural networks are a very effective tool for the mathematical representation of structural responses needed for the formulation of performance functions, particularly when they cannot be given explicitly. These functions are required for reliability analysis, and the neural network representation facilitates the efficient calculation of nonperformance probabilities by direct simulation. These probabilities, in turn, are required by the optimization process implied in performance-based design. Obtaining an optimal design becomes then a straightforward process, involving first the development of a deterministic response database, using appropriate software, and then using that database in the training of the networks for use in the optimization. The search for the optimum design implies that the input vectors to the networks include the design parameters themselves. One advantage of this process is its clarity, something which is very useful in certain applications like earthquake engineering. The calculation of probability of nonperformance, for example, just implies a direct simulation using the trained neural networks, an advantage over more involved methods relying on integration of fragilities over all possible values of a hazard. In addition, Performance-Based design, using the fragility approach, would require the availability of fragility functions for different sets of design parameters, and the determination of these functions could be a computationally intensive task.

The development of the deterministic databases could also be computationally intensive, particularly if the calculations involve nonlinear behaviour. However, after this is done, the corresponding neural network remains applicable unless the structural form is changed. Thus, in the previous example of the concrete frame, the neural network obtained is valid for all frames of the same geometric configuration, and performance-based design can be readily carried out for any changes in the statistics of the intervening variables. Availability of neural networks for different structural forms could thus permit the development of very efficient software, as an engineering design tool, for direct performance-based design. This would only require the definition of the performance criteria and the corresponding reliability levels with which they must be met. The availability of computer power and the clarity of the approach would facilitate the future use of these direct methods by the design engineer. Today, there is still an emphasis on codified approaches. Codes normally specify a design procedure and appropriate modification factors. For example, a codified seismic design normally starts with an elastic dynamic analysis, followed by the introduction of reduction factors to modify elastic forces in order to account for ductility. In the process, simplifying assumptions are usually made, as for example the equality of displacements between elastic and a nonelastic structure. Any code procedure should be calibrated against the more complete, direct method. This is to ensure that, over the set of situations for which the codified approach will be used, the reliability levels achieved with the code approximately match the reliabilities desired. The calibration process is itself one of optimization, and the same steps discussed here can be followed by considering the modification factors as instances of the design parameter vectors d_d and/or d_c.

Other neural networks applications to dynamics and earthquake engineering could be envisioned and, in fact, have already received consideration (Bento, Ndumu, & Dias, 1997; Yi, Lee, Lee, & Kim, 2000). These applications refer to a recurrent network that could be used for the development of hysteretic behaviour in structural members or connections, representing the current output force when an input is given by the current displacement and the immediate prior history of forces and associated displacements. Such representation should be more robust than the usual method of calibrating a hysteretic algorithm to an experimental, nondynamic, result obtained for a cyclic displacement history. There is no guarantee, following such an approach, that the output from the adopted hysteretic algorithm can be extrapolated to any other excitation beyond that used for its calibration.

Finally, neural networks have been used here to represent responses in terms of input variables and design parameters. However, carrying the analysis a step further, reliability levels β can themselves be represented by a network in terms of the same input. This further application would facilitate even more the optimization implied by performance-based design.

Acknowledgments

The collaboration of Dr. Jiansen Zhang in reviewing this chapter and the development of some of the data presented here is gratefully acknowledged. Research grant support to the author, from NSERC, the National Scientific and Engineering Research Council of Canada, is also gratefully acknowledged.

References

Bento, J., Ndumu, D., & Dias, J. (1997). Application of neural networks to the seismic analysis of structures. In H. Dajian & M. Haihong (Eds.), *Proceedings, EPMESC VI (Enhancement and Promotion of Computational Methods in Engineering and Science)* (Vol. 1, pp.72-81). China: South China University of Technology Press.

Bertero, R. D., & Bertero, V. V. (2002) . Performance-based seismic engineering: the need for a reliable conceptual comprehensive approach, *Earthquake Engineering and Structural Dynamics, 31,* 627-652.

Breitung, K., & Faravelli, L. (1996). Response surface methods and asymptotic approximations. In F. Casciate & J. B. Roberts (Eds.), *CRC mathematical modelling series: mathematical models for structural reliability analysis* (CRC Mathematical Modelling Series, pp. 227-285).

Bucher, C. G., & Bourgund, U. (1990). A fast and efficient response surface approach for structural reliability problems, *Structural Safety, 7,* 57-66.

Clough, R. W., & Penzien, J. (1975). *Dynamics of structures.* McGraw-Hill.

Deng, J., Gu, D., Li, X., & Qi Yue, Z. (2005). Structural reliability analysis for implicit performance functions using artificial neural network. *Structural Safety, 27,* 25-48.

Ditlevsen, O., & Madsen, H. O. (1996) . *Structural Reliability Methods.* Wiley.

Fajfar, P., & Krawinkler, H. (1997). *Seismic design methodologies for the next generation of codes.* Balkema.

Faravelli, L. (1989). Response-surface approach for reliability analysis. *Journal of Engineering Mechanics, ASCE, 115*(2), 2673-2781.

Foschi, R. O., Li, H., & Zhang, J. (2002) . Reliability and performance-based design: A computational approach and applications. *Structural Safety, 24,* 205-218.

Foschi, R.O. (2005). Performance-based design and application of neural networks, *Proceedings, International Conference on Structural Safety and Reliability (ICOSSAR-05),* Rome, Italy.

Ghaboussi, J., & Lin, C. C. J. (2000). Performance-based design using structural optimization, *Earthquake Engineering and Structural Dynamics, 29,* 1677-1690.

Koehler, J. R., & Owen, A. B. (1996). Computer Experiments. In S. Ghosh & C. R. Rao (Eds.), *Handbook of statistics* (pp. 261-308). New York: Elsevier Science.

Krawinkler, H. (1999). Progress and challenges in performance-based earthquake engineering. In *Proceedings of the Seismic Engineering for Tomorrow Congress—In Honour of Professor Hiroshi Akiyama*, Tokyo, Japan.

Melchers, R. E. (1987). *Structural reliability: Analysis and predictions.* Chichester, UK: Ellis Horwood series in civil engineering.

Möller, O., & Foschi, R. O. (2003). Reliability evaluation in seismic design: A response surface methodology. *Earthquake Spectra, 19*(3), 579-603.

Patterson, D. W. (1996). *Artificial neural networks: Theory and application.* Prentice Hall.

Sacks, J., Welch, W. J., Mitchell T. J., & Wynn, P. (1989). Design and analysis of computer experiments. *Statistical Sciences, 4*(4), 409-435.

Schüeller, G. (Ed.). (2006). Structural reliability software. *Structural Safety, 28,* 1-216.

Schüeller, G. I., Bucher, C. G., & Bourgund, U. (1989). On efficient computational schemes to calculate structural failure probabilities. *Probabilistic Engineering Mechanics, 4*(1), 10-18.

Wen, Y. K. (2001). Reliability and performance-based design. *Structural Safety, 23,* 407-428.

Yi, W-H., Lee, H-S., Lee, S-C., & Kim, H-S. (2000). *Development of artificial neural networks-based hysteretic model.* Paper presented at the 14th Engineering Mechanics Conference, University of Texas, Austin, TX.

Zhang, J., & Foschi, R. O. (2004). Performance-based design and seismic reliability analysis using designed experiments and neural networks. *Probabilistic Engineering Mechanics, 19*(3), 259-267.

Chapter III

Evolutionary Seismic Design for Optimal Performance

Arzhang Alimoradi, John A. Martin & Associates, Inc., USA

Shahram Pezeshk, University of Memphis, USA

Christopher M. Foley, Marquette University, USA

Abstract

The chapter provides an overview of optimal structural design procedures for seismic performance. Structural analysis and design for earthquake effects is an evolving area of science; many design philosophies and concepts have been proposed, investigated, and practiced in the past three decades. The chapter briefly introduces some of these advancements first, as their understanding is essential in a successful application of optimal seismic design for performance. An emerging trend in seismic design for optimal performance is speculated next. Finally, a state-of-the-art application of evolutionary algorithms in probabilistic performance-based seismic design of steel moment frame buildings is described through an example. In order to follow the concepts of this chapter, the reader is assumed equipped with a basic knowledge of structural mechanics, dynamics of structures, and design optimizations.

Introduction and Background

About 10 years ago, the rupture of a major blind trust fault in southern California drifted the path of developments in the area seismic design research and practice. The Northridge earthquake of January 1994 (M_w = 6.7) and its consequences appeared to be a story of coincidental success and failure. Prior to Northridge, almost all seismic design provisions were set forward based upon the philosophy of minimization of loss of life. During Northridge, only about 57 people lost their lives in an area that is highly developed. Yet, the direct and indirect financial losses in the aftermath of Northridge were estimated in the range of $40 billion[1]. Shortly after, numerous steel moment frame buildings were discovered with a peculiar type of brittle damage in their moment connections in the form of cracking of the complete joint penetration welds that was often propagated into the columns (see Figure 1). The repair was costly but necessary because of a significant loss of lateral strength and stiffness these moment frames had experienced. The problem lead to a large research program on seismic behavior of steel moment frames in the United States (SAC Joint Venture[2]). At the same time, Pacific Earthquake Engineering Research Center (PEER)[3] was developing methodologies for design and evaluation of structures that could perform desirably and predictably under multiple scenarios of seismic attack. The extensive investigations at the PEER center, the SAC project, and other initiatives in California has led to a set of important methodologies and tools (e.g., OpenSees,[4] FEMA-350, FEMA-356, and FEMA-440) for performance-based seismic design and evaluation—many of them currently being practiced in engineering offices (American Society of Civil Engineers [ASCE], 2000; Applied Technology Council [ATC], 2005; Federal Emergency Management Agency [FEMA], 2000a). These advancements make it possible to probabilistically and reliably estimate the structural response parameters (such as interstory drifts, story shear, etc.) under various seismic design scenarios.

Implementation of the new analytical developments and design provisions into reliable and practical computational tools appears to be of essential value to the success of post-Northridge investigations. Optimal seismic design for multiple performance-levels is computationally expensive. We propose and implement a formulation for design automation of structural systems for multiple levels of seismic performance using a naturally-inspired computational agent. The proposed formulation utilizes the advancements in the areas of second-order nonlinear dynamic response evaluation and evolutionary computation. The objective is to develop a general framework, with respect to model and mathematical formulation, as well as mechanical behavior of structural systems and components, in an optimal design formulation. The principles of optimal seismic design for performance, implementation of a genetic algorithm, and an example are explained in this chapter so that they could be used, programmed, and/or practiced by engineers, students, and researchers equally. More examples and a thorough coverage of the methodology can be found in references (Alimoradi, 2004; Foley, 2002; Pezeshk, 1998).

Problem Statement

A Big Picture

How is seismic performance defined and calculated? The most logical definition of seismic performance (which is perhaps the most useful to the society) is through calculating seismic risk, that is, damage, death, and downtime (these are decision variables—DVs—in performance-based design, also referred to as 3Ds). One problem in defining seismic performance this way is that these DVs will not be readily available from structural analysis and investigation of engineering demand parameters such as interstory deformations, story shear, modal properties, and such. Fragility curves that relate engineering demand parameters to damage measures should be used for this purpose. However, seismic performance is and most probably will always remain a function of engineering demand parameters. Therefore, optimal design for seismic performance could be conveniently carried out by seismic design for optimal response.

Designing a structure to have "the best" seismic response is possible by finding an optimal set of optimization variables for the structure such that its measure of seismic performance is optimized. Structural damage is often correlated with the amount of interstory deformations (drift ratio) and so minimizing the drift ratios can enhance the seismic performance. However, reducing the deformations is also analogous to larger size members and in turn

Figure 1. Typical pre-Northridge type of damage to moment resisting frame connections: (a) Fracture at a fuzed zone, (b) column flange "divot" fracture (Adapted from FEMA, 2000b)

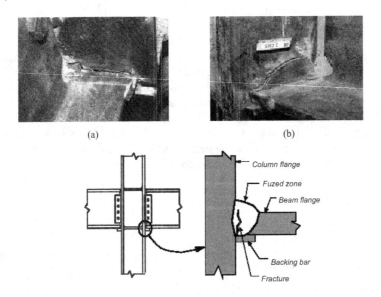

(a)   (b)

heavier and more expensive designs. The problem of optimal design for seismic performance, therefore, is of a multiobjective nature.

Optimization variables are usually taken in member-level (cross sectional geometric properties and member's elastic and inelastic properties), but the objective function may be defined in system-level (maximizing the collapse capacity [Ibarra, Medina, & Krawinkler, 2002], minimizing the ductility demand, maximizing the lateral strength, maximizing the post-yield strength, etc.) or in member-level (minimizing the plastic hinge rotations at specific elements, bounding the demand-to-capacity ratios at various elements, maximizing columns capacity against buckling, etc.). Tools to achieve the objectives of design may be obtained by combining structural analysis methods and design optimization algorithms. Several structural analysis methods for seismic performance assessment have been proposed (nonlinear static procedures such as modal pushover, and adaptive pushover analysis [Chopra & Goel, 2002; Elnashai, 2001]). However, the advancements in computational power make it possible to practically incorporate nonlinear response history analysis for sets of ground motion records in an optimization algorithm. Structural design can be conveniently eliminated in the process of advanced analysis when proper modeling of structural members (in the elastic and inelastic regime), components' behavior, and system response is considered. This is because the process of design (the change of structural members and section properties) becomes automated in the numerical trial-and-error iterations of the optimization algorithms in achieving a design objective.

Structural design is the art of creating beautiful and functional facilities that are economical. The necessity for a good design to be economical and safe implies application of trial and error over many viable configurations. When it comes to design for seismic performance, structural analysis and the process of trail and error could be time consuming. Generalization of design rules by prior experience is difficult due to the complexity of nonlinear dynamic response. Although a number of off-shelf commercial software systems offer utilities for both detailed earthquake analysis and design optimization, integration of nonlinear response history-based design and optimization is yet to receive broad applications. The preliminary version of such system was developed and tested successfully in 2004 and will be explained here (Alimoradi, 2004). Besides providing practical ease in engineering practice, such a tool can help answering the following questions:

- Are structural performance objectives, in a given geographical location, independent of each other or correlated?

- How does variation of performance in one specific level change the behavior of the structure under other performance levels?

- Does optimal performance for ductility (displacement) necessarily create viable designs for strength (forces)?

- How does the economy of designs change in the process of seismic design?

The theory that is developed to meet the aforementioned objectives is general. The application however, is illustrated on a number of benchmark steel moment frames. Various levels of detail in modeling (such as partially restrained connections) and complexity of the analysis could be considered.

Theory

A newly developed method of probabilistic performance-based seismic design for steel moment frames provides a framework for estimation of the levels of confidence on the probabilities of certain structural behavior (limit-states or performance-levels) under different earthquake scenarios (FEMA, 2000a). The computation of a design's levels of confidence is explained in this section as well as the way these levels of confidence could be enhanced.

Our optimal design problem is formulated through the following statement:

$$\min\left[W = \sum_{i=1}^{NM} A_i l_i \rho_i \right] \qquad \max\left[q^{CP} \right] \qquad \max\left[q^{IO} \right] \tag{1}$$

subject to:

$$0.60 - q_{Drift}^{Global\_CP} \leq 0.0 \tag{2}$$

$$0.60 - q_{CCF}^{Global\_CP} \leq 0.0 \tag{3}$$

$$0.40 - q_{Drift}^{Global\_IO} \leq 0.0 \tag{4}$$

$$0.40 - q_{CCF}^{Global\_IO} \leq 0.0 \tag{5}$$

in which W is the total weight of the structure; q^{IO} and q^{CP} are the confidence levels for immediate occupancy (IO) and collapse prevention (CP) performance computed through the provisions of FEMA-350, respectively. It is observed that when the total weight of the structure goes down, the confidence levels in attaining IO and CP performance may go down as well. The confidence levels are to be obtained by consideration of various structural response parameters such as the maximum over time and space of the median interstory drifts (subscript *Drift* in equation (1)) and column compression forces (subscript *CCF* in equation (1)). The minimum confidence obtained from different response quantities in one performance level controls the confidence of the design in that level.

The computation of the levels of confidence follows the formulation derived after extensive statistical analyses of nonlinear seismic response at Stanford University, and has a form similar to the more common method of LRFD (Cornell, Jalayer, Hamburger, & Foutch, 2002):

$$\lambda = \frac{\gamma \cdot \gamma_a \cdot D}{\varphi \cdot C} \tag{6}$$

which is simply the ratio of structural demand to capacity (D/C) modified for uncertainties.

It is important to notice that optimization of a structure's seismic performance at a specific level (IO or CP) could be essentially done in the following ways:

1.   Minimizing demand (through seismic isolation and energy dissipation technologies, as one alternative)

2.   Maximizing capacity (by appropriate considerations of members elastic and inelastic properties), and

3.   Reducing uncertainties (through better estimation of hazard, accurate modeling of the structure and its behavior, and using advanced analytical techniques for response estimation).

David Carroll's micro genetic algorithm (GA) is used to perform the optimization (Carroll, 2005). Genetic algorithms, developed through the pioneering work of Holland (1975) and Goldberg (1989) based on the theory of natural evolution, perform optimization using stochastic search. Adaptation, natural parent selection, and survival of the fittest in nature, when simulated on a computer, become a very powerful optimization method. Genetic algorithms work with sets of assumed solution variables (so-called population of chromosomes in GA jargon) rather than a single starting point for iterations. GAs also have stochastic mechanisms such as mutation that prevent the algorithm from focusing the search around local optima in the objective space. Beside, there is usually no need to define an objective function and its gradient explicitly as the measure of convergence can be simply defined by a fitness function. GA components and processes are as follow:

•   **Population of chromosomes:** An assumed set of optimization variables is used in the iterative processes of GAs. This set is commonly called a population of chromosomes whereas each iteration is called a generation, stemming from the processes of natural evolution in nature. A chromosome consists of a string of decimal or binary values (genes) that represent the variables in the system. Naturally, the best genes in a chromosome represent the fittest individual in a competitive environment, and are what an evolutionary algorithm is looking for. The nature's solution to enhancement of species is through the process of natural parent selection and recombination in a genetic pool.

•   **Fitness function, natural parent selection, and fitness penalties:** Fitness function is a mathematical expression that defines how fit an individual chromosome is within an environment. Fitness function should yield higher absolute values for individuals that are fitter. Once the fitness of individual chromosomes is evaluated in a generation, pairs of parents are selected based on the value of their fitness. The fitter the individual is, the higher the chances are for that individual to be selected as a parent of the next generation. Therefore, fitter parents would be able to pass on their good genetic characteristics to the forthcoming offspring generations.

•   **GA operators:** These are mechanisms such as crossover and mutation. Although different schemes are proposed and applied (such as single-point and multiple-point mutation and crossover), they all ensure diversity in a population by introducing some random variations in the solution set. Crossover is the reproduction of two mating chromosomes in order to create offspring chromosomes by sharing pieces of genetic information and is done by switching parts of the chromosomes. The location along the length of chromosome strings where switching of the pieces and genes would take place may be determined randomly. The chances of two chromosomes mating are controlled by the probability of crossover. Mutation (which usually has a low probability) changes the value of a gene by flipping a randomly selected bit of information along its length and hence creating a form of information block (commonly referred to schema) that may be located in an area of the search space not looked before.

A novel radial fitness evaluation formulation is developed for the multiobjective problem of equations (1)-(5):

$$fitness_{jk} = \frac{R_{jk}}{R_{max}} \cdot \left[\frac{1}{1 + \frac{d_{jk}^{CP} - d_{limit}^{CP}}{d_{limit}^{CP}}} \right] \cdot \left[\frac{1}{1 + \frac{d_{jk}^{IO} - d_{limit}^{IO}}{d_{limit}^{IO}}} \right] \cdot \left[\frac{1}{1 + \left| \frac{1.2 \cdot \sum M_p^{girders} - \sum M_p^{Columns}}{\sum M_p^{Columns}} \right|} \right]$$

$$R_{jk} = \sqrt{(d_{limit}^{IO} - d_{jk}^{IO})^2 + (d_{limit}^{CP} - d_{jk}^{CP})^2 + (W_{max} - W_{jk})^2}$$

$$R_{max} = \sqrt{(d_{limit}^{IO})^2 + (d_{limit}^{CP})^2 + (W_{max})^2} \qquad (7)$$

in which, an optimal design is obtained by maximizing the distance of the designs (chromosomes) in a GA population from a worst case design scenario in the objective space. The radial ordinates of the worst case scenario are identified by limit subscripts. Interstory drift is d and the total weight of the design is W. Mp represents the plastic moment capacity of the girders and columns.

Several example frames are investigated in this study using the second-order inelastic response history analysis of two sets of ground motion records representing IO and CP level events. Only a typical portal frame will be presented here.

Implementations and the Software System

The example frame is shown in Figure 2. The analytical model for this frame consists of lumped masses at the beam-to-column nodes; pinned supports; and partially to fully restrained connections consisting of zero-length springs. Input ground motions are applied in the form of acceleration time histories at the base. The dimensions for the frame are also provided in the figure. Rigid offsets were considered. The material properties for the members in the frame are: the elastic modulus $E = 29000$ ksi and $F_y = 50$ ksi. Lumped mass magnitudes are: 0.085 $k - s^2 / in$.

The genetic algorithm is carried out with the design represented using two optimization variables: The first represents the cross section of both columns and the second represents the beam's cross section. The design variables are taken from a set of 236 possible cross sections of standard AISC beam and column shapes. No special grouping according to beams and columns was done. Structural designs are represented by binary chromosomes that decode to the configuration and cross sections in the frames. In general, a population size of 30 chromosomes with probability of crossover of 60% and probability of mutation of 3.5% were found to produce stable optimization trajectories toward an optimal design solution. Chromosomes of parents reproduce two children subsequent in generations with 4.0% probability of creep. The new generations were reproduced until there is no improvement in the maximum fitness over the past 20 generations or the program stops at the maximum number of generations of 300.

The portal frame example considered two fully restrained (FR) beam-end connection configurations as part of special moment resisting frames (SMRFs), and one partially restrained (PR) connection configuration within an ordinary moment resisting frame (OMRF). Additional details regarding the frame configurations denoted using Analysis Case Numbers (ACNs) are given below:

1.   **ACN-1:** FR connections with strong-column weak-beam (SCWB) criterion;

2.   **ACN-2:** FR connections with the omission of SCWB criterion;

3.   **ACN-3:** PR moment connections with SCWB criterion based on the connection's plastic moment capacity. The connection's hysteretic behavior includes inelastic unloading with gap or pinching. The parameters needed to define this connection behavior are provided in the references (Prakash, Powell, & Campbell, 1993).

The initial stiffness and yield moment capacities for the connections are:

$$K_c = 10EI/L \tag{8}$$

and

$$M_{cy}^+ = M_{cy}^- = 0.66M_{pb} \tag{9}$$

The hardening stiffness parameter is defined using a connection rotation of 0.03 radians and a connection moment capacity equal to $1.4M_{cy}$, L is the length of the beam; M_{pb} is the plastic moment of the beam; and I is the second moment of area for the beam.

The example frame will be subject to two sets of SAC ground motion records, for the IO and CP level design, shown in Figure 2.

Fitness trajectories obtained during execution of the GA are shown in Figure 3. No significant epistatic behavior was observed in the fitness trajectories (each run of the GA resulted in consistent convergence to higher fitness without stagnation). Therefore, the GA parameters appear to have been appropriately chosen for this problem. The average fitness trajectories illustrate that the genetic algorithm is indeed performing search of the design space as exhibited by the fluctuation in the average fitness of the populations with increasing generations.

Figure 2. Portal frame analytical model used in the design example and the input ground motion records (Adapted from Alimoradi, 2004)

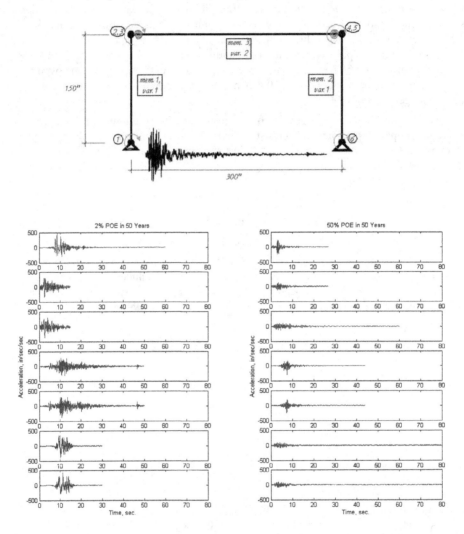

Figures 4a and 4b show two slices through objective space (corresponding to objective pairs) for ACN-1. Since 30 chromosomes compete in over 1,000 generations, these figures represent thousands of unique designs. As expected, when the total weight of the structural system decreases, the median interstory drift angle (ISDA) demands increase for immediate occupancy (IO) and collapse prevention (CP). These figures illustrate the competing nature of weight and median drift objectives. From Figure 4c, the median estimates of

Figure 3. Fitness trajectories for portal frame with various connections (Adapted from Alimoradi, 2004)

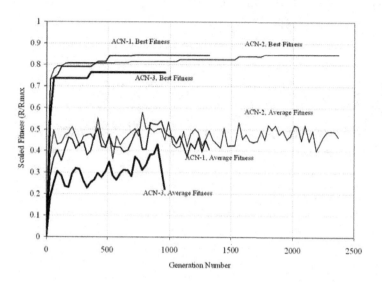

ISDA demand under CP and IO input records appear to be correlated at lower interstory drift ratios (e.g., 1% for IO and 2.5% for CP). Higher dispersions were usually observed at higher ranges of nonlinear response. This indicates that for a given structural configuration and IO and CP–level ground motion records, a certain performance level may be controlling the multiple-level design.

A radial fitness formulation was used in the design of the frame (Alimoradi, 2004). The impact of using this method is shown in Figures 4a and 4b through the distribution of candidate designs along the Pareto front. The Pareto front can be used by the owner and engineer to understand how confidence in meeting IO and CP performance affects the volume (cost) of the structural system. Obviously, the portal frame is highly simplistic. However, the concept of using multiple-objective optimization algorithms in structural engineering design is highly beneficial.

To understand the difference between three designs (ACN-1 through ACN-3) three frames were chosen from the Pareto front for these frame design runs that have roughly equal weight. Pertinent details from these three designs are shown in Table 1.

In this table, M_{pb} / M_p is the ratio of plastic moment capacity of columns to that of the beam, $ISDA^{PO}_{median}$ is the median interstory drift angle demand at a performance objective (PO) – CP or IO, q^{PO} is the confidence level in meeting a defined performance objective, $d_{residual}$ is the permanent interstory drift at the end of a component of Tabas (1978) M7.8 ground motion record, f_1 is the first mode natural frequency, ξ_1 is the first mode equivalent damping ratio, and G is the generation number.

Figure 4. Distribution of designs in objective space for portal frame with ACN-1: (a) Collapse prevention performance and weight; (b) immediate occupancy performance and weight; (c) correlation between IO and CP performance objectives

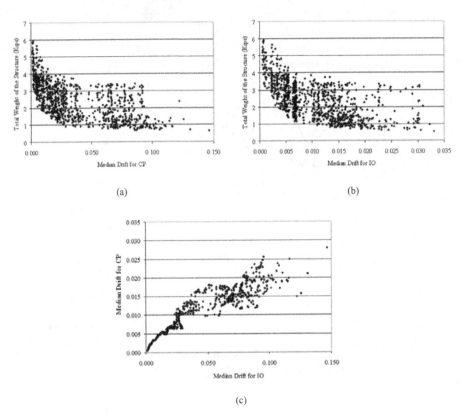

(a)   (b)

(c)

These three designs are different with respect to the ratio of plastic moment capacity of their columns to their girder, but interestingly all of them have very close first-mode natural frequencies, which perhaps suggests a strong correlation between displacement responses of such single-mode dominant structures and their first-mode natural frequency. ACN-3 has a slightly shorter first mode frequency that may be attributed to its PR connections. The low IO confidence level of ACN-3 might also be the result of a lower lateral resistance under IO level input motion that can cause lateral deformations to approach the IO capacity faster than CP. Longer natural periods may be advantageous under severe input ground motions provided IO performance can be met with the added interstory drift that is likely to occur.

To evaluate nonlinear static lateral behavior of the near-equal weight designs (Table 1), pushover curves (displacement control) were generated. These curves are shown in Figure 5. A lateral force equivalent to 0.10g PGA multiplied by the total mass at roof level was applied horizontally. Gravity loads (companion actions) corresponding to the mass magni-

Table 1. GA-generated designs and performance information for the three portal frame configurations considered

Parameter (1)	ACN-1 (2)	ACN-2 (3)	ACN-3 (4)
Volume (in^3)	14,010	13,830	13,890
Column Section	W27×102	W24×55	W27×84
Beam Section	W21×57	W21×101	W21×73
$\dfrac{M_{po}}{M_{pb}}$	2.36	0.53	1.42
$ISDA^{CP}_{median}$ (%)	2.06%	2.04%	3.71%
q^{CP} (%)	99.99	99.99	96.78
$ISDA^{IO}_{median}$ (%)	0.70	0.69	0.78
q^{IO}(%)	99.97	99.97	42.10
$d_{residual}$ (%)†	0.12	1.09	0.08
$\dfrac{R}{R_{max}}$	76.3 % $G = 44$	76.6% $G = 79$	76.4% $G = 32$
f_1 (Hz)	2.64	2.62	2.40
ξ_1 (%)	4.64	4.61	2.23

Note: † – Residual ISDA demand at the end of Tabas (1978) ground motion record

tude were present. The frame was pushed to a collapse mechanism, or 10 inches of lateral displacement. Since strong column-weak beam constraint is active in cases ACN-1 and ACN-3, the final designs exhibit beam-type collapse mechanisms with P-Δ effect after a drift ratio of about 1.3%—the drift angle capacity of low-rise ordinary moment frames at IO performance level is 1% according to SAC (2000; FEMA, 2000a). The lack of available redundancy in the simple portal framework when strong column-weak beam (SCWB) criteria are not considered is shown in the response as well. The pushover curve for ACN-2 has very limited ductility after the initial plastic hinge formation.

Study of response of near-equal weight designs with various connection types and properties in Table 2 reveals the intuitive assumption that stiff and strong connections can reduce lateral deformations in seismic response of regular frames is valid. The reduction in median ISDA (columns 7 and 8 in the table) can be seen as one moves down Table 2 from ACN-2 to ACN-4.

Similar behavior is seen with ACN-5 to ACN-7. By using stiff and strong inelastic PR connections (ACN-7) as opposed to flexible PR pinching connections (ACN-6), CP response is reduced by 72%, while IO response is down by 36%. This is a significant improvement for virtually the same amount of structural material used. The SCWB constraint seems to create heavier designs for the same level of performance and as a result, near-equal weight

Figure 5. Lateral force-displacement behavior (pushover response) of equal weight designs

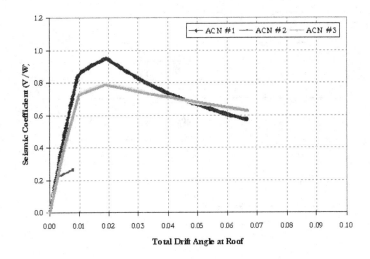

structures without SCWB criterion tend to have a lower displacement response. This can be seen by comparing: ACN-2 with ACN-5; and ACN-3 with ACN-6. Comparison of ACN-4 and ACN-7 reveals that when flexible connections are utilized, the SCWB criteria have an unpredictable impact on the volume of the optimal design. It should be noted that lower displacement response, in general, might not be taken exclusively as a good measure of seismic performance, as it may be associated with undesirable patterns of plastic hinge formation in the structure and loss of postyield stability.

The aforementioned interpretations of the PR frames' behavior may also be extended to the response of FR special moment frames in ACN-1 and ACN-8. Here, omission of the SCWB criterion (ACN-8) reduces the CP and IO responses by as much as 32% and 2%, respectively. This response reduction in FR special moment frames is not as significant as in PR ordinary moment frames, which can be attributed to a higher level of integrity of FR special moment frames compared to PR ordinary moment frames.

ACN-9 and ACN-10 utilize the connection's yield moment capacity to evaluate the SCWB criteria rather than the beam's plastic moment capacity as done in the preceding PR analysis cases. The optimal structural design corresponding to ACN-10 lies somewhere between systems with and without SCWB criterion. From Table 2, the responses of Cases 9 and 10 are slightly better than Cases 2 and 3 (with SCWB) and worse than Cases 5 and 6 (No SCWB).

Figure 6a illustrates the variation in normalized interstory drift demand (ISDA) with variation in normalized connection elastic stiffness. Normalization is done with respect

Figure 6. Variation of interstory drift with (a) connection stiffness, and (b) strength

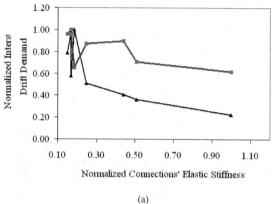

(a)

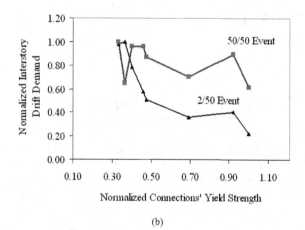

(b)

to largest value observed in the analysis set. It is obvious that stiffer moment connections can reduce the lateral deformation response drastically, although IO response seems to be less sensitive to the connections stiffness. Figure 6b illustrates the variation in normalized ISDA with variation in normalized connection yield strength. As in Figure 6a, it appears that 50/50 response for the optimal designs does not show the same consistent reduction as 2/50 response. The ISDA response did not appear to be sensitive to the connections' hardening ratio in this study.

Table 2. Summary of the analyses results from the study of optimal response sensitivity to connections' behavior (α is the hardening ratio of the connection)

ACN (1)	Vol. (in^3) (2)	Nonlinear Connection Parameters				Median ISDA (%)		R/R_{max} (9)	First Mode	
		$K_c \times 10^6$ (lb/in.) (3)	α (%) (4)	$M_{cy} \times 10^3$ (lb-in.) (5)	Gap (6)	CP (7)	IO (8)		f_1 (Hz) (10)	ξ_1 (%) (11)
1	15570	463.5	0.050	5.350	NO	2.50	0.64	0.736	2.51	4.46
2	16470	1.745	0.979	4.670	YES	4.60	0.78	0.446	2.57	3.37
3	16260	1.547	5.575	5.676	YES	3.70	0.75	0.724	2.38	3.23
4	15600	5.307	2.624	9.800	NO	1.70	0.55	0.736	2.86	4.42
5	16230	2.591	0.948	6.732	YES	2.40	0.68	0.725	2.34	3.63
6	16230	4.601	4.171	13.04	YES	1.90	0.70	0.725	2.70	4.21
7	16230	10.47	1.887	14.15	NO	1.05	0.48	0.725	3.51	5.39
8	15390	773.3	0.015	8.600	NO	1.70	0.63	0.740	2.76	4.80
9	16170	1.929	0.977	5.148	YES	4.70	0.51	0.725	2.94	3.47
10	16380	1.769	5.552	6.468	YES	2.71	0.75	0.722	2.54	3.38

Results and Conclusion

The multiobjective problem of equation (1), when solved, will result decision making curves for the cost of construction and levels of confidence on meeting particular limit states under various earthquake scenarios. A sample of such decision making curves in 3D is shown in Figure 7. It is apparent from this curves that the structural performance under one performance-objective, in general, may be a function of other performance objectives

It was observed that the median drift demand and the structural performance under CP could be enhanced by improving the behavior under IO (and visa versa), whereas improving the behavior for either of the structural performance levels usually results in heavier (and more expensive) designs. Practical considerations in developing a Pareto decision-making surface were discussed. The algorithm is capable of presenting designs with minimum weight that satisfy predefined ranges of preferred seismic performance. A full nonlinear time-history analytical engine embedded in the program executes the designs using sets of input ground motion records for maximum accuracy. The computational platform is flexible, relatively fast, and easy to use for practical purposes.

Figure 7. The decision surface obtained for a three-story moment frame through optimizing equation (1)

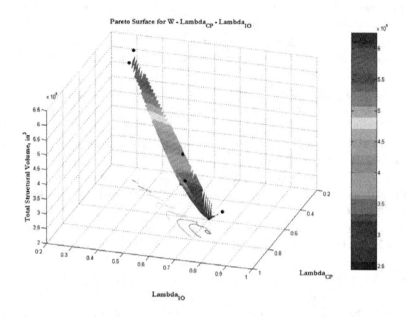

References

Alimoradi, A. (2004). *Probabilistic performance-based seismic design automation of non-linear steel structures using genetic algorithms.* Unpublished doctoral dissertation, The University of Memphis, TN.

American Society of Civil Engineers. (2000). *Pre-standard and commentary for the seismic rehabilitation of buildings* (FEMA-356). Washington, DC: Federal Emergency Management Agency.

Applied Technology Council. (2005). *Improvements of nonlinear static seismic analysis procedures* (FEMA-440 preprint ed.). Redwood City, CA.

Carroll, D. L. (2005). FORTRAN genetic algorithm (GA) driver [Computer software]. Retrieved November 2005, from http://cuaerospace.com/carroll/ga.html

Chopra, A. K., & Goel, R. K. (2002). A modal pushover analysis procedure for estimating seismic demands for buildings. *Earthquake Engineering and Structural Dynamics, 31,* 561–582.

Cornell, A. C., Jalayer, F., Hamburger, R. O., & Foutch, D. A. (2002). Probabilistic basis for 2000 SAC Federal Emergency Management Agency steel moment frame guidelines. *Journal of Structural Engineering, 128*(4), 526–533.

Elnashai, A. S. (2001). Advanced inelastic static (pushover) analysis for earthquake applications. *Structural Engineering and Mechanics, 12*(1), 51–69.

Federal Emergency Management Agency. (2000a). *Recommended seismic design criteria for new steel moment-frame buildings* (FEMA-350). Washington, DC: Author.

Federal Emergency Management Agency. (2000b). *State of the art report on systems performance of steel moment frames subject to earthquake ground shaking* (FEMA-355C). Washington, DC: Author.

Foley, C. M. (2002). Optimized performance-based design for buildings. In S. A. Burns (Ed.), *Recent advances in optimal structural design* (pp. 169–240). American Society of Civil Engineers.

Goldberg, D. (1989). *Genetic algorithms in search, optimization, and machine learning.* Addison-Wesley Professional.

Holland, J. H. (1975). *Adaptation in natural and artificial systems: An introductory analysis with applications to biology, control, and artificial intelligence.* University of Michigan Press.

Ibarra, L., Medina, R., & Krawinkler, H. (2002). Collapse assessment of deteriorating SDOF systems. In *Proceedings of the 12th European Conference on Earthquake Engineering.*

McKenna, F., Fenves, G. L., Scott, M. H., & Jeremić, B. (2000). *Open system for earthquake engineering simulation.* Available online from http://opensees.berkeley.edu

Pezeshk, S. (1998). Design of framed structures: An integrated nonlinear analysis and optimal minimum weight design. *International Journal for Numerical Methods in Engineering, 41,* 459–471.

Prakash, V., Powell, G. H., & Campbell, S. (1993). *DRAIN-2DX base program description and user guide: Version 1.10* (Rep. No. UCB/SEMM-93/17-18). Berkeley: University of California - Berkeley.

SAC. (2000). *Suites of earthquake ground motions for analysis of steel moment frame structures.* Sacramento, CA: Task 5.4.1., SAC Joint Venture.

Endnotes

[1] Data courtesy of the USGS Pasadena office (http://Pasadena.wr.usgs.gov/north/) and Southern California Earthquake Center (http://www.data.scec.org/chrono_index/northrEquationhtml), accessed on November 19, 2005.

[2] More information at http://www.sacsteel.org; accessed on November 19, 2005.

[3] The Pacific Earthquake Engineering Research Center at the University of California - Berkeley, http://peer.berkeley.edu/), accessed on November 19, 2005.

[4] More information at: http://opensees.berkeley.edu/ accessed on November 19, 2005; please also see McKenna et al. (2000).

Chapter IV

Optimal Reliability-Based Design Using Support Vector Machines and Artificial Life Algorithms

Jorge E. Hurtado, Universidad Nacional de Colombia, Colombia

Abstract

Reliability-based optimization is considered by many authors as the most rigorous approach to structural design, because the search for the optimal solution is performed with consideration of the uncertainties present in the structural and load variables. The practical application of this idea, however, is hindered by the computational difficulties associated to the minimisation of cost functions with probabilistic constraints involving the computation of very small probabilities computed over implicit threshold functions, that is, those given by numerical models such as finite elements. In this chapter, a procedure intended to perform this task with a minimal amount of calls of the finite element code is proposed. It is based on the combination of a computational learning method (the support vector machines) and an artificial life technique (particle swarm optimisation). The former is selected because of its information encoding properties as well as for its elitist procedures that complement those of the a-life optimisation method. The later has been chosen du to its advantages over classical genetic algorithms. The practical application of the procedure is demonstrated with earthquake engineering examples.

Introduction

Since its initial developments in the middle of the last century, earthquake engineering incorporated the modelling of uncertainties in its procedures. This was due to the fact that, much more than gravity loads, earthquake loads are rather unpredictable. In addition, account must be taken of the randomness of shaking duration, ground amplification and structural parameters to have an adequate picture of the probabilistic behaviour of structures subject to strong earthquakes. The introduction of performance-based design represents an improvement over classical design procedures which use equivalent static forces and reduction factors that take into account nonlinear energy dissipation. In practical analysis, performance-based design consists in doing a nonlinear static analysis of a predesigned structure with increasing lateral loads in order to determine a unique load-displacement curve (the so-called capacity spectrum) and comparing it to the plot of spectral acceleration vs. spectral displacement (the so-called composite spectrum) (Reinhorn, 1997). However, the question about the seismic safety of the structure remains unsolved while uncertainties are not incorporated into this kind of design process.

Despite the developments of probabilistic modelling, specifically those of random vibration analysis and structural reliability, practical seismic design as prescribed in design codes has incorporated the uncertainties only in load modelling. In fact, codes declare that the design is performed for a maximum acceleration with a certain probability of exceeding a given threshold, but nothing is required about the probability of structural failure, i.e., the probability of surpassing one or more undesirable performance limits. This is defined as:

$$P_f = P[g(x) \le 0] = \int_{F=\{x:g(x)\le 0\}} p_X(x)dx \tag{1}$$

where X is the vector of random variables whose deterministic realisation is represented by x, F is the failure region determined by mapping the surpassing a dangerous threshold (such as a critical floor acceleration or drift) onto the x-space, $P_X(x)$ is the joint probability density function of the random vector and $g(x)$ is a function describing the performance of the structure with respect to the critical threshold, such that the contour $g(x) = 0$ divides the x-space into two subdomains: safe and failure. The former is defined as $S=\{x: g(x)>0\}$ and the latter as indicated in equation (1). The reliability is defined as the complementary probability $R = 1 - P_f$.

The difficulty to accomplish a full probabilistic analysis incorporating all uncertainties explains why only the maximum acceleration uncertainty is included in code procedures. In fact, to assess the failure probability of a certain structure under earthquake loads in an accurate manner it is necessary to perform a Monte Carlo simulation, implying the generation of a large number of random accelerograms corresponding to the spectral model of the site and then to analyse an equal number of structural models differing from each other in the structural parameters, which are randomly generated from their respective probability distributions. This is evidently an expensive computational task.

While the main goal of structural reliability analysis is to estimate failure probabilities for critical thresholds defined for serviceability or collapse conditions (Melchers, 1999), structural optimisation is oriented towards the assessment of a set of design variables that define a structural model exhibiting a minimal cost while satisfying performance constraints (Haftka, Gurdal, & Kamat, 1990). The combination of these two kinds of analysis leads to the field of reliability-based optimization, which aims at finding a structural solution satisfying two normally conflicting purposes: minimal cost and maximum reliability (Frangopol, 1995). Evidently, this combination overcomes the outstanding deficiencies of both fields of research, which are as follows:

1. Structural optimization is normally performed without considering the uncertainties of structural loads and material parameters. In some cases, however, the uncertainties are so high that the result of an optimization conducted in a deterministic way is a matter of concern. This if, for instance, the case in designing structures liable to be excited by strong earthquakes.

2. Structural reliability methods have reached an important level of maturity. However, their incorporation into structural design practice is very modest (Sexmith, 1999). It is argued that this is partly due to educational reasons. However, it is probable that as long as reliability concepts are not directly incorporated into design practices and codes the structural design will continue to be performed with safety factors and other simple techniques that hide uncertainties "under the rug" (Elishakoff, 2005).

These ideas motivated Rosenblueth and Mendoza (1971) to propose the introduction of structural reliability concepts and tools into the design process. Nowadays, there are several formulations of the reliability-based optimisation problem, the selection among which depends on the purpose at hand. In this chapter we are interested in the solution of the following problem:

Π:Minimise $C(y)$

subject to $P[g_i(x, y) \leq 0] \leq Q, \quad i = 1, 2, \ldots$ (2)

where $C(y)$ is the initial cost of the structure, $P[A]$ is the probability of random event A, Q is its given upper bound and y is a vector of the design variables, i.e., those which are the target of the design as they determine its cost (usually member sizes). This is a practical approach to considering uncertainties in design practice in such a way that the random variables x are different from the design variables y. Other formulations are also possible. For instance, that the mean of the random variables are the design variables (Frangopol, 1995), or that the expected cost associated to different failure modes are involved in the cost function (Gasser & Schuëller, 1997; Royset & Polak, 2004).

The solution of Problem $\prod$ faces the following difficulties: (a) Functions $g_i(x, y)$ are normally given in implicit form; (b) The failure probabilities are very low and their estimation by some standard procedures such as the first-order reliability method (FORM) may be inaccurate in many cases; (c) There is a need of solving an entire reliability problem in each step of the optimisation procedure.

In this chapter it is shown that the use of artificial intelligence techniques can be of avail for facing problems of this kind. Specifically, the proposed approach employs two methods, one developed in the area of computational learning (the support vector machines) and the other in the field of artificial life algorithms (the particle swarm optimisation), due to reasons that are exposed in the sequel.

Support Vector Machines

This is a computational (or statistical) learning technique for solving three main tasks (Vapnik, 1998): pattern recognition, function approximation and probability density estimation. Despite some similarities to neural networks, the method does not have a biological inspiration.

Two tasks involved in the subject of present chapter can be identified as pattern recognition problems, namely the separation of safe and failure domains in the reliability context and the discrimination between feasible and infeasible domains in optimization analysis. In both cases it is necessary to identify an implicit separating function. This is done with some training samples, whose class is evaluated with a numerical solver of the structural model. Among the numerous pattern classification methods (see Duda & Hart, 2002), support vector machines have been selected because they provide explicit equations for ancillary margin functions that constitute a search domain for new samples. This allows refining the current estimate of the discriminating function. For this reason they have proved to be very useful in reliability analysis (Hurtado, 2004a, 2004b; Hurtado & Álvarez, 2003).

Figure 1 shows the basic elements of a support vector classifier. White and black samples belong to classes A and B, respectively. The method yields an estimate of the unknown separating function (solid line) together with two ancillary margin functions (dashed lines) passing through the samples lying the closest to the discriminating function. These special samples are the support vectors.

The approximation of the discrimination function reads:

$$\hat{g}(x) = \sum_{i=1}^{S} \alpha_i c_i K(x, x_i) - b$$

(3)

where x_i are the so-called support vectors, $\alpha_i > 0$ are Lagrange multipliers found in a quadratic optimisation procedure, $K(x, x_i)$ is a kernel function and b is a threshold.

The main features of the support vector method for classification are as follows:

Figure 1. Elements of a support vector machine for classification

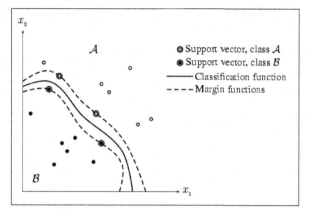

1. According to a theorem of pattern recognition, the probability of perfect separation of samples in two classes with a hyperplane grows with the number of dimensions. The SVM method exploits this fact by means of the kernel function in equation (3), because it can be implicitly expressed as a linear product of vector in a space of virtually infinite dimensions.

2. From the information viewpoint, the SVM performs an important information compression (or encoding) properties, represented by the small number of support vectors, in comparison with the samples needed for training ($s < < n$).

3. The optimisation mentioned above consists in maximising the margin shown in Figure 1. The support vectors are the only samples with Lagrange multipliers greater than zero. Thus, the method is not oriented by the probability structure of the two populations (as in Bayesian methods) but by the samples lying in the boundary between them. This allows the calculation of the discriminating functions with a small number of samples located in the vicinities of the discrimination function, without regard to their probabilistic structure

The latter feature suggests a simple sequential approach for updating the classifier: If a new random sample lies in the margin band, it is accepted for updating the margin; otherwise it is discarded. The condition for being inside the margin is:

$$\left| \sum_{i=1}^{s} \alpha_i c_i K(x, x_i) - b \right| \leq 1$$

(4)

Accordingly, the importance of the explicit margin functions is that they allow checking if a sample is worth testing, that is, if it should be evaluated with the finite element solver either

Figure 2. Instances of convergence of support vector machine to a classification function Top left: First approximation with two samples Top right: New discrimination function updated with samples lying inside the margin in the previous stage. Bottom left and right: Further refinements obtained in like manner.

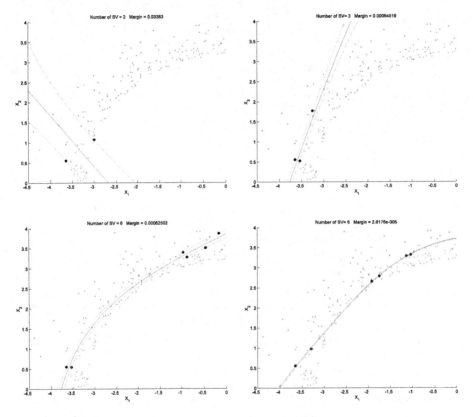

for updating the implicit limit state function or an optimisation constraint. As an illustration, Figure 2 shows instances of the approximation of function:

$$g(x) = 4 - x_2 - 0.16(x_1 - 1)^2.$$

In the first stage the algorithm selects one sample in each class from the pool shown in the Figure and fits a first SVM trial function. Then in each step selects another sample lying inside the margin band an updates the classifier. As the approximation progresses the margin band becomes narrower.

Support vector machines perform similarly to neural networks in both classification and regression tasks needed in structural engineering (see Hurtado, 2004b). However, the former method is preferred herein because of the advantages provided by the availability of special samples (the support vectors), which concentrate all the information about the two classes.

Particle Swarm Optimisation

Since the first proposal for developing optimization algorithms with inspiration in biological processes in the 1960s, there has been a large amount of research on this area (see Forbes, 2004). This has conducted to the consolidation of at least the following families of optimization methods: genetic algorithms (Goldberg, 1989), evolutionary strategies (Bäck, 1996) and artificial life methods. While the first two imitate the reproduction and evolution of species, the last group finds its inspiration in the social life of species such as birds, bees or ants in which a kind of distributed intelligence is found. Among them are the ant colony optimization (Dorigo & Stützle, 2004) and the particle swarm optimization (Kennedy & Eberhart, 2001). In the former case, the algorithm imitates the back-and-forth trajectories of ants that allow them finding minimal length routes. This makes this algorithm suitable for optimization problems such as task assignation, routing, scheduling, and so forth.

The second case is more interesting for our purposes. The particle swarm optimisation algorithm imitates the associative motion of flocks of birds or schools of fishes, which proceed

Figure 3. Example function for particle swarm optimisation

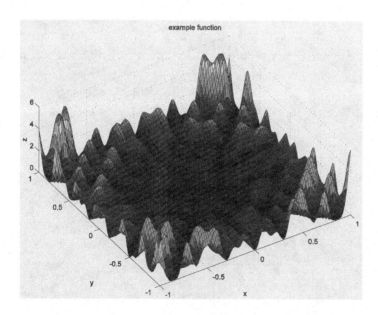

according to a feedback communication of the findings of individuals in such a way that the cohesion of the group is maintained. The algorithms make the individuals to follow those of them that found the best results in the history of their survey.

Particle swarm optimisation is especially useful for nonlinear problems involving many local minima or multi-objective programming, both of which are common in structural engineering. As an illustration, Figure 3 is a plot of function $z = (x\ sin(20y)+y\ sin(20x))^2$ $cosh(sin(10x)x)+(x\ cos(10y)\ y\ sin(10\ x))^2\ cosh(cos(20\ y)\ y)$, whose global minimum is at $(0,0)$. It can be noted that it has multiple local minima, so that the application of conventional gradient procedures is not granted to lead to the global solution. On the contrary, PSO easily finds the solution after some iterations, as illustrated by Figure 4.

Figure 4. Three instances of convergence of particle swarm optimization. Top panels: initial population (left) and population after 5 iterations (right) appearing over the contours of the function. Bottom panel: population after 10 iterations. Notice the clustering about the global minimum located at $x = 0$, $y = 0$.

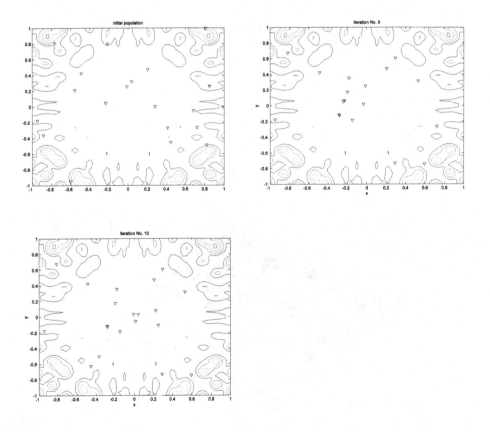

Table 1. Comparison of performance of genetic algorithms (GA) and particle swarm optimisation (PSO) for Rosenbrock's function

Number of iterations	GA	PSO
20	(0.55, 0.31)	(0.90,0.79)
40	(0.68,0.46)	(0.77,0.61)
60	(0.95,0.90)	(0.92,0.82)
80	(0.36,0.12)	(0.98,0.95)
100	(0.56,0.31)	(0.92,0.89)

Particle swarm optimisation has been selected as the optimisation method in this chapter because it offers distinct advantages over classical genetic algorithms (Goldberg, 1989). In order to illustrate this, let us consider the so-called Rosenbrock's function, given by:

$$f = \sum_{i=1}^{n} 100(x_{i+1} - x_i^2)^2 + (1 - x_i)^2$$

for two dimensions ($n = 2$) in the rectangular domain $(-2, 2) \times (-2, 2)$. This function is characterised by a wide and flat valley that makes very difficult to find the minimum, which is located at $(1, 1)$. Table 1 shows the results given by genetic algorithms (GA) and PSO. It can be seen that, as the number of iterations increases, the results given by the later method are better and more stable than those of the former.

Earthquake Engineering Applications

After the above brief summaries of the pattern recognition and optimisation methods used in the research reported in this chapter, it is important to describe how their combination can be of avail for solving reliability, optimisation and reliability-based optimisation problems in earthquake engineering.

To be specific, the advantages of this combination will be illustrated for the following cases:

1.   Approximating an implicit discriminating function, such as those arising in structural optimisation or reliability analysis. In fact, in the former case constraints formulated in terms of displacements or stresses are implicit, because they are only known through a finite element (or similar) solution of the model. On the other hand, in reliability analysis the limit state function for large structures is normally formulated as a critical threshold minus a response whose value can only be obtained with a numerical code.

Figure 5. Generating training samples for a support vector machine with particle swarm optimization. From top to bottom: Notice that the samples sequentially move towards the target limit state function. It them becomes possible to approximate it with a SVM.

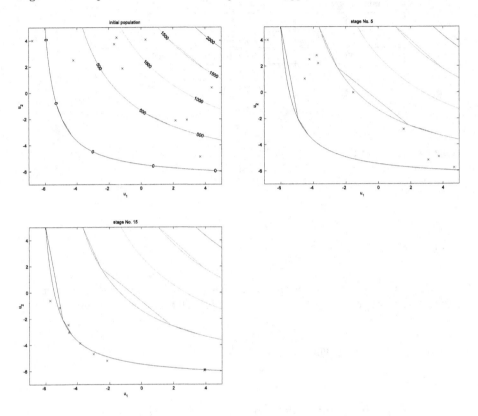

2.   Reliability-based optimisation with active probabilistic constraints of a basic seismic design calculation. This is a specific case of Problem Π described above, in which the solution is located on those constraints. Such a situation is common case in reliability-based optimisation, because the reduction of construction costs implies increasing the failure probability and, therefore, the solution is to be found on the constraint threshold.

3.   Reliability-based optimisation of a nonlinear building model using random vibration analysis. This is also an instance of Problem Π. However, instead of working with samples of the random variables, the uncertainties are managed with methods of nonlinear random vibration analysis.

Problem No. 1: Sample generation for approximating an implicit discrimination function as an optimisation problem.

Neural networks, support vector machines and other soft-computing methods are very suitable to approximating classification functions due to the fact that they are made up with several functions which are active in small domains. If the functions to approximate are implicit, the performance of these soft computing techniques is the better, the closer the samples are to the separating boundary. In this section it is shown that this task can be performed with particle swarm optimisation by formulating the sample preparation as an optimisation problem.

Consider, for instance, the problem of approximating an implicit limit state function is a structural reliability problem, especially in the zone where there is a higher probability mass in the failure domain. Assume first that the problem is formulated in a space where the random variables have been either normalised by subtracting the mean and dividing by the standard deviation. Alternatively, the variables can be transformed to standard Normal random variables by means of Rosenblatt, Nataf or Polynomial Chaos transformations (see Hurtado, 2004b). In this u-space the important zone is located about the so-called design point, that is, the point of the limit state function lying the closest to the origin. For solving this problem use is commonly made of the popular FORM (first-order reliability method), consisting in solving the following optimisation problem:

FORM: Minimise $\beta = u^T u$

$\quad$ subject to: $G(u) = 0$

where $G(u)$ is the limit state function in the u-space. The failure probability is estimated as the normal distribution evaluated at $-\beta$. however, there are several problems in which there are multiple design points, because the curvature of the limit state function makes it close to a hypersphere portion. Also, if the probability density function is very flat in the failure region, the failure probability can no longer be tied to a single point. Besides, the function to approximate can be highly nonlinear, so that the assessment of the failure probability on the basis of the reliability index β can be rather inaccurate. In such situations it is preferable to approximate the function in an ample region and to perform a Monte Carlo simulation using this approximation instead of the structural finite element code. In order to do this, it is suggested to generate the training samples for a neural network or a support vector machine by solving the following unconstrained optimisation problem:

Minimise $u^T u + \alpha\, G^2(u)$ $\hspace{6cm}$ (5)

where α is a penalty factor. The combined minimisation of the distance to origin and the square of the limit state function means that, at a difference to FORM, no single global solution is sought. Figure 5 illustrates the convergence of the particle swarm to the function (Kuschel, Rackwitz, & Pieracci, 1999):

$$G(u) = (u_1\sigma_1 + \mu_1)(u_2\sigma_2 + \mu_2) - 146.14$$

with $\sigma_1 = 11,710$, $\mu_1 = 78,064$, $\sigma_2 = 0.00156$, $\mu_2 = 0.0104$. This function has been selected because it shows at least three design points. The samples appearing in the last panel of the Figure can be used to train a support vector machine to approximate the function. Also, a more sophisticated algorithm can be devised to reduce the number of iterations by exploiting the search space reduction properties of the margin band.

Problem No. 2: Structural optimisation with an active probability constraint.

Consider now Problem $\prod$ formulated above. The rationale of the proposed procedure for solving it is as follows: Since all random variables are all contained in vector x, it is possible to associate the limit failure probability Q to a set of support vectors that, due to the compression properties of the SVM method, represent in some sense the probabilistic constraint. To this end generate a large set of N random numbers after the joint probability density function of x in a normalised space (as explained in the preceding problem) and assign to the fraction QN of them the failure label and the safe one to the rest. Then fit a support vector machine. (Notice that to this purpose the large set of numbers is not classified, i.e., it is not processed with the finite element solver). The support vectors $\tilde{u}$ and they score (defined as the difference between the number of support vectors in the safe and failure classes) can then be used for testing the feasible or infeasible solutions in y-space (the structural model space). This is done by evaluating the limit state function $g(u, y)$ with each trial solution y and all the support vectors. If the trial model displays more positive or negative values of the limit state function for the support vectors in the reference problem in u-space, the model can be labelled as feasible or infeasible, respectively. Notice that this corresponds to substituting the original problem by:

$\hat{\prod}$:Minimise $C(y)$

subject to
$$g_i(\tilde{u}_i, y) \leq 0, \quad i = 1,2,...S^-$$
$$g_i(\tilde{u}_i, y) \geq 0, \quad i = 1,2,...S^+$$

where $S^-(S^+)$ is the number of negative (positive) support vectors in u-space. Finally, a support vector machine in the y-space is trained with feasible are infeasible samples thus labelled. The classifier can be updated in a sequential manner with new samples generated with particle swarm optimisation, whose fitness function must include a penalisation term for trespassing the current SVM margin.

To be specific, the PSO is called to solve the following problem, formulated according to standard theories of penalty techniques (see, e.g., Chong & Zak, 2001):

$$\text{Minimise } J(y) = C(y) + \eta \sum_{i=1}^{2} \theta(r_i(y)) r_i(y)^{\gamma_i} \tag{7}$$

where $r_1(y) - 1$, $r_2(y) = -(d(y) + 1)$, $d(y)$ is the distance of sample y to the current separating SVM function; $\theta(r_i(y))$ equals 30, but it equals 10, 2 or 1 if its argument is respectively less than or equal to 1, 0.1 or 0.001, for both i = 1, 2. Finally, for both $i = 1, 2$ the exponent $\gamma_i = 1$ if $r_i(y)$ is less than one; otherwise it is equal to 2. On the other hand, $\eta = k\sqrt{k}$, where k is the iteration counter.

In detail the algorithm is as follows:

1. Generate a small set of K structural model y and a large set of N random numbers obeying the joint probability density function of the random variables x. Normalise them by subtracting the mean and dividing by the standard deviation of each. This lead us to a normalised space u.

2. Sort the transformed variates in the critical orthant (i.e., that in which failure is most likely to occur, normally corresponding to low capacity and high demand values) by their distance to the origin.

3. Select the last $M = QN$ samples in the sorted set, i.e., the farthest from the origin in the critical orthant of the transformed space. Label them as failure samples and the rest as safe ones.

4. Fit a support vector classifier to these two classes in u-space. Let S be the number of support vectors and T the difference between the number of support vector on the safe side minus the corresponding number at the failure side. Evaluate the fitness for all $k = 1, 2,..., K$ models.

5. Calculate the limit state function at all the support vectors $\tilde{u}$ for each model and the corresponding score. If the score $T(k) > T$, this means that the model has a failure probability higher than Q and therefore it is infeasible. Otherwise it is feasible.

6. Fit a support vector classifier in y-space with samples in the feasible and infeasible classes thus obtained. Apply particle swarm optimization to the current population in order to generate a new one. In this procedure, besides the fitness function, the cost function must include a penalization term for trespassing the margin band in y-space.

7. If the optimum fitness stabilizes, stop. Otherwise, return to step 3.

Before illustrating the practical application of this method it is important to compare its computational effort to that needed by a reliability-based optimisation analysis using an optimisation technique combined with advanced Monte Carlo simulation. Assume the optimisation method requires K steps. In each of them it is necessary to calculate the failure probability in order to check if the solution is feasible or infeasible. Therefore, M solver calls are necessary, where $M = O(100/Q)$ if the crude Monte Carlo method is applied and $M = O(10/Q)$ if a variance-reduction technique such as importance sampling is used (Melchers, 1999). If, for instance, $Q = 0.01$, which is a common value in earthquake engineering applications, $1,000K$ samples will be necessary for a reliability-based optimisation using importance sampling. This figure is much larger than the SK solver calls required by the proposed method, because the number of support vectors resulting in a perfectly separable

Figure 6. Frame model for Problem No. 2

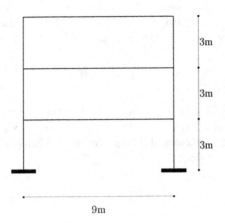

Figure 7. Design spectrum for Problem No. 2

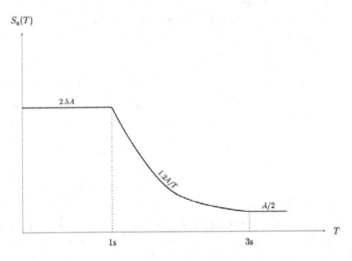

classification problem (such as that of structural reliability or constrained optimisation) is normally very low (see Hurtado, 2004b; Vapnik, 1998).

Consider the one-bay, three-story RC frame subject to seismic action shown in Figure 6. The frame is to be designed with a code acceleration spectrum (see Figure 7). The elastic response spectrum at low periods is specified as *2.5A* and the inelastic structural displacements are considered equal to the elastic displacements evaluated with the elastic spectrum

Figure 8. Initial population of structural models for Problem No. 2

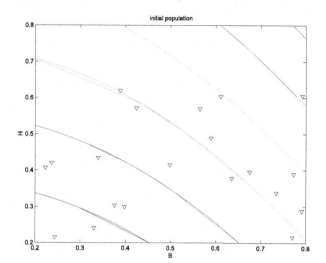

without any reduction factor. Therefore, since no use is made of artificial accelerograms, the random variables are only (a) the maximum base acceleration A, defined as a lognormal variable with mean 0.25 g and a large coefficient of variation (0.5); (b) the drift capacity of the frame in its first floor, specified as a Normal variable with mean 0.005 m and a coefficient of variation equal to 0.15. The purpose of the problem is to find the structural model with minimal weight with the following constraints: a probability of failure lower than or equal to 0.05 and member sizes B and H (supposed to be equal for all beams and columns) greater than 0.3 m.

This problem has been designed in two dimensions in y-space in order to facilitate visualisation. Figure 8 displays the initial PSO population, while Figure 9 displays the population at iterations No. 7. It can be observed that the end PSO population concentrates in the region defined by the current support vectors. This indicates that individuals lying beyond the region they mark are infeasible because the probability of failure associated to them is either too small or too large. From the particles appearing in Figure 9, that showing the minimal weight is the pair $B = 0.3382$, $H = 0.3768$, which can be taken as the final solution.

Problem No. 3: Optimisation of a nonlinear building model with an active probability constraint using random vibration analysis.

Figure 10 shows one of the five equal frames of a four-story RC building subject to earthquake actions. The building will be analysed as a nonlinear shear beam model, using a lumped mass and a single lateral stiffness for each floor. The purpose of the calculation is

Figure 9. End instance of convergence of the proposed reliability-based optimisation algorithm in Problem No. 2. Notice that the population clusters about the support vectors.

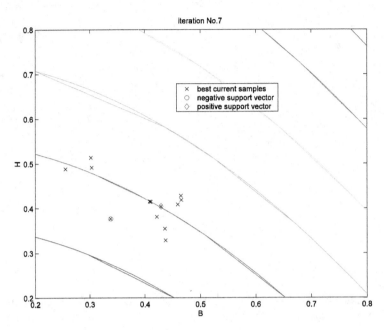

to maximise the flexibility of the frame under probabilistic constraints imposed over the story drifts. To be specific, the problem is as follows:

Find $y = [k_1 \, k_2 \, k_3 \, k_4]$
that minimises $C(y) = k_1 + k_2 + k_3 + k_4$
subject to $P[\delta_i(y) \geq 0.02 H_i] \leq Q_i, \quad i = 1,...,4$ (8)

where δ_i is the story drift, H_i is the story height and Q_i are the maximum admissible probabilities for lateral deformation of the building, equal to 0.001, 0.005, 0.01 and 0.01. In words, the tolerance of a large story drift is ten times more restrictive for the first story than for the top one.

The analysis is carried out within a random vibration analysis context using the method of stochastic equivalent linearisation. The shear beam springs are modelled with the well-known Bouc-Wen hysteretic model (Wen, 1976), in which the restoring force is given by:

$$f = \alpha k \delta + (1-\alpha)kz$$ (9)

Figure 10. Frame for Problem No. 3

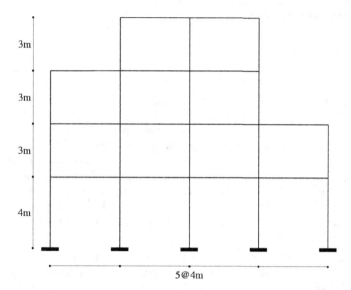

where α is the ratio between post-yielding stiffness to the preyielding stiffness, given by k, and z is a nonlinear auxiliary variable (with displacement units), which is specified by a differential equation:

$$\dot{z} = A\dot{\delta} - \beta \left| \dot{\delta} \right| \left| z \right|^{n-1} z - \gamma \dot{\delta} \left| z \right|^{n}$$

(10)

in which A, β, γ, n are parameters that in this specific problem take the values $A = 1$, $\beta = 288$, $n = 1$. Use is made of the relationship $\beta = -3\gamma$ as recommended by Sues, Wen, and Ang (1983) for RC frames. The lumped masses are equal to 300 metric tons for the first two floors, 225 for the third and 150 for the fourth. The horizontal seismic excitation is modelled as a Kanai-Tajimi power spectral density given by:

$$G(\omega) = \frac{\omega_g^2 + 4\zeta_g^2 \omega^2}{(\omega_g^2 - \omega^2)^2 + 4\zeta_g^2 \omega^2} G_0$$

where G_0 stands for the intensity of the driving white noise and ω_g, ξ_g are parameters. The values used in present example are 0.00743 m^2 / s^3, 16.5 rad / s and 0.8, respectively. This corresponds to the compound response $2\xi_g \omega_g \dot{u}_g + \omega_g^2 u_g$ of a SDOF oscillator subject to white noise with one-sided power spectral density equal to G_0. The stochastic equivalent linearisation procedure consists in substituting equation (10) by:

$$\dot{z} = C_e \delta + K_e z \tag{11}$$

where the linearisation coefficients are found by minimising the expected error, given by the difference between the two expressions. Gaussian distribution is assumed for calculating this expected value, because it facilitates the computation of the coefficients (Atalik & Utku, 1976). Other, non-Gaussian strategies are, however, possible (Hurtado & Barbat, 1996, 2000; Pradlwarter, Bucher, & Schuëller, 1991). In any case the coefficients are given in terms of second-order statistical moments of the drift, its velocity and the auxiliary variable z. These values can be drawn form the covariance matrix of the state vector $q = [\delta_1 \quad \dot{\delta}_1 \quad z_1 \quad \cdots \quad \delta_4 \quad \dot{\delta}_4 \quad z_4 \quad u_g \quad \dot{u}_g]$ which can be obtained by solving an algebraic Lyapunov equation in the stationary case or a differential Lyapunov equation in the nonstationary case.

Once the system has been linearised and the solution for the covariance matrix has been found, it is possible to estimate the probability of exceeding a certain threshold by the drift displacement by means of the level-crossing theory of stochastic processes and assuming that the drift obeys a Gaussian distribution. In the present case the following equation was used for this estimation (Nigam, 1983):

$$P(T, \bar{\delta}) = 1 - (1 - P(0, \bar{\delta}) \quad \exp(-\eta T) \tag{12}$$

where η is a factor depending on the statistical moments of the response, T is the strong-motion duration where the maximum displacement is likely to occur and $\bar{\delta}$ is the maximum allowable interstory drift. The latter two values are respectively equal to 7 sec. and $0.02H$ in present example.

The algorithm applied is therefore as follows:

1. For each individual of the current population, identified by a vector of story stiffnesses, calculate the covariance matrix by solving the Lyapunov equation and then the linearisation coefficients.

2. Calculate the probabilities of exceeding the drift thresholds using equation (12).

3. Classify the samples in two classes: (a) samples for which all level crossing probabilities are lower than or equal to the maximum allowable values (feasible samples), and (b) samples for which at least one probability is higher than its respective maximum (unfeasible samples).

4. Fit a support vector machine with the labelled samples.

5.   Generate a new population within the current margin using particle swarm optimisation.

6.   If convergence has not been achieved go to step 1. Exit otherwise.

The analysis showed that this algorithm converges very rapidly. After five iterations the following optimal solution vector was found:

$$y = [k_1 \quad k_2 \quad k_3 \quad k_4 \quad] = [1.090 \quad 1.195 \quad 1.491 \quad 0.437] \times 1e5$$

in kN/m. On this basis adequate frame member sections yielding the above stiffness values can then be selected.

Conclusion

Reliability, optimisation and reliability-based optimisation are structural problems that imply deciding whether samples are either safe or unsafe, feasible or infeasible. To this end soft-computing pattern recognition techniques such as Neural Networks or support vector machines are very useful. In this chapter, the latter method has been selected because of its information compression and search space reduction properties. Its application in combination with an artificial life programming method, such as particle swarm optimisation, allows solving complicated problems such as the minimisation of structural cost subject to probabilistic constraints. Algorithms proposed to such a purpose have been illustrated with earthquake engineering examples.

References

Atalik, T. S., & Utku, S. (1976). Stochastic linearization of multi-degree of freedom non-linear systems. *earthquake engineering and Structural Dynamics, 4,* 411-420.

Bäck. T. (1996). Evolution strategies: An alternative evolutionary algorithm. In J. M Alliot, E. Lutton, E. Ronald, M. Shoenhauer, & D. Snyers (Eds.), *Artificial evolution* (pp. 3-20). Berlin, Germany: Springer.

Chong, E. K. P., & Zak, S. H. (2001). *An introduction to optimization.* New York: Wiley.

Dorigo, M., & Stützle, T. (2004). *Ant colony optimization.* Cambridge, MA: MIT Press.

Duda, R. O., Hart, P. E., & Stork, D. G. (2001). *Pattern classification.* New York: Wiley.

Elishakoff, I. (2005). *Safety factors and reliability: Friends or foes?* New York: Kluwer.

Forbes, N. (2004). *Imitation of Life*. Cambridge, MA: MIT Press.

Frangopol, D. (1995). Reliability-based structural design. In C. Sundararajan (Ed.), *Probabilistic structural mechanics handbook* (pp. 352-387). New York: Chapman & Hall.

Functions from the viewpoint of statistical learning theory. (n.d.). *Structural Safety*, 26, 271-293.

Gasser, M., & Schuëller, G. I. (1997). Reliability-based optimization of structural systems. *Mathematical Methods of Operations Research*, *46*(3), 287-307.

Goldberg, D. E. (1989). *Genetic algorithms in search, optimization and machine learning*. New York: Addison-Wesley.

Haftka, R. T., Gurdal, Z., & Kamat, M. P. (1990). *Elements of structural optimization*. Dordrecht, Holland: Kluwer Academic.

Hurtado, J. E. (2004a). *An examination of methods for approximating implicit limit state*.

Hurtado, J. E. (2004b). *Structural reliability. Statistical learning perspectives*. Heidelberg, Germany: Springer.

Hurtado, J. E., & Álvarez, D. A. (2003). A classification approach for reliability analysis with stochastic finite element modelling. *Journal of Structural Engineering, 129*, 1141-1149.

Hurtado, J. E., & Barbat, A. H. (1996). Improved stochastic linearization method using mixed distributions. *Structural Safety, 18*, 49-62.

Hurtado, J. E., & Barbat, A. H. (2000). Equivalent linearization of the Bouc-Wen hysteretic model. *Engineering Structures, 20*, 1121-1132.

Kennedy, J., & Eberhart, R. C. (2001). *Swarm intelligence*. San Francisco: Morgan Kaufmann.

Kuschel, N., Rackwitz, R., & Pieracci, A. (1999). Multiple β points in structural reliability. In P. D. Spanos (Ed.), *Computational stochastic mechanics* (pp. 181-190). Rotterdam, Holland: Balkema.

Melchers. R. E. (1999). *Structural reliability: Analysis and prediction*. Chichester, UK: Wiley.

Nigam, N. C. (1983).*Introduction to random vibrations*. Cambridge, MA: MIT Press

Pradlwarter, H. J., Bucher, C. G., & Schuëller, G. I. (1991). Nonlinear systems. In G. I. Schuëller (Ed.), *Structural dynamics: Recent advances* (pp. 146-213). Berlin, Germany: Springer-Verlag.

Reinhorn, A. M. (1997). Inelastic analysis techniques in seismic evaluations. In P. Fajfar & H. Krawinkler (Eds.), *Seismic design methodologies for the next generation of codes* (pp. 277-287). Rotterdam, Holland: Balkema.

Rosenblueth, E., & Mendoza, E. (1971). Reliability optimization in isostatic structures. *Journal of the Engineering Mechanics Division ASCE, 97*, 1625-1640.

Royset, J. O., & Polak, E. (2004). Reliability-based optimal design using sample average approximations. *Probabilistic Engineering Mechanics, 19*, 331-343.

Schölkopf, B., & Smola, A. (2002). *Learning with kernels.* Cambridge, MA: MIT Press.

Sexmith, R. (1999). Probability-based safety analysis: Value and drawbacks. *Structural Safety, 21,*303-310.

Sues, R. H., Wen, Y. K., & Ang, A. H. S. (1985). Stochastic evaluation of seismic structural performance. *Journal of Structural Engineering, ASCE, 3*(6), 1204-1218.

Vapnik, V. N. (1998). *Statistical learning theory.* New York: Wiley.

Wen, Y. K. (1976). Method for random vibration of hysteretic systems. *Journal of Engineering Mechanics, 102,* 249-263.

Chapter V

Optimum Design of Structures for Earthquake Induced Loading by Wavelet Neural Network

Eysa Salajegheh, University of Kerman, Iran

Ali Heidari, University of Shahrekord, Iran

Abstract

Optimum design of structures for earthquake induced loading is achieved by a modified genetic algorithm (MGA). Some features of the simulated annealing (SA) are used to control various parameters of the genetic algorithm (GA). To reduce the computational work, a fast wavelet transform is used. The record is decomposed into two parts. One part contains the low frequency of the record, and the other contains the high frequency of the record. The low-frequency content is used for dynamic analysis. Then using a wavelet neural network, the dynamic responses of the structures are approximated. By such approximation, the dynamic analysis of the structure becomes unnecessary in the process of optimisation. The wavelet neural networks have been employed as a general approximation tool for the time history dynamic analysis. A number of structures are designed for optimal weight and the results are compared to those corresponding to the exact dynamic analysis.

Introduction

Optimum design of structures is usually performed to select the design variables such that the weight or cost of the structure is minimized, while all the design constraints are satisfied. The external loads on the structures can be static (Salajegheh, 1996a, 1996b; Salajegheh & Salajegheh, 2002) or dynamic (Papadrakakis & Lagaros, 2000). In the present study, the design variables are considered as the member cross-sectional areas, which are chosen as discrete variables. The design constraints are bounds on member stresses and joint displacements. Optimum design problem is formulated as a mathematical nonlinear programming problem and the solution is obtained by the MGA (Salajegheh et al., 2005).

For problems with large number of degrees of freedom, the structural analysis is time consuming. This makes the optimal design process inefficient, especially when a time history analysis is considered. In order to overcome this difficulty, a discrete wavelet transforms (DWT) and a fast wavelet transforms (FWT) are used. Using these transformations, the main earthquake record is modelled as a record with a very small number of points. Thus, the time history dynamic analysis is carried out with fewer points. In Refs. (Salajegheh, & Heidari 2002, 2005) the DWT and FWT are used for the dynamic analysis of structures. These transformations are powerful means for the dynamic analysis and the time required is far less than the classical methods. Therefore, the FWT are used for optimisation of structures with earthquake loading (Salajegheh, Heidari, & Saryazdi, 2005). Despite substantial reduction in the dynamic analysis, optimisation process requires a great number of time history dynamic analyses; thus, the overall time of the optimisation process for earthquake record is very long.

In this work, in order to overcome this difficulty, using a wavelet neural network (WNN) (Thuillard 2001; Zhang & Beveniste, 1992) the dynamic responses of the structures are approximated. By such approximation, the dynamic analysis of the structure is not necessary during the optimisation process. Both feedforward neural networks and wavelet decompositions inspire this network. An algorithm of backpropagation type is proposed for training the network. In this network, the input is the damping ratio and the angular natural frequency of the structure and the output is the dynamic responses of a single degree of freedom structure against these reduced points. After training the network, using inverse wavelet transform (IWT) the results of the dynamic analysis is obtained for the original earthquake record from the output of the network. The numerical results of optimisation show that this approximation is a powerful technique and the required computational effort can be substantially reduced.

In the following, first a brief discussion of the MGA is presented. Then a brief discussion of the FWT and WNN are outlined. The details of the optimisation approach with approximation concepts are discussed and some numerical examples for optimum design of structures are presented. The computational time is compared to those of the exact optimisation method.

Modified Genetic Algorithm

In this research, the optimisation is based on the genetic algorithm but the probability of the genetic algorithm's operations namely crossover and mutation is controlled with Boltzmann probability distribution criterion (Bennage & Dhingra, 1995). The crossover and mutation operations are accepted, if the offspring's objective function values are less than their parents or the criterion specified is satisfied. When the number of generation increases, the control parameter imaginary temperature is reduced according to cooling progress and the chance of acceptance of the offspring with large value of objective function is reduced. In fact, by employing the features of the simulated annealing, we create an adaptive probability rate, which is different in each generation of the genetic algorithm. The numerical results show that by employing this type of adaptive control probabilities for crossover and mutation, a smooth convergence is achieved in the process of genetic algorithm. The combination of the GA and SA methods is referred to as the MGA.

Decomposition of Earthquake Record by FWT

The wavelet transform (WT) is being increasingly applied, in the fields ranging from communications to engineering, to analyse signals with transient or nonstationary components (Farge, 1992). *Nonstationary* means that the frequency content of the signal may change over time and the onset of changes in the signal cannot be predicted in advance. Earthquake records, which are transient-like and have very short duration, fit the definition of nonstationary signals. The analysis of nonstationary signals often involves a compromise between how well sudden variations can be located, and how well long-term behaviour can be identified. Choosing a basis function well suited for the analysis of nonstationary signals is an essential step in such applications. Wavelets are mathematical functions that cut up data or function into different frequency components, and then study each component with a resolution matched to its scale. They have advantages over traditional Fourier methods in analysing physical states where the signal contains discontinuities and sharp spikes. Unlike the Fourier transform (FT), the wavelet transform has dual localization, both in frequency and in time (Grossmann & Morlet, 1984; Mallat, 1989). These characteristics make wavelets an active subject with many exciting applications. There are various types of wavelet transforms, continuous wavelet transforms (CWT), discrete wavelet transforms (DWT) and fast wavelet transforms (FWT) (Farge, 1992; Mallat, 1989). In this chapter, both the FWT and DWT are used. The FWT is used to reduce the number of points of the earthquake record. The DWT is used to construct wavelet neural network.

It turns out that the wavelet transform can be simply achieved by a tree of digital filter banks. The main idea behind the filter banks is to divide a signal into two parts: the first is the low-frequency part and the other is the high-frequency part. This idea can be achieved by a set of filters. A filter banks, consists of a low-pass and a high-pass filter, which separate a signal into different frequency bands. A filter may be applied to a signal to remove or enhance certain frequency bands of the signal (Oppenheim, Schafer, & Buck, 1999). By applying a low-pass filter to a signal $s[t]$ of length N (number of points), the high-frequency bands of the signal are removed and an approximate version of the original signal is obtained. A high-pass filter removes the low-frequency components of the original signal, and the result is a signal containing the details of the main signal. By combining these two filters into a filter bank, the original signal is divided into an approximate and a detail signal.

A multilevel decomposition of the original signal is performed, by repeating the decomposition process. In the next stage, the low-pass filtered output signal is used as input to the filter bank (Strang & Nguyen, 1996). If the computation of the wavelet transform is reduced to the FWT, then the resulting implementation is very efficient. Several the FWT algorithms have been devised for computation of wavelet transform coefficients. In this chapter, the Mallat algorithm is used for dynamic analysis of structures for earthquake induced loading (Salajegheh & Heidari, 2005). In this method the original record is decomposed into two records. Detail record D_j and approximate record A_j which are obtained as follows:

$$D_j(t) = \sum_t s(t) h_j^*(t - 2^j k) \qquad j = 1, 2, ..., J \quad k = 1, 2, ..., K \qquad (1)$$

$$A_j(t) = \sum_n s(t) g_j^*(t - 2^j k) \qquad j = 1, 2, ..., J \quad k = 1, 2, ..., K \qquad (2)$$

where j and k are integer numbers. The symbol * denotes complex conjugate, the value of 2^j is the dilation factor; the value of k is shifting factor, and [.] is used for a discrete notation. $h_j[t-2^j k]$ is the discrete wavelet analysis and $g_j[t-2^j k]$ is the scaling function analysis, defined as (Mallat, 1989):

$$h_j[t - 2^j k] = \frac{1}{\sqrt{2^j}} \psi(\frac{t - 2^j k}{2^j}) \qquad (3)$$

$$g_j[t - 2^j k] = \frac{1}{\sqrt{2^j}} \varphi(\frac{t - 2^j k}{2^j}) \qquad (4)$$

Different functions can be chosen for ψ and φ. In the present study the following functions are selected. In the wavelet literature, these functions are referred to as the Harr functions (Strang & Nguyen, 1996) that are easy to implement and the numerical results indicate that the Harr functions are appropriate.

$$\psi(t) = \begin{cases} 1 & 0 \leq t \leq 0.5 \\ -1 & 0.5 < t \leq 1 \end{cases} \qquad \varphi(t) = \begin{cases} 1 & 0 \leq t \leq 1 \\ 0 & otherwise \end{cases} \qquad (5)$$

Wavelets and scaling functions must be deduced from one stage to the next as follows:

$$g_{j+1}[t] = \sum_k g_j[k] g_l[t - 2k] \qquad (6a)$$

$$h_{j+1}[t] = \sum_k h_j[k]g_l[t-2k] \tag{6b}$$

The number of points of the earthquake record is reduced by the FWT. The decomposition starts from the original record of the earthquake ($s[t]$), and produces two sets of records, D_j, and A_j. These vectors are obtained by convolving $s[t]$ with the low-pass filter for A_1 by equation (4), and with the high-pass filter for D_1 by equation (3). Then from the record A_1, the two records A_2 and D_2 are evaluated and the process is continued until A_J and D_J are evaluated for the Jth stage. For the earthquake record the approximation record (A_j) with low-frequency components is the effective part. The detail record (D_j) with high-frequency components is not effective. Therefore A_j is used for the dynamic analysis. For the all earthquake records, the low-frequency content is the effective part, because most of the energy of the record is in the low-frequency part of the record. On the other hand, for a record, the shape and the effects of the entire low-frequency component are similar to those of the main record. For an earthquake record if we remove the high-frequency components, the record is different, but we can still distinguish the pattern of the record. The numerical results of the dynamic analysis show that this approximation is a powerful technique and the required computational work can be reduced greatly. The errors in the proposed methods are small. In Figure 1, the El Centro earthquake record (S-E 1940) for A_j signal is shown.

In this chapter, this process is repeated in three stages, and the number of points of the original record is reduced to 0.125 of the primary points. The error is negligible, in particular in the first three stages of decomposition (Salajegheh & Heidari, 2005). In this method, the process can be inversed and the original record can be computed. This process is named as *inverse wavelet transform* (IWT). The original signal $s[t]$ can be achieved through the IWT process, by using the D_j and A_j as:

$$s(t) = \sum_{j=1}^{J} \sum_k D_j(t)\tilde{h}_j(t-2^j k) + \sum_{j=1}^{J} \sum_k A_j(t)\tilde{g}_j(t-2^j k) \tag{7}$$

where $\tilde{h}_j(t-2^j k)$ is called the synthesis wavelet and $\tilde{g}_j(t-2^j k)$ is called the synthesis scaling function (Salajegheh & Heidari, 2005). The $\tilde{h}$ and $\tilde{g}$ are used for high and low-pass filters, respectively. In the present study, the A_j record is only used in the process of the IWT, because the effects of the D_j are negligible as explained.

WNN for Dynamic Response Approximation

Artificial neural network (ANN) is a mathematical system that mimics the way in which the brain works. It consists of fully interconnected layers of processing units called "neurons." There are always one input layer, one output layer and a number of hidden layers. Each layer of nodes receives its input from previous layer or from the network input. The output of each node feeds the next layer or the output of the network. Internal or hidden layers provide the interconnection between input and output layers. The output of neuron is related to the summed inputs by a linear or a nonlinear transfer function. Different the ANN

Figure 1. (a) The El Centro record; (b) decomposition of the El Centro record (A_1); (c) decomposition of the El Centro record (A_2); (d) decomposition of the El Centro record (A_3); (e) decomposition of the El Centro record (A_4); (f) decomposition of the El Centro record (A_5)

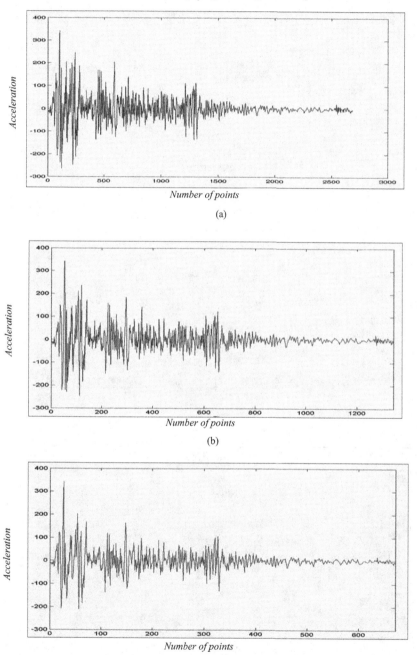

(a)

(b)

(c)

Figure 1. continued

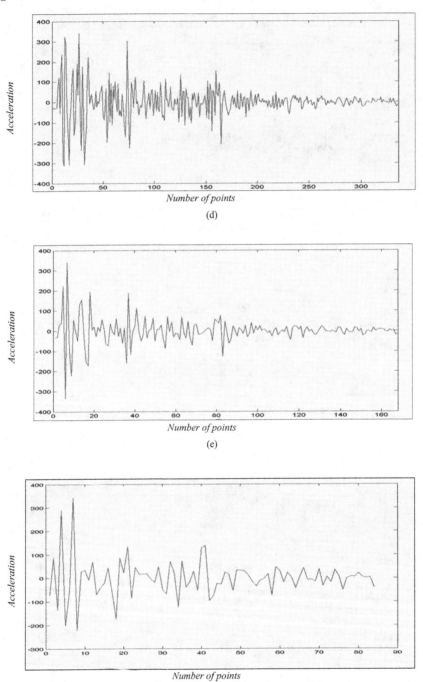

(d)

(e)

(f)

classes use different definition of activation functions and different training algorithms. The most common training network algorithm is the backpropagation algorithm, and a sigmoid function is a popular activation function. The ANN is capable of representing functional relations. To create such a representation, it suffices to train a neural network with a set of known input-output pairs. During the training, interconnection weights between neurons are set to produce the correct input-output relation. After this initial stage, upon presentation of a new input vector, an approximation to an output vector is returned.

Two neural networks that can be used in function approximation are backpropagation neural network and counterpropagation neural network. These two networks are not capable to approximating the output of the dynamic analysis. In this chapter, to overcome this difficulty, a novel ANN is used.

The wavelet neural network (WNN) (Xu & Ho, 2002; Zhang et al., 2001) is inspired from the backpropagation. The activation function in the WNN is mother wavelet. Based on discrete wavelet transform, the WNN has been proposed as a novel universal tool for functional approximation, which shows surprising effectiveness in solving the conventional problems of poor convergence or even divergence encountered in other kinds of neural networks. It can dramatically increase convergence speed. In spite of having great potential applicability of the WNN, there are only a few papers on the WNN theory.

The WNN to approximate the dynamic responses of the structure consist of three layers: input layer, hidden layer, and output layer. Each layer has one or more nodes. Figure 2 depicts the schematic diagram of the three layers of the WNN. As illustrated in Figure 2, the input data vector X is connected to the input nodes of the networks. The connections between input units and hidden units, and between hidden units and output units are called weights U and W, respectively. The modified training steps of the WNN are as follows:

1. The number of nodes in the three layers is defined. These numbers in input, hidden and output layers are Q, R and M, respectively.

2. The activation function of the hidden layer is chosen. Different functions can be employed. In this chapter the mother wavelet is used as follows [24]:

$$\psi(x) = -x\exp(-\frac{x^2}{2}) \tag{8}$$

Then, the function ψ_{a_i,b_i} can be calculated from the mother wavelet with dilation a_i and translation b_i as:

$$\psi_{a_i,b_i}(x) = \sqrt{2^{a_i}}\,\psi(2^{a_i}x - b_i) \tag{9}$$

where a_i and b_i are integer numbers. Now, the activation function of the wavelet nodes by substitution of (9) into (10) has the following form:

$$\psi_{a_i,b_i}(x) = -\sqrt{2^{a_i}}\,(2^{a_i}x - b_i)\exp(\frac{(2^{a_i}x - b_i)^2}{2}) \tag{10}$$

Figure 2. Wavelet neural network

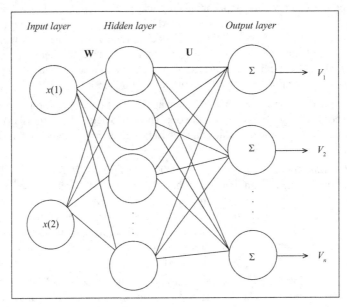

3.   The weights, u_{qr} and w_{rm} (components of U and W), are endowed with random values.

4.   Input learning samples $X_n(q)$ and the corresponding target output values $V_n(m)^T$, are chosen, where n is the number of learning samples, T stands for the target value, q and m are the appropriate components of vectors X_n and V_n, respectively.

5.   The output value of the sample V_n is calculated as:

$$V_n(m) = \sum_{r=1}^{JK} \frac{w_{rm}}{\sqrt{2^j}} \psi \left(\frac{1}{2^j} \sum_{q=1}^{Q} u_{qr} X_n(q) - k\right) \tag{11}$$

where $JK = (J+1)(K+1)$ and the values of j and k are computed as:

$$j = r / (2K + 1) - J \tag{12}$$

$$k = rem(r, 2K + 1) - K \tag{13}$$

in which $rem(r, 2K + 1)$ is the remainder of $r / (2K + 1)$.

6. The instantaneous gradient vectors are computed as follows:

$$\delta w_{rm} = \frac{\partial E}{\partial w_{rm}} = -\sum_{n=1}^{N} \frac{[V_n(m)^T - V_n(m)]}{\sqrt{2^j}} \psi \left(\frac{1}{\sqrt{2^j}} \sum_{q=1}^{Q} u_{qr} X_n(q) - k\right) \tag{14}$$

$$\delta u_{qr} = \frac{\partial E}{\partial u_{qr}} = \sum_{m=1}^{M} \left(-\sum_{n=1}^{N} \frac{V_n(m)^T - V_n(m)]}{\sqrt{2^j}} w_{rm} \frac{\partial \psi}{\partial q'} X_n(q)\right) \tag{15}$$

where $q' = \sum_{q=1}^{Q} \sum_{r=1}^{JK} u_{qr} X_n(q)$. If $p'_n = 2^{-j} q' - k$, then the derivative of ψ with respect to q' is:

$$\frac{\partial \psi}{\partial q'} = \frac{1}{2^j} p_n'^2 \exp\left(-\frac{p_n'^2}{2}\right) - \frac{1}{2^j} \exp\left(\frac{p_n'^2}{2}\right) \tag{16}$$

The error function is the mean square error function, that is:

$$E = \sum_{n=1}^{N} \sum_{m=1}^{M} \frac{[V_n(m)^T - V_n(m)]^2}{2} \tag{17}$$

In this chapter, steepest descent method is used to minimize E.

7. The modified weight in backpropagation (BP) is calculated by:

$$w_{rm}^{new} = w_{rm}^{old} + \Delta w_{rm}^{new} \tag{18}$$

$$u_{qr}^{new} = u_{qr}^{old} + \Delta u_{qr}^{new} \tag{19}$$

in which the values of Δw_{rm}^{new} and Δu_{qr}^{new} are computed as:

$$\Delta w_{rm}^{new} = -\eta \frac{\partial E}{\partial w_{rm}^{old}} + \alpha \Delta w_{rm}^{old} \tag{20}$$

$$\Delta u_{qr}^{new} = -\eta \frac{\partial E}{\partial u_{qr}^{old}} + \alpha \Delta u_{qr}^{old} \tag{21}$$

where η is the learning rate factor, and α is momentum factor.

8. If the output error falls below a setting value, the learning procedure of the WNN is stopped, otherwise it will return to step (4).

In fact, the method of evaluation of the errors in the WNN is similar to the BP but the node activation function is the mother wavelet with varying translation and scale values. This modification results in better output than the standard BP as the activation function in the standard BP is a simple function and this is not adequate for approximating the functions such as the earthquake records. The results of the training process are the approximate dynamic responses (displacements) of a structure with single degree of freedom at all the time intervals as explained further in the subsequent sections.

Simulation of Dynamic Responses by WNN

One of the methods used for the evaluation of the dynamic responses of a linear analysis of a structure is the modal superposition. In modal superposition method it is shown that multi-degrees of freedom system (MDOF) can be converted into multisystems of single degree of freedom (SDOF). The solution of each SDOF system may be found by Duhamel's integral. After calculating the response of each SDOF system, the response of MDOF system can be found by the superposition method. The Duhamel's integral is shown as (Paz, 1997):

$$y(t) = \frac{1}{\varpi_D} \int_{t_i}^{t_{i+1}} \ddot{y}(t) \exp(-\xi\varpi(t-\tau)) \sin \varpi_D(t-\tau) d\tau \tag{22}$$

where ξ is damping ratio, ϖ is angular natural frequency, $\ddot{y}$ is acceleration record, and ϖ_D is equal to $\varpi(1-\xi^2)^{0.5}$. In Duhamel's integral for a specific earthquake record, $\ddot{y}$, t_i and t_{i+1} are constants. Therefore, two independent variables exist and the input variables of WNN are as ξ and ϖ. The target vector is the response of SDOF system. In this method, the responses of the SDOF are computed at all time intervals of the earthquake record. The number of nodes of input layer, hidden layer and output layer are chosen as 2, 12 and 336, respectively. If we use $a = 12$, the value of 2^a is equal to 4096. This number is grater than the number of points of the earthquake record. According to wavelet theory choosing at least 12 values can show the variant of the record suitability (Thuillard, 2001). In fact, the values of J and K in equations (3) and (4) should be chosen such that $(J+1)(K+1) = a$. In this chapter, $J = 3$ and $K = 2$ are used as different values of these parameters would not affect the final results. Total time intervals of the El Centro earthquake are 2688 and the number of reduced intervals by FWT (third stage of decomposition) is evaluated as 336.

In fact, the analysis of the SDOF structure by the standard Duhamel's integral is not required. By changing the values of ξ and ϖ, the trained network provides the complete time history dynamic analysis of any SDOF against the A_3 record with less computational efforts. The efficiency of the dynamic analysis of the SDOF structure is due to two factors. The major factor is that the number of the points of the record is reduced, and the other factor is that the computational time of the dynamic analysis of the structure against the reduced points by the WNN is less than that of Duhamel's integral (Heidari, 2004). Then by modal superposition, the MDOF system can be converted into a number of SDOF systems, each of which has different ξ and ϖ. Thus the dynamic time history analysis of the MDOF system under investigation can be obtained by using the results of the WNN.

Main Steps of Optimisation with FWT and WNN

The main steps in the optimisation process employing FWT and WNN for earthquake loading are as follows:

1. The functions ψ and φ are defined. In this study, equation (7) is employed.

2. The number of stages for decomposition of the record is chosen. Here three stages are used (Salajegheh & Heidari, 2005).

3. The FWT of the earthquake record in three stages is computed.

4. The approximate version of the earthquake record in the third stage (A_3) is used for dynamic analysis.

5. The dynamic responses of a SDOF structure against A_3 are calculated. This process is repeated for a number of the SDOF structures with different ξ and ϖ for training the network.

6. The number of nodes in the input layer, hidden layer, and output layer are chosen. In this chapter, 2, 12, and 336 nodes are used, respectively.

7. The activation function in wavelet layer is chosen by using equation (12). In this chapter, 12 activation functions are employed.

8. The weights of the wavelet neural network are optimised. This process is repeated until the network is converged.

9. For each ξ and ϖ, the trained network is used to calculate the dynamic responses of any SDOF structure against A_3 record.

10. The responses of the SDOF structure against the original earthquake record are evaluated by equation (9). Now, the dynamic responses of the structure under investigation are determined by modal superposition method.

11. MGA is used for optimisation. The optimisation convergence is checked, if convergence is satisfied, the process is stopped, otherwise the member cross-sections are updated and the process is repeated from step (9).

It can be observed that in the process of optimisation, the direct dynamic analysis of the structure is not required. In fact, the necessary responses are found by the trained WNN. In addition, the time history analysis of the structure required for the training is achieved for a record with less number of time intervals. For the El Centro earthquake, the best ratio of the reduced intervals with respect to the original record is chosen as 0.125 (Heidari, 2004; Salajegheh & Heidari, 2005). It is to be mentioned that the training process should be carried out once for the specified earthquake record. Thus, for optimum design of any structure against this earthquake, the time history dynamic analysis is not required. The numerical results show that the evaluation of the structural responses by trained network is faster than employing the Duhamel's integral (Heidari, 2004).

In summary, the FWT and WNN are used in two different ways to enhance the efficiency of the optimisation process. The first application is to reduce the cost of the time history dynamic analysis by the FWT. By the FWT, the original record is filtered and the number of points in the resulting record (A_3) is about 0.125 of the original record. The second aspect is to reduce the computational cost of the overall optimisation process by the WNN. This is achieved through training a special network; the result of which is that the analysis of the structures is not necessary during the optimisation process. In addition, the WNN was applied to the A_3 record. The WNN was not used for the original record, because the number of output nods would be increased and training the system would be impossible. In fact, the main application of the FWT is to reduce the number of points of the original earthquake accelerogram for the process of the WNN and the application of the WNN is to omit the time history dynamic analysis of the SDOF systems during the optimisation process which is faster than the Duhamel's integral (Heidari, 2004).

It should be noted that the procedure outlined in the present study is for structural optimisation under earthquake loads when the dynamic analysis is carried out by the modal superposition method. However, the investigation is under progress for other methods of dynamic analysis as the training process is different.

Numerical Examples

Three examples are optimised for minimum weight against the El Centro Earthquake record (S-E 1940). The computer time for analysis are obtained in clock time by a personal Pentium 2. The number of points of the El Centro is 2688. The time interval for the record was 0.02 seconds. The number of points in A_3 was 336. The optimisation is carried out by the following methods:

1. MGA with the original earthquake record (MGE)
2. MGA using the FWT and WNN (MFW)

In all the examples, allowable stress is taken as 1,100 kg/cm², Young's modulus is 2.1×10^6 kg/cm², weight density is 0.0078 kg/cm³, and damping ratio for all modes are 0.05. The members are pipes for examples 1 and 2, with radius to thickness less than 50. The problems are designed with stress, the Euler's buckling and horizontal displacement constraints. The set of available discrete values considered for the cross-sectional areas of the members are given in Table 1 for examples 1 and 2. The training time for the El Centro earthquake for a SDOF system is 47 minutes.

Example 1. The 10 bars truss shown in Figure 3 is designed with height and span of $3m$. The truss is simply supported at the joints 1 and 4. The mass of 5,000 kg is lumped at each free node. The horizontal displacement at joint 6 is considered to be less than 10 cm.

The results of optimisation are given in Table 2. In the cases MGE and MFW the final weights are 708.32 and 769.23 kg, respectively. The time of computation for MGE and MFW are 136 and 4 minutes, respectively.

Table 1. Available member areas (cm²)

No.	Area	No.	Area	No.	Area	No.	Area	No.	Area
1	0.8272	9	3.789	17	10.57	25	25.11	33	68.35
2	1.127	10	4.303	18	12.99	26	27.54	34	70.7
3	1.427	11	4.479	19	13.66	27	29.69	35	71.4
4	1.727	12	5.693	20	15.11	28	33.93	36	73.0
5	2.267	13	6.563	21	17.13	29	40.14	37	80.7
6	2.777	14	7.413	22	18.74	30	43.02	38	85.2
7	3.267	15	8.229	23	19.15	31	51.03	39	87.4
8	3.493	16	9.029	24	21.15	32	63.6	40	90.5

Figure 3. Ten-bar truss

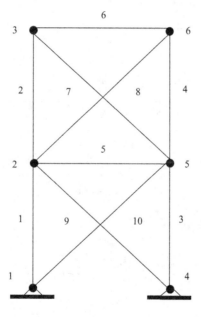

Example 2. A double layer grid of the type shown in Figure 4 is chosen with dimensions of 10×10 m for top layer and 8×8 m for bottom layer. The height of the structure is 0.5 m and is simply supported at the corner joints 1, 5, 21 and 25 of the bottom layer. The mass of 300 kg is lumped at each free node. The vertical displacement of joint 13 at the centre of bottom layer must be less than 10 cm. The members are grouped into 13 different types as shown in Table 3.

Table 2. Results of optimization for 10- bar truss

Member no.	Areas (cm²)	
	MGE	*MFW*
1	25.11	29.69
2	13.66	18.74
3	25.11	33.93
4	13.66	15.11
5	0.827	0.827
6	0.827	0.827
7	27.54	29.69
8	27.54	29.69
9	51.03	51.03
10	51.03	51.03
Weight (kg)	708.32	769.23
Time (min.)	136	4

Table 3. Member grouping for double layer grid

No.	Member no.	No.	Member no.	No.	Member no.
1	1-4; 37-40	6	7; 16; 25; 34	11	47; 50; 58; 61; 69; 72; 80; 83; 91; 94
2	10-13; 28-31	7	41-45; 96-100	12	48; 49; 59; 60; 70; 71; 81; 82; 92; 93
3	19-22	8	52-56; 85-89	13	All diagonal members
4	5; 9; 14; 18; 23; 27; 32; 36	9	63-67; 74-78		
5	6; 8; 15; 17; 24; 26; 33; 35	10	46; 51; 57; 62; 68; 73; 79; 84; 90; 95		

For this example, the results of maximum displacements are compared for the original earthquake record and those of the FWT. Results of analysis of maximum displacements of joints 2, 6, 13, 16, 23, 33, 40, 49, 51, and 57 in the X, Y, and Z directions, for the original earthquake record (OER), A_1, A_2, A_3, A_4 and A_5 for the El Centro record for the optimal results of the MFW method are given in Tables 4 to 6, respectively. In this study, the allowable error for the average maximum displacement is considered to be less than 10%.

The results show that the maximum displacement of joints in A_4 and A_5 are not suitable, therefore A_3 is used for dynamic analysis and optimisation of the structures.

Figure 4. (a) Double-layer grid, (b) top layer, (c) bottom layer

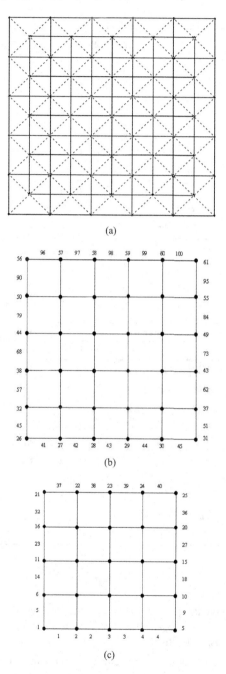

(a)

(b)

(c)

Table 4. Results of maximum displacement of Example 2 in X direction

Joints No.	Maximum dynamic displacement $*10^{-2}$						$|OER-A_j| / OER*100$				
	OER	A_1	A_2	A_3	A_4	A_5	A_1	A_2	A_3	A_4	A_5
2	117	117	108	105	97	90	0.3	7.8	10.3	17.1	23.1
6	117	117	108	107	98	91	0.4	8.2	8.5	16.2	22.2
13	114	115	105	103	93	87	0.2	7.9	9.6	18.4	23.7
16	116	117	106	105	95	86	0.8	8.4	9.5	18.1	25.9
23	104	105	96	94	85	75	1.6	7.7	9.6	18.3	27.9
33	125	129	119	117	103	90	3.4	4.9	6.4	17.6	28.0
40	91	89	85	83	75	65	3.0	6.7	8.8	17.6	28.6
49	82	80	77	73	67	56	2.4	6.1	10.9	18.3	31.7
51	97	95	89	88	80	72	1.8	8.2	9.3	17.5	25.8
57	89	87	82	81	73	63	1.8	8.0	8.9	18.0	29.2

Table 5. Results of maximum displacement of Example 2 in Y direction

Joints No.	Maximum dynamic displacement $*10^{-2}$						$|OER-A_j| / OER*100$				
	OER	A_1	A_2	A_3	A_4	A_5	A_1	A_2	A_3	A_4	A_5
2	112	116	122	123	135	147	3.6	8.9	9.8	20.5	31.3
6	104	107	108	99	91	80	3.3	3.6	4.5	12.5	23.1
13	101	103	104	108	120	132	2.2	2.6	6.9	19.3	30.7
16	68	65	61	67	81	91	4.2	10.3	2.5	19.3	33.8
23	46	44	42	44	55	63	4.3	8.7	4.1	19.6	36.9
33	72	69	66	65	60	51	5.2	8.3	9.7	16.7	29.2
40	66	65	60	59	54	46	2.5	9.1	10.6	18.5	30.3
49	58	55	53	52	46	39	5.2	8.6	10.3	20.7	32.8
51	74	70	68	67	59	48	6.2	8.1	9.5	20.2	35.1
57	43	41	39	38	34	29	4.7	9.3	11.6	20.9	32.6

The results of optimisation are given in Table 7. In the cases of MGE and MFW, the final weights are 5383.6 and 5730.21 kg, respectively. The time of computation in the MGE and MFW are 323, 10 minutes, respectively.

The results show that the maximum displacement of joints in A_4 and A_5 are not suitable, therefore A_3 is used for dynamic analysis and optimisation of the structures.

Table 6. Results of maximum displacement of Example 2 in Z direction

Joints No.	Maximum dynamic displacement *10^{-2}						$|OER-A_j| / OER*100$				
	OER	A_1	A_2	A_3	A_4	A_5	A_1	A_2	A_3	A_4	A_5
2	349	369	386	389	415	433	5.7	10.6	11.5	18.8	24.1
6	340	359	374	380	332	414	5.7	9.9	11.9	2.3	21.8
13	319	335	348	353	390	399	5.1	9.1	10.7	22.1	25.1
16	281	294	300	307	332	360	4.6	6.5	9.1	18	27.9
23	261	272	282	291	307	336	4.1	8.3	11.3	17.6	28.9
33	289	300	311	315	341	361	3.7	7.4	8.9	17.9	24.7
40	321	338	352	359	389	411	5.2	9.6	11.7	21.2	27.9
49	249	259	269	273	300	331	4.2	8.3	10.1	20.8	33.3
51	295	308	322	328	358	372	4.4	9.2	11.1	21.3	26.1
57	273	282	302	305	323	340	3.5	10.7	11.9	18.5	24.8

Table 7. Results of optimization for double layer grid

Group no.	Areas (cm²)	
	MGE	MFW
1	68.35	68.35
2	12.99	13.66
3	3.493	4.479
4	10.57	12.99
5	10.57	12.99
6	17.13	18.74
7	10.57	13.66
8	10.57	12.99
9	10.57	13.66
10	12.99	12.99
11	5.693	10.57
12	18.74	21.15
13	25.11	21.15
Weight (kg)	5383.6	5730.21
Time (min.)	323	10

Figure 5. Shear building of seven stories

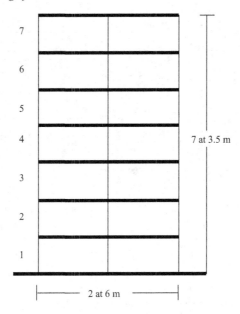

Example 3. The plane shear building model of 7 storys shown in Figure 5, is designed, whose floor masses move only horizontally within a vertical plane. It is assumed that the mass of each rigid floor of the model includes the effect of masses of all the structural elements adjacent to the floor of the prototype building. The mass of each floor is 90 tons. The set of available discrete values for the members are considered from the standard I sections. The horizontal displacement at the seventh floor is considered to be less than 15 cm.

In the cases MGE and MFW the final weights are 64.3 and 64.9 tons, respectively. The time of computation for MGE and MFW are 169 and 8 minutes, respectively.

Conclusion

An efficient method is presented for discrete optimum design of structures for earthquake load. The main goal is to reduce the computational cost of optimum design procedure. Numerical optimisation techniques, such as modified genetic algorithms, require the evaluation of the objective function and the design constraints at a great number of design points. In each design point the structure should be analysed to evaluate the necessary information. It can be seen that the number of structural analyses is excessive and for large-scale structures with many degrees of freedom, the optimisation is difficult; in particular, when a time history dynamic analysis is required. In the present study to reduce the overall cost of optimisation, some attempts have been made to reduce the cost of dynamic analysis as well as the cost of optimisation process. As far as the dynamic analysis is concerned, the idea of signal processing together with wavelets and filter banks are employed. By this method, the acceleration record is filtered and the number of points of the record is reduced. The structure is analysed dynamically with less number of time intervals. To reduce the cost of optimisation, a neural network type is employed. A wavelet neural network is presented to approximate the dynamic analysis of the structure during the optimum design process. Thus during optimum procedure the dynamic analysis of the structure is not required.

The FWT is used in two different aspects to enhance the efficiency of the optimisation process; the first application is to reduce the cost of time history dynamic analysis. The second aspect is to reduce the computational cost of the overall optimisation process. The numerical results show that in the proposed method, the time of optimisation is reduced to about 0.05 of the time required for exact optimisation. But the error is increased by a factor of about 8%.

References

Bennage, W. A., & Dhingra, A. K. (1995). Single and multiobjective structural optimisation in discrete-continuous variables using simulated annealing. *International Journal for Numerical Methods in Engineering, 38,* 2553-2573.

Farge, M. (1992). Wavelet transforms and their application to turbulence. *Annual Review of Fluid Mechanics,* 24, 395-457.

Grossmann, A., & Morlet, J. (1984). Decomposition of Hardy function into square integrable wavelets of constant shape. *SIAM Journal of Mathematics Analysis, 15,* 723-736.

Heidari, A. (2004). *Optimum design of structures against earthquake by advanced optimisation methods.* Unpublished doctoral dissertation, University of Kerman, Iran.

Mallat, S. (1989). A theory for multiresolution signal decomposition: the wavelet representation. *IEEE Tranactions Pattern Analysis and Machine Intelligence, 11,* 674-693.

Oppenheim, A. V., Schafer, R. W., & Buck, J. R. (Eds.). (1999). *Discrete-time signal processing.* NJ: Prentice Hall.

Papadrakakis, M., & Lagaros, N. D. (2000). Advances in computational methods for large-scale structural optimisation. In B. H. V. Topping (Ed.), *Computational mechanics for the twenty-first century* (pp. 431-449). Edinburgh, UK: Saxe-Coburg.

Paz, M. (Ed.). (1997). *Structural dynamics: Theory and computation.* New York: McGraw Hill.

Salajegheh, E. (1996a). Approximate discrete variable optimisation of frame structures with dual methods. *International Journal for Numerical Methods in Engineering, 39,* 1607-1617.

Salajegheh, E. (1996b). Discrete variable optimisation of plate structures using dual methods. *Computers and Structures, 58,* 1131-1138.

Salajegheh, E., & Heidari, A. (2002). Dynamic analysis of structures against earthquake by combined wavelet transform and fast Fourier transform. *Asian Journal of Civil Engineering, 3,* 75-87.

Salajegheh, E., & Heidari, A. (2005). Time history dynamic analysis of structures using filter banks and wavelet transforms. *Computers and Structures, 83,* 53-68.

Salajegheh, E., Heidari, A., & Saryazdi, S. (2005). Optimum design of structures against earthquake by a modified genetic algorithm using discrete wavelet transform. *International Journal for Numerical Methods in Engineering, 62,* 2178-2192.

Salajegheh, E., & Salajegheh, J. (2002). Optimum design of structures with discrete variables using higher order approximation. *Computers Methods in Applied Machines and Engineering, 191,* 1395-1419.

Strang, G., & Nguyen, T. (Eds.). (1996). *Wavelets and filter banks.* New York: Wellesley-Cambridge Press.

Thuillard, M. (Ed.). (2001). *Wavelets in soft computing.* New York: World Scientific.

Xu, J., & Ho, D. W. C. (2002). A basis selection algorithm for wavelet neural networks. *Neurocomputing, 48,* 681-689.

Zhang, Q., & Beveniste, A. (1992). Wavelet networks. *IEEE, Transactions on Neural Networks, 3,* 889-898.

Zhang, X., Qi, J., Zhang, R., Liu, M., Hu, Z., Xue, H., & Fan, B. (2001). Prediction of programmed-temperature retention values of naphthas by wavelet neural networks. *Computers and Chemistry, 25,* 125-133.

Chapter VI

Developments in Structural Optimization and Applications to Intelligent Structural Vibration Control

Sk. Faruque Ali, Indian Institute of Science, India

Ananth Ramaswamy, Indian Institute of Science, India

Abstract

The chapter introduces developments in intelligent optimal control systems and their applications in structural engineering. It provides a good background on the subject starting with the shortcomings of conventional vibration control techniques and the need for intelligent control systems. Description of a few basic tools required for intelligent control such as evolutionary algorithms, fuzzy rule base, and so forth, is outlined. Examples on vibration control of benchmark building and bridge under seismic excitation are presented to provide better insight on the subject. The chapter provides necessary background for a reader to work in intelligent structural control systems with real-life examples. Current trends in the research area are given and challenges put forward for further research.

Introduction

Civil engineering structures (e.g., tall buildings, long-span bridges, nuclear power plants), are an integral part of a modern society. Present trend in civil engineering is to build more flexible structures like long-span bridges and high-rise buildings. Dynamic loads (e.g., earthquake, wind gusts, wave forces, blasts), can cause severe vibratory motion especially for very flexible long span bridges and slender tall buildings (Soong, 1990). Protection of these structures, their material content and human occupants, against damage induced by large environmental loads (e.g., earthquake, strong winds) is no doubt a worldwide priority.

The control of structural vibration can be achieved by various means such as modifying structural rigidities, masses, damping, and by attaching external devices, known as control devices, either to dissipate vibrational energy of the structure or to impart a restoring force to the structure so as to minimize vibration. These control devices may be grouped into three broad categories (Soong & Spencer, 2002). They are passive energy dissipation system, active control system and combination thereof, (i.e., hybrid and semiactive systems). Each of these control systems has their own merits and demerits as discussed later. Parameters of control devices deployed in active systems are driven by control algorithm based on measured structural responses. An accurate structural model is important. Performance of control devices and algorithms decreases with uncertainties and nonlinearities in external load and sudden change in material properties (Soong & Spencer).

Major effort has been devoted in recent years to develop new unconventional control techniques to handle uncertainties and nonlinearities in external load and material properties with ease. These techniques incorporate knowledge assimilated from diverse area such as neurology, psychology, operation research, conventional control theory, computer science, and communication theory. These methods are collectively known as soft computing techniques. Intelligent control is a derivative of soft computing techniques, which focus on stochastic, vague, empirical, and associative situations. It establishes functional relationship (linear and nonlinear) between input and output space from empirical data without using an explicit model for the control plant (King, 1999). Intelligent control seeks solution to the problem of controlling plants from the viewpoint of replacing human operators. This is the point where intelligent control departs from conventional control. The near reproduction of human intelligence and the mechanism for inferring, decision making for an appropriate control action, and strategies or policy that must be followed are embedded in these tools. Computational intelligence includes expert systems; fuzzy logic; artificial neural network, and evolutionary computing like genetic algorithm; simulated annealing; and swarm optimization and their derivatives.

This chapter traces concurrent developments in areas of optimization, fuzzy logic and control theory leading to the present state of the art in intelligent optimal structural control and future directions. The chapter dwells on the basic approaches of intelligent control and offers examples from benchmark exercises in control of building and bridge vibration under seismic excitation.

Background of Conventional Control Techniques

The structural engineering community first embraced the notion of structural control in the 1960s (Zuk, 1980). Since then, a number of new techniques and devices have emerged and have been installed in different structures.

Conventional Control Techniques

Most civil engineering structures have low damping characteristics. Thus, there is a need for devices that can enhance structural damping and/or stiffness properties to mitigate excessive vibrations. This characteristic may be achieved by passive energy dissipating devices. Passive systems do not need external power while they enhance energy dissipation of the system. They have an inherent property of mitigating the structural responses within a particular frequency range, tuned at the time of installation. Passively controlled systems show undesirable responses when the frequency content of the external forces is out of the bandwidth for which the passive system is preset.

An active control system uses an external power source to alter the dynamic characteristics of the structure continuously so as to decrease the structural responses. It has the capability of changing energy imparted to the structure in real time to suppress structural vibration based on the feedback of the responses of the structure. The problem with these systems lies in the substantial power requirement for large-scale structures. Moreover there is a concern that power sources may fail during severe earthquake rendering active control systems to be unreliable (Soong & Spencer, 2002).

Combination of active and passive control (i.e., hybrid and semiactive systems) alleviate the shortcomings of active and passive systems deployed alone. A hybrid control system is defined as one that employs a combination of a passive and an active device. Its reliability is higher than that of an active system as under a power failure the passive component of the system still controls the primary structure. Research in hybrid system has been mostly centered on *hybrid mass damper* (HMD) *systems* and *hybrid base isolation systems*. Semi-active control systems comprise a class of active control systems for which the external energy requirements are orders of magnitude less than typical active systems. A semiactive control device does not add mechanical energy to the structural system, therefore bounded input-bounded output stability is guaranteed. The development in the area of semiactive system is attributed to the fact that semiactive control devices offer the adaptability of active system without requiring the associated large power. Commonly used semiactive devices include variable orifice damper, semiactive tuned mass dampers (SATMD), controllable fluid dampers and smart base isolation systems. Applications of aforementioned control systems to buildings and bridges over the world can be found in Soong and Spencer (2002) and Spencer and Nagarajaiah (2003).

There have been several benchmark problem defined on buildings and bridges to make systematic comparison of various control strategies. Three generations of benchmark building problem under seismic and wind induced excitations have been studied since 1996 (Ohtori, Christenson, Spencer, & Dyke, 2004; Spencer, Dyke, & Deoskar, 1998; Yang, Agrawal, Samali, & Wu, 2004). Base-isolated benchmark building problems have been defined and studied by Narasimhan, Nagarajaiah, Johnson, and Gavin (2003), Nagarajaiah and Narasimhan (2003). Dyke et al. (2003) and Caicedo et al. (2003) have defined a benchmark cable-stayed bridge problem. The recently proposed benchmark highway bridge vibration control problem (Agrawal, Tan, Nagarajaiah, & Zhang, 2005) is studied and presented in this chapter.

Problems in Conventional Control Techniques and Advantages in Use of Fuzzy Logic Control

When designing for safety of civil structures, active control should identify the changes in the current state due to modifications of systems parameters; identify changes in external disturbances, such as earthquakes or winds; compensate for internal and external disturbances, such as noisy measurements; and be able to provide nonlinear control mitigating excessive structural responses. In order to provide a structure with such capabilities, a number of sensors must be incorporated into the structure to monitor the deflection and acceleration at various locations. Actuators, tendons or other control devices can be used to actively modify the stiffness and damping characteristics of the structure.

The first stage in the analysis of control systems is the development of an adequate mathematical model for the structural system. The model can be derived either from physical laws or can be obtained using system identification. Thereafter, control force required is determined based on a control algorithm (e.g., pole assignment method, optimal control, and independent model-space control; Soong, 1990). A control unit then uses the sensor measurements to manipulate the actuators so as to modify in real time the behavior of the structure subjected to a dynamic action based on a reduced order structural model. This introduces errors due to control and observation spillovers. In real time implementation, time delays in processing sensor measurements, online computation, and executing the control force as required (i.e., driving the actuators) is present. In addition to these there are hardware related constraints such as saturation and resolution of the sensor, analog-to-digital converter (ADC) and digital-to-analog converter (DAC), which lead to saturation and quantization errors. Furthermore, limited number of sensors can only provide limited number of measurements. Therefore, active control techniques are prone to instability under structural nonlinearity, modeling errors, parameter uncertainty, sensor and observer disturbances, and unknown excitations. Unless the natural frequency of the sensors and actuators are well above the structural dominant frequencies, classical control becomes very sensitive to sensor and actuator dynamics. For detailed discussion on these aspects, interested readers are referred to Soong (1990), Soong and Spencer (2002) and Housner et al. (1997).

Some of the above discussed limitation of conventional control techniques can be alleviated using an intelligent control system such as fuzzy logic control that can treat the nonlinearities in structures, sensitivity to sensor and actuator dynamics, uncertainties in input excitations effectively and easily (Ahlawat & Ramaswamyl, 2002a). An accurate model of the structural dynamics is not required for the design of an intelligent controller.

Intelligent Control System

In a broad perspective, intelligent systems underlie what is called *soft computing* techniques. Intelligent control considers the integration of the computational process, reasoning, and decision making along with the level of precision or uncertainty in the available information as the design parameters. Therefore, an intelligent control system is more realistic and often has multiple solutions. Eventually, this demands that the designer make a selection from a suit of nondominated solutions. The principal tools in such a consortium are fuzzy logic, neural network computing, genetic algorithms and probabilistic reasoning. Furthermore, these methodologies, in most part, are complementary rather than competitive (Ali & Mo, 2001). Increasingly, these approaches are also utilized in combination, referred to as *hybrid*. Presently, the most well-known systems of this type are neuro-fuzzy and genetic-fuzzy systems.

The concept of intelligent control was first given by Fu to enhance and extend the applicability of automatic control systems (Housner et al., 1997). The increase in complexity due to design of more and more flexible civil engineering structures have led to the development of controllers that unlike conventional control system do not require complex models for its functioning, rather these methods rely on a knowledge base together with a rule base so as to decide and replicate human thought and decision making process. Intelligent controller systems are supported by four basic operations namely, fuzzy logic, neural network, neuro-fuzzy interface, and evolutionary computing. Unlike active control intelligent systems do not handle the tedious mathematical models of the controlled structure. They need only to set a simple controlling method based on engineering experience. Therefore, they are particularly useful in complicated structural control systems. Fuzzy controller does not involve complicated mathematical calculations and the whole fuzzy controller can be easily implemented on parallel fuzzy digital signal processing (DSP) boards (Patyra & Mlynek, 1996), which guarantees immediate reaction times resulting in reduced time-delays. The next section presents a discussion on the different methods of intelligent control.

Fuzzy Logic Control

Zadeh (1965) introduced fuzzy set theory to treat imprecision and uncertainty that is often present in implementation of problems in real world. Mamdani (1974), by applying Zadeh's theories of linguistic approach and fuzzy inference, successfully used the *if-then* rule on the automatic operating control of steam generator. Since then fuzzy control theory has been applied on a number of linear and nonlinear systems. In civil engineering, the fuzzy set theory has been applied by Faravelli and Yao (1997), Teng, Peng, and Chuang (2000), Ahlawat and Ramaswamy (2002a, 2002b, 2003, 2004a, 2004b), Wang and Lee (2002), and so forth.

Fuzzy logic control is a simulation of logical reasoning of human brain; it maps an input space to a corresponding output space based on fuzzy rules specified in *if–then* format known as *knowledge base*. Fuzzy logic–based control includes a fuzzification interface, an inference engine and a defuzzification interface as shown in Figure 1. Definitions of a few terms are provided here to facilitate the ensuing discussions:

Figure 1. Fuzzy control system

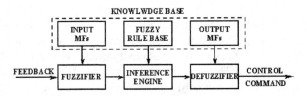

- **Domain of discourse:** The range over which input and output variable spaces are defined is known as the domain of discourse. If the domain is not known properly, -1 to +1 is taken as the domain of discourse and prescaling and postscaling is applied to the input and output variables. Examples given in the chapter takes acceleration and pseudo-velocity as input and actuator voltage is given as output. The inputs are scaled down to 1 using prescaling depending upon sensor sensitivity and control voltage is scaled up based on actuator voltage capacity.

- **Membership function:** Unlike crisp set fuzzy variable can take a value or *measure of the membership* between 0 and 1. A measure of the degree to which a variable belongs to a particular set is determined using membership function. There are various kinds of membership functions available and their shape depends upon the definition and the work they are to perform. The most frequently used membership functions are triangular, trapezoidal, generalized bell shaped and gaussian membership function.

A generalized bell shaped function can approximate other membership functions with appropriate choice of its parameters. Figure 2 shows the shape of generalized bell shaped function describing parameters *a*, *b*, and *c*, which, as an example are taken as 2, 2.5, and 0 respectively. The membership grade μ_x for input *x* in a generalized bell shaped membership function is given by equation (1) (MATLAB, 2004):

$$\mu_x = \frac{1}{1 + \left| \dfrac{x - c}{a} \right|^{2b}} \tag{1}$$

The following steps provide a simple design for fuzzy logic controller:

- **Defining input, output variables:** Decision on what responses of the system are subject to observation and measurement, leading to choice as input variables is the first step. Choice of control functions needed, results in the choice of output variables. In the examples provided pseudo-velocity and absolute acceleration data have been considered to be input variables and actuator voltage as an output variable.

Figure 2. Fuzzy membership function

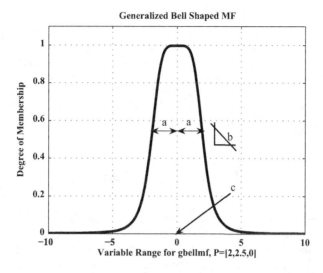

- **Fuzzification of input variables:** The fuzzification interface maps the measurable input variables in the form of a crisp set to a fuzzy linguistic value based on their membership grade in the domain of discourse. Here, the values of the state variables to be monitored during the process are fuzzified using linguistic terms defined by membership functions of the fuzzy set. Usually, within a domain of discourse the number of partitions for linguistic variables should be odd in number. This number will decide the status of a partition of the space. The more the linguistic variables, the more sophisticated is the partition of spaces. Example 1 gives the fuzzified form of the inputs and the output and the variables are shown in Figure 6.

- **Inference engine:** The inference engine has a dual role in fuzzy control theory. It maps the input fuzzified variables to the output variables based on user-defined rules known as *knowledge base*. It also provides a *decision* based on the results obtained from implementation of these rules. The most commonly used inference methods and the operators used for inference of these strategies are (Faravelli & Yao, 1997):

  ○ **Mamdani's strategy:** This fuzzy inference method is based on fuzzy MAX-MIN operator.

  ○ **Larsen's strategy:** It is based on a fuzzy PRODUCT operator.

  ○ **Takagi and Sugeno's strategy:** Takagi and Sugeno's inference method characterizes the fuzzy outputs as the functions of fuzzy input set.

Usually, the control rule base of the fuzzy controller is formed from operator experience and expert knowledge. The more the control rules, the more the efficiency of the control system. Control rules are usually in the form of *if–then* rules to link input and output variables (*if* is called "antecedent"; *then* is called "consequence").

For example, if x is A_i; and y is B_i; then z is C_i; $i = 1 \ldots n$; where n represents the number of control rules:

- **Defuzzification of output variables:** Defuzzification describes the mapping from the space of fuzzy outputs to a crisp set output. Defuzzification operation takes most of the processing time in a fuzzy control algorithm. A large number of defuzzification methods are available but only few are practically amenable for fuzzy control systems. They are: *center of area* (COA); *center of gravity* (COG); *height defuzzification* (HD); *center of largest area* (COLA); *mean of the maximum* (MOM), and so forth (King, 1999).

Neural Networks and Neuro-Fuzzy Controller

Fuzzy systems are powerful in representing linguistic and structured knowledge by means of fuzzy set theory, but it is up to the experts to establish the knowledge base to be used in the system. Therefore, formation of appropriate fuzzy *if–then* rules and membership functions remains a thorny issue. Furthermore, static fuzzy rules and membership function are vulnerable to changes in system parameters. One way to get rid of sensitivity to parameter changes is to combine a fuzzy system with an artificial neural network (ANN) referred to as hybrid neuro-fuzzy control (Faravelli & Yao, 1997; Jang, 1993). ANNs are capable of learning the changes in the given system input and accommodate these changes in the output. The basic elements of a neural network are nodes, the layers and the activation function. Commonly used ANN structure is multiple layer perceptron and radial basis function network. A two-layer feedforward network is shown in Figure 3. The output from the network is given as equation (2):

$$y_i = \sigma \left(\sum_{j=1}^{N_2} (w_{ij} \ \sigma (\sum_{k=1}^{N_1} v_{jk} + b_{vj}) + b_{wi})) \right) \tag{2}$$

where W_{ij}'s, V_{ij}'s are connection weights, σ is the activation function, X_i's are input, and Y_i's are output. Therefore, ANN creates or modifies the fuzzy rules to adopt with the changes in parameters. An example of hybrid fuzzy-neural network based model is an adaptive neuro fuzzy inference system (ANFIS) developed by Jang (1993). ANFIS has been deployed in a number of applications successfully (e.g., nonlinear system identification, nonlinear control, function approximation). A brief introduction to the subject is given here, but detailed coverage of ANFIS can be found in Jang (1993), King (1999), and Ali and Mo (2001).

An adaptive network, as shown in Figure 4, is a multilayer feedforward network consisting of rectangular and circular nodes and relational links. Each node performs a particular function on incoming signals based on a set of parameters pertaining to that node to achieve optimal control. Rectangular nodes are adaptive and represent input and output fuzzy membership function, while circular nodes are fixed-type nodes. Each output depends on nodal parameters that are adjusted by the ANN to minimize the overall error in the output measured from the desired one. The architecture of ANFIS can be fitted to different infer-

Figure 3. Two-layer perceptron network

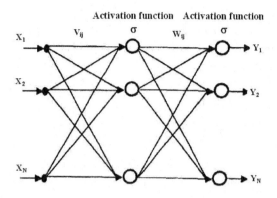

Figure 4. Adaptive neural networks

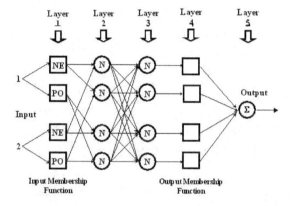

ence schemes, but the Takagi and Sugeno type is computationally more efficient (Nauck, 1999; MATLAB, 2004).

As shown in Figure 4, each node of the first layer is connected to exactly one of the n input variables (here only two input variables, velocity and acceleration data considered and taken for Example 1) and stores the parameters of the input membership function (as shown in Figure 2). The nodes of the second layer are antecedents of the fuzzy rules. Third-layer nodes compute the degree of fulfillment for each rule. The nodes in layer four compute the output values of the rules. The node in the fifth layer is the output nodes and computes the crisp output value. ANFIS requires a predefined structure of the fuzzy system that it trains for the given input, output, and the shape of the membership functions. For Layer 1, every node is associated with a node membership function or the input membership function seen in equation (3).

$$F_i^1 = \mu_{A^i}(x_m) \tag{3}$$

where x_m is the input to node i, μ_{A^i} is the membership function, given by equation (1), A^i is the linguistic label for the membership function (e.g., 'PO' = positive), and F_i^1 is the output from i^{th} node of Layer 1.

In Layer 2, each node computes the weights, given by equation (4), of the input with respect to the given rule base and sends to Layer 3.

$$w_i = \mu_{A^i}(x_m) \times \mu_B(y_m), \quad i = 1, 2. \tag{4}$$

where $\mu_{A^i}(x_m)$ and $\mu_B(y_m)$ are the membership grades of the inputs x_m, y_m in the fuzzy subset A^i and B^i. The outputs of the Layer 3 are normalized, given by equation (5), with respect to the input weights to Layer 3.

$$\overline{w}^i = \frac{w^i}{w^1 + w^2}, \quad i = 1, 2. \tag{5}$$

In Layer 4, output of each rule is compared using a linear combination, given by equation (6), as given by Takagi and Sugeno (Nauck, 1999; Nguyen, Sugeno, Tong, & Yager, 1995).

$$F_i^4 = \overline{w}^i f^i = \overline{w}^i (p_i x_m + q_i y_m + r_i) \tag{6}$$

where p_i, q_i, r_i are parameters of Sugeno type fuzzy inference. The final crisp output (i.e., output of Layer 5) is the sum of all the outputs of Layer 4, as given by equation (7).

$$F_i^5 = \sum \overline{w}^i f^i = \frac{\sum w^i f^i}{\sum w^i}, \quad i = 1, 2. \tag{7}$$

In Example 1, three bell-shaped MFs are taken for each input. The desired output is taken as provided by the LQG. The hybrid learning procedure considering back-propagation and least square is considered (MATLAB, 2004) for error (difference in desired output and trained output) minimization. In Example 1, six earthquake records considering both the horizontal components have been considered for training and testing of the ANFIS.

Evolutionary Optimization Algorithm

Global optimization algorithms emulating certain principles observed both in nature and other man-made physical processes have proved their usefulness in various domains of complex engineering applications. Examples of such phenomena can be found in the annealing processes in metallurgy, central nervous systems, and biological evolution in humans, which in turn have lead to various global optimization methods, such as, *simulated annealing* (SA), *artificial neural networks* (ANNs) and the field of *evolutionary computation* (EC) (Ali & Mo, 2001). EC and related techniques are based upon biological observations that date back to Charles Darwin's discoveries in the 19th century of the means of natural selection, the survival of the fittest, and theories of evolution. The inspired algorithms are termed *evolutionary algorithms* (EAs).

An evolutionary algorithm is a collective name used to describe computer simulation of evolution processes observed in nature and includes genetic algorithms (GAs), evolutionary programming, evolutionary strategy, and genetic programming.

The seminal book, *Adaptation in Natural and Artificial Systems,* published in 1975 by J. H. Holland, laid the foundation of the technique and introduced the GA. During the 1980s rapid progress in computer technology permitted the use of GA in difficult large-scale problems with complicated design space and the method rapidly diffused into the engineering community. A two-branch tournament algorithm for multi-objective optimization using GA used in Example 2 has been discussed.

All EA share a common concept of simulating the individual structural evolution, and selecting superior individuals based on the relative fitness of the individual in a population as happens in a biological evolution process. The *population*, a cluster of candidate solutions of the optimization process called *individuals*, evolves in accordance with the law of natural selection. The more and random the initial population size the faster and better is the optimization. After initialization, the populations are kept for *selection, recombination* and *mutation* repeatedly. Iteration, termed as a *generation*, is continued until some stopping criteria are met. The individuals that undergo recombination and mutation are named *parents* reproducing *offspring*. The selection is made based on the value of fitness of the population obtained from the *fitness function* (objective function). The objective functions ϕ_1, ϕ_2 (equations (8a), (8b)) considered in Example 2 were to *minimize* the maximum of the nondimensionalized peak interstory drift $d_i(t) / d_{max}$ and the maximum of the nondimensionalized peak acceleration $\ddot{u}_{ai}(t) / \ddot{u}_{amax}$ due to the given set of known earthquake excitations (EQ rec) (defined later in equations (9a), (9b)). The objective functions are given by:

$$\phi_1 = \min \left\{ \max_{EQ\ rec} \left[\max_{t,i} \left\{ \frac{|d_i(t)|}{d_{max}} \right\} \right] \right\} \tag{8a}$$

$$\phi_2 = \min \left\{ \max_{EQ\ rec} \left[\max_{t,i} \left\{ \frac{|\ddot{u}_{ai}(t)|}{\ddot{u}_{amax}} \right\} \right] \right\} \tag{8b}$$

A two-branch tournament algorithm is considered for selection of best-fit individuals. The procedure for this selection is shown in Figure 5 and outlined in the following steps:

1. Place the entire population of the current generation in the pot

2. Select two individuals randomly from the pot without replacement

3. Compute the first fitness and copy the better performing individual to the parent pool

4. Repeat the steps 2 and 3 till the pot is empty

5. Refill the pot with population from the current generation

6. Select two individuals randomly from the pot without replacement

7. Compute the second fitness and copy the better performing individual to the parent pool

8. Repeat the steps 6 and 7 till the pot is empty

Figure 5. Two-branch tournament algorithm (Source: Ahlawat & Ramaswamy, 2002; © 2002 by John Wiley & Sons, Ltd., reproduced with permission)

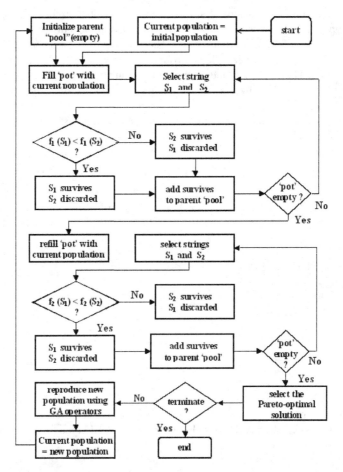

At the end of the above steps, the parent pool is full. A new population is generated from this population of the parent pool using crossover and mutation operations. The random partial exchange of genetic information of successful parental traits occurs through *recombination (cross-over)*. *Mutation* is introduced to input random changes in the specification of the trait in an individual. Mutation aims at introducing new characteristics in the individual that is not there in the parents, in a population by randomly changing one particular string in the gene. Exploration of the search space is achieved by the operation of recombination and mutation. *Exploitation* and *exploration* are the two basic operations that make the GA more powerful than conventional optimization techniques. In a generalized optimization problem, the objective is to search for vector $x \in M$ such that the objective function $\varphi_i(x)$ is minimized (as seen in equations (8a), (8b)). In practice the optimization variables are restricted by constraint functions of the form of $g_j \leq 0$. Analytical solution of the problem are available only when φ and g_j are simple enough.

Moreover, conventional numerical optimization methods require simplification of the original problem (e.g., linearization of the objective and constraint functions) and their derivative to exist. Therefore, solution obtained is locally optimal to the simplified problem. Conventional techniques do not provide global solution. This is the primary motivation for adopting non-conventional optimization methods like GAs for complex nonlinear structures built today.

In intelligent control, issues concerning the use of best form of fuzzy sets in fuzzy controller, that is, to obtain the optimum values of membership function variables (Lee, 1990a, 1990b) or the best topology of the neuron structure, or number of neurons or number of hidden layers in case of neural control are being optimized using evolutionary optimization. The benefits of EA are in its robustness, flexibility, adaptability, and ability to produce near global solutions.

Furthermore, discontinuities, noise and unpredictable phenomena have little effect on the application of GA for multi-dimensional optimization problems.

Numerical Examples

A number of numerical examples are available in the literature, where fuzzy logic based control is used to design controller for building and bridge structures developed in a benchmark exercise.

Example 1. As a first example, we discuss the effectiveness of ANFIS driven hydraulic actuator based control strategy on currently proposed benchmark exercise on a highway bridge control problem provided by ASCE structural control committee (Agrawal, Tan, Nagarajiah & Zhang, 2005; Tan & Agrawal, 2005). A two-span, prestressed concrete box-girder bridge on 91/5 over crossing located in Orange County of southern California forms the benchmark problem.

In the present benchmark, highway bridge problem full order nonlinear model of the bridge has been used as an evaluation model to preserve the effects of column nonlinearity and realistic implementation of control systems. Bilinear force deformation relationship is

Table 1. Inference rules for FLC used in the study

		Acceleration				
		NL	NE	ZE	PO	PL
Velocity	**NL**	NL	NE	NS	NS	ZE
	NE	NE	NS	ZE	ZE	ZE
	ZE	NS	ZE	ZE	ZE	PS
	PO	ZE	ZE	ZE	PS	PO
	PL	ZE	PS	PS	PO	PL

considered for the nonlinear analysis of center columns and isolation bearings. The effect of soil-structure interaction at the end abutments and approach embankments are modeled as spring-damper. The bridge is analysed for six earthquakes considering both near source and far source excitations. The ground motions are considered to be applied to the bridge in both longitudinal and lateral directions. Eight control devices are located between each abutment-end and deck of the bridge. Sensor and actuator dynamics are not considered for the design. A set of 21 evaluation criteria is defined to consider the effectiveness of control design. For complete problem description one can see (Agrawal, Tan, Nagarajiah, & Zhang, 2005; Tan, & Agrawal, 2005).

Multilinear regression analysis is done to determine the effect of states on the control variable. It was found that the velocity and acceleration data have highest regression co-efficient. Cross-correlation analyses revealed it is necessary to feed the ANFIS network with accelerometer data from east and west ends of the bridge. Three bell-shaped MFs for each input is taken and the network is trained for desired output same as LQG control given in the benchmark problem. Of the 16 actuators deployed for longitudinal and transverse vibration response control, 8 actuators (4 on each end), longitudinally attached, are trained using ANFIS to control the longitudinal component of vibration. Of the remaining eight actuators deployed for mitigating the transverse vibration response, four of them (two in each side) are tuned using ANFIS, while the other four are tuned using the FLC rule based shown in Table 1. Acceleration and velocity feedback from the central bent columns are given as input to fuzzy logic controller (FLC) based on rule base shown in Table 1. The velocity and acceleration feedback components help in generating the initial inference rule base capturing the fundamental mode of the motion of a simple pendulum (e.g., if velocity is zero and acceleration is high), the structure is at its extreme position and control action is not needed because it is going to return to its neutral position due to the restoring force.

The control output based on FLC drives actuators transversely attached to the deck. The FLC has been designed using five generalized bell shaped membership functions for each of the input variables (acceleration and velocity) and seven similar membership functions for the output variable (control command) as shown in Figure 6. The input subsets are: NL = negative large, NE = negative, ZE = zero, PO = positive, PL = positive large. The output subsets are: NL = negative large, NE = negative, NS = negative small, ZE = zero, PS = positive small, PO = positive, PL = positive large.

Figure 6. Input and output membership functions

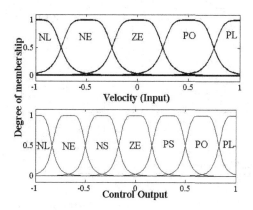

Figure 7. Comparisons of LQG and ANFIS

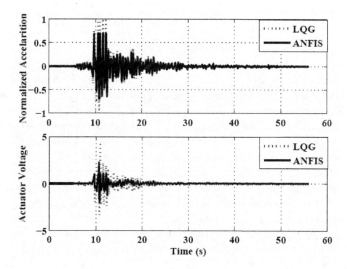

ANFIS is trained separately for both near source earthquakes like Chichi in 1999, Northridge in 1994, and Turkey-Bolu in 1999, and far source earthquake excitations, namely, North Palm Springs in 1986, El Centro in 1940, and Kobe in 1995. Instead of eight accelerometer data, ANFIS needs only four accelerometer data for efficient control with responses comparable to LQG (see Table 2).

Two ANFIS trained FLC are used, on each end of the deck, to control vibration along transverse as well as longitudinal direction with inputs from both ends and central accelerometer. The

Table 2. Comparisons of LQG and ANFIS for Benchmark Highway Bridge

Performance Index	LQG	ANFIS	Performance Index	LQG	ANFIS	Performance Index	LQG	ANFIS
Max Base Shear	0.4861	0.35182	Max Dissipated Energy	0	0	Norm Abutment Displ	0.4512	0.38185
Max Base Moment	0.2880	0.20559	Max Plastic Connections	0	0	Norm Ductility	0.0169	0.01239
Max Midspan Displ	0.1277	0.09009	Norm Base Shear	0.6326	0.46333	Max Control Force	0.0138	0.01687
Max Midspan Accel	0.6771	0.59206	Norm Base Moment	0.1758	0.12845	Max Device Stroke	0.5783	0.6212
Max Abutment Displ	0.5834	0.48209	Norm Midspan Displ	0.1479	0.10703	Max Power	0.0401	0.0220
Max Ductility	0.0634	0.04529	Norm Midspan Accel	0.7556	0.61194	Total Power	0.0082	0.0046
Devices	16	16	Sensors	12	4	Computational Resources	28	20

hybrid control system is found to be effective in reducing the peak values of the responses. Comparison of acceleration and control force at west end for LQG control and ANFIS under Turkey-Bolu earthquake (transverse direction) is shown in Figure 7. Comparison of LQG and ANFIS based control design for 21 performance indices is shown in Table 2. It is evident from Table 2 that hybrid system reduces most of the peak responses by about 30% in comparison to corresponding LQG controller based response and nearly 12% and 20 % respectively for abutment displacement and mid span acceleration at the cost of increase in peak control force by 20% without reaching the actuator maximum limit. Moreover, the present method needs less number of sensors than the LQG based control system. Therefore, probability of failure of control scheme due to sensor failure is less than that of LQG. It is expected that even better results can be obtained by simultaneous optimization of cost function using EA along with FLC. Research on issues such as choice of parameters for optimization and efficient optimization schemes, treatment of seismic excitations accounting for spatial variability in long span bridges that contain many closely spaced modes is presently just underway.

Example 2. FLC driven HMD control of ten-story shear building is presented as second example.

•   **Problem formulation:** A ten-story shear building with HMD at the top floor has been used to demonstrate the multiobjective optimal design procedure for a FLC driven HMD. The mass, stiffness and damping of each floor is taken as m = 3.6×105 kg, k =

Figure 8. (a) Interstory drift vs. peak floor acceleration; (b) stability of FLC (Source: Ahlawat & Ramaswamy, 2002; © 2002 by John Wiley & Sons, Ltd., reproduced with permission)

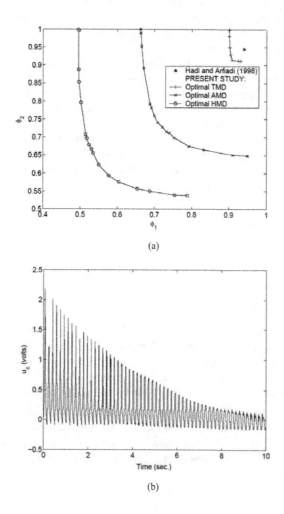

(a)

(b)

650 MN/m, $c = 6.2$ MN-s/m. El Centro, Hachinohe, Northridge, and Kobe earthquake records are considered as input excitation.

- **FLC:** The top floor absolute acceleration and the top floor pseudo-velocity have been used in this study as the input variables. FLC has been designed similar to FLC in the previous example, that is, with five input membership functions and seven out membership functions. The adopted inference rules in this study have been shown in Table 1. In this example a two-branch tournament genetic algorithm (Ahlawat &

Ramaswamy, 2002a) has been employed for multi-objective optimization of the HMD and FLC parameters.

- **Optimization problem formulation:** The objectives function in the optimization problem in this example was to *minimize* equations (8a) and (8b), where the interstory drift and $d_i(t)$ absolute acceleration $\ddot{u}_{ai}(t)$ due to the given set of known earthquake excitations are given in equations (9a), and (9b). The multiobjective optimization of FLC driven HMD has resulted in a set of nondominated (Pareto-optimal) designs. Variation of the objective function ϕ_1 with the corresponding value of the objective function ϕ_2 has been shown in Figure 8a. This variation has been shown for three different cases (1) for HMD, (2) for TMD alone (i.e., if there is no active control device) and (3) for AMD alone (i.e., there is no passive control device). The result available from the literature was for an optimal TMD installed at the top floor for N-S component of El Centro earthquake excitation. For comparison, results for the optimal TMD designed in the same study have been computed for all the four earthquakes to determine the performance of ϕ_1 and ϕ_2.

$$d_i(t) = \left[\begin{array}{cc} u_i(t) & for \ i=1 \\ u_i(t) - u_{i-1}(t) & for \ i \neq 1 \end{array} \right] \tag{9a}$$

$$d^{max} = \max_{t,i} \{|d_i(t)|\} \qquad \ddot{u}_a^{max} = \max_{t,i} \{|\ddot{u}_{ai}(t)|\} \tag{9b}$$

It is shown (Figure 8a) that the HMD performs better than the TMD and AMD acting alone. The stability of the FLC has been examined for each of the Pareto-optimal design using the extreme (worst) initial conditions (Ahlawat & Ramaswamy, 2002a) and has been found to be stable. Results for the worst case of stability has been shown in Figure 8b.

Optimal FLC, using GA for multiple objective, have been applied to various other problems including minimization of seismically excited torsionally coupled flexural building (Ahlawat & Ramaswamy, 2002a, 2002b, 2003), vibration control of third generation benchmark nonlinear building (Ahlawat and Ramaswamy, 2004a) and tall buildings under wind load (Ahlawat & Ramaswamy, 2004b).

Future Directions

Increase in computational facility in recent years and performance expectation in future has paved the way for implementing soft computing techniques to large and complex civil structures. Future is the age of computational intelligence, when structures will think and can take decisions like humans accounting for uncertain events like the magnitude and frequency content of earthquakes. Applications of Fuzzy based control systems; ANN based control systems and hybrid strategies in mitigating the responses of civil structures like buildings have been explored. These tools were found to be efficient and relatively simple to deploy. Computationally hybrid control systems using combination of neural networks, fuzzy systems and evolutionary algorithm have potential application in civil structures due its ability to control nonlinear structural responses and their insensitivity to noise in measurements. Application of intelligent control techniques to vibration control of long span cable supported bridges, which are strong nonlinear systems with many closely spaced modes, remains a demanding field. Incorporation of spatial variation of earthquake excitation leading to differential support movements in intelligent control of structures spread over long spans is a new challenge to structural control engineers.

Conclusion

Hybrid intelligent control techniques are presented in connection with the recent benchmark bridge and building problems. Example 1 shows the efficiency of ANFIS based hybrid intelligent systems over the LQG. A 30% improvement in performance parameters of the structure and controller is obtained using ANFIS approach compared to the LQG controllers. Example 2 gives a comparative study of active, passive and GA optimized FLC. FLC is shown to provide better control over other conventional techniques (LQG, TMD).

Introductory discussion on the related topics of soft computing is provided so that the examples are self-explanatory. Conventional control techniques and their shortcomings have been discussed. It has been shown that the hybrid intelligent control techniques provide better vibration control of structures with complex models and multiple objectives, where linear models are no longer valid. Therefore, linearization of structural systems and identification of parameters introduces errors. Conventional control systems designed for structures based on linear models do not provide optimal control due to the model approximation and related errors. Therefore, controlling structures intelligently without introducing accurate mathematical modeling of the system is shown to be a viable alternative to structural engineers.

References

Agrawal, A. K., Tan, P., Nagarajaiah, S., & Zhang, J. (2005). *Benchmark structural control problem for a seismically excited highway bridge, Part 1: Problem definition* (pp. 1–37). Retrieved June 14, 2005, from http://www-ce.ccny.cuny.edu/People/Agrawal/Highway%20Benchmark%20Problem.htm

Ahlawat, A. S., & Ramaswamy, A. (2002a). Multiobjective optimal design of FLC driven hybrid mass damper for seismically excited structures. *Earthquake Engineering and Structural Dynamics, 31*(7), 1459–1479.

Ahlawat, A. S., & Ramaswamy, A. (2002b). Multiobjective optimal FLC driven hybrid mass damper systems for torsionally coupled, seismically excited structures. *Earthquake Engineering and Structural Dynamics., 31*(12), 2121–2139.

Ahlawat, A. S., & Ramaswamy, A. (2003). Multiobjective optimal absorber system for torsionally coupled, seismically excited structures. *Engineering Structures, 25*(7), 941–950.

Ahlawat, A. S., & Ramaswamy, A. (2004a). Multiobjective optimal fuzzy logic controller driven active and hybrid control systems for seismically excited nonlinear buildings. *Journal of Engineering Mechanics, ASCE, 130*(4), 416–423.

Ahlawat, A. S., & Ramaswamy, A. (2004b). Multiobjective optimal fuzzy logic controller system for response control of wind excited tall buildings. *Journal of Engineering Mechanics, ASCE, 130*(4), 524–530.

Ali, Z., & Mo, J. (Eds.). (2001). *Intelligent control systems using soft computing methodologies*. CRC Press.

Caicedo, J. M., Dyke, S. J., Moon, S. J., Bergman, L. A., Turan, G., & Hague, S. (2003). Phase II benchmark control problem for seismic response of cable-stayed bridge. *Journal of Structural Control, 10*(3), 137–168.

Dyke, S. J., Caicedo, J. M., Turan, G., Bergman, L. A., & Hague, S. (2003). Phase I benchmark control problem for seismic response of cable-stayed bridge. *Journal of Structural Engineering, ASCE, 129*(7), 857–872.

Faravelli, L., & Yao, T. (1997). Elements of fuzzy structural control. In B. M. Ayyub, A. Guran, & A. Haldar (Eds.), *Uncertainty modeling in vibration and fuzzy analysis of structural systems* (pp.147–165). Singapore: World Scientific..

Holland, J. H. (1975). *Adaptation in natural and artificial systems*. Ann Arbor: The University of Michigan Press.

Housner, G. W., Bergman, L. A., Caughey, T. K., Chassiakos, A. G., Claus, R. O., Masri, R. E., et al. (1997). Structural control: Past, present and future. *Journal of Engineering Mechanics, ASCE, 123*(9), 897–970.

Jang, J. S. R. (1993). ANFIS: Adaptive-network-based-fuzzy-inference-system. *IEEE Transactions on System, Man and Cybernetics, 23*(3), 665–685.

King R. E. (1999). *Computational intelligence in control engineering*. Marcel Dekker.

Lee, C. C. (1990a). Fuzzy logic in control system: Fuzzy logic controller: Part 1. *IEEE Transactions on System, Man and Cybernetics, 20*(2), 404–418.

Lee, C. C. (1990b). Fuzzy logic in control system: Fuzzy logic controller: Part 2. *IEEE Transactions on System, Man and Cybernetics, 20*(2), 419–435.

Mamdani, E. H. (1974). Application of fuzzy logic algorithms for control of simple dynamic plant. In *Proceedings of the Institute of Electrical Engineers* (pp. 1585–1588).

MATLAB. (2004). *The software for Numerical Computing Version 7.0.1. (R14)*. The Math Works. Retrieved November 27, 2006 from http://www.mathworks.com/products/matlab/

Nagarajaiah, S., & Narasimhan, S. (2003, July 16–18). Smart base isolated benchmark building, Part II: Sample controller for linear and friction isolation. In *Proceedings of the 16th ASCE Engineering Mechanics Conference*, University of Washington, Seattle, WA.

Narasimhan, S., Nagarajaiah, S., Johnson, E. A., & Gavin, H. P. (2003). Benchmark problem for control of base isolated building. In *Proceedings of the Third World Conference on Structural Control,* Como, Italy (Vol. 2, pp. 349–354).

Nauck, D. (1999). Neuro-fuzzy methods. In H. B. Verbruggen & R. Babuska (Eds.), *Fuzzy logic control: Advances in applications* (pp. 65–86). Singapore: World Scientific.

Nguyen, H. T., Sugeno, M., Tong, R. M., & Yager, R. R. (Eds.). (1995). *Theoretical aspects of fuzzy control.* New York: Wiley.

Ohtori, Y., Christenson, R. E., Spencer, B. F., & Dyke, S. J. (2004). Benchmark control problem for seismically excited nonlinear buildings. *Journal of Engineering Mechanics, ASCE, 130*(4), 366–385

Patyra, M. J., & Mlynek, D. M. (1996). *Fuzzy logic: Implementation and applications.* Chichester, UK: Wiley & Teubener.

Soong T. T. (1990). *Active structural control: Theory and practice.* New York: Wiley.

Soong, T. T., & Spencer, B. F. (2002). Supplemental energy dissipation: State of the art and state of the practice. *Engineering Structures, 24*(3), 243–259.

Spencer, B. F., Dyke, S. J., & Deoskar, H. S. (1998). Benchmark problem in structural control-Part I: Active mass driver system. *Earthquake Engineering and Structural Dynamics, 27*(11), 1127–1139.

Spencer, B. F., & Nagarajaiah, S. (2003). State of the art of structural control. *Journal of Structural Engineering, ASCE, 129*(7), 845–856.

Tan, P., & Agrawal, A. K. (2005). *Benchmark structural control problem for a seismically excited highway bridge–Part II: Sample control designs.* Retrieved June 14, 2005, from http://www-ce.ccny.cuny.edu/People/Agrawal/Highway%20Benchmark%20Problem.htm

Teng, T. L., Peng, C. P., & Chuang, C. (2000). A study on the application of fuzzy theory to structural active control. *Computer Methods in Applied Mechanics and Engineering, 189*(2), 439–448.

Wang, A. P., & Lee, C. D. (2002). Fuzzy sliding mode control for a building structure based on genetic algorithms. *Earthquake Engineering and Structural Dynamics, 31*(6), 881–895.

Yang, J. N., Agrawal, A. K., Samali, B., & Wu, J. C. (2004). Benchmark problem for response control of wind excited tall buildings. *Journal of Engineering Mechanics, ASCE, 130*(4), 437–446.

Zadeh, L. A. (1965). Fuzzy sets. *Information and Control, 8,* 338–353.

Zuk, W. (1980). The past and future of active control systems. In *Proceedings on the First International IUTAM Symposium on Structural Control*, University of Waterloo, Canada, North Holland, Amsterdam.

Section II

Structural Assessment Applications

Chapter VII

Neuro-Fuzzy Assessment of Building Damage and Safety After an Earthquake

Martha L. Carreño, Universidad Politécnica de Cataluña, Spain

Omar D. Cardona, Universidad Nacional de Colombia, Colombia

Alex H. Barbat, Universidad Politécnica de Cataluña, Spain

Abstract

This chapter describes the algorithmic basis of a computational intelligence technique, based on a neuro-fuzzy system, developed with the objective of assisting nonexpert professionals of building construction to evaluate the damage and safety of buildings after strong earthquakes, facilitating decision-making during the emergency response phase on their habitability and reparability. A hybrid neuro-fuzzy system is proposed, based on a special three-layer feedforward artificial neural network and fuzzy rule bases. The inputs to the system are fuzzy sets, taking into account that the damage levels of the structural components are linguistic variables, defined by means of qualifications such as slight, moderate or severe, which are very appropriate to handle subjective and incomplete information. The chapter is a contribution to the understanding of how soft computing applications, such as artificial neural networks and fuzzy sets, can be used to complex and urgent processes of engineering decision-making, like the building occupancy after a seismic disaster.

Introduction

After an earthquake, it is necessary to answer some urgent questions, such as: How many buildings were affected? What is the geographic distribution of damage? What was the degree of damage? Are the buildings habitable, and what is the level of safety? Must people be evacuated? What type of alternative actions should be immediately taken? Are there trapped people? What structures represent danger for neighbours and pedestrians? What types of buildings were affected? In order to answer all these questions, it is necessary to carry out an accurate process of damage evaluation that requires the participation of professional experts in the field. Unfortunately, the number of professionals who fulfil that expertise is always insufficient and, therefore, the evaluation process becomes even more difficult. By one hand, for nonexperts, the impact caused by seeing damage is so great that they tend to describe it as more severe than it really is. By the other hand, non experts can underestimate cases of severe damage because of their innocuous appearance. There is no doubt that the information obtained during the evaluation process is highly subjective and that it depends on the conception and the impression that the inspectors have about each case. In all evaluation methods, the damage levels are defined with linguistic qualifications such as light, minor, moderate, average, severe, etc. These definitions can have a remarkable variation in their meaning according to the person who uses them.

Soft computing can be used to overcome these difficulties of damage evaluation. Neural networks have been used to face complex problems simulating the function of the human neural system, imitating the adaptive and cognitive mechanisms of human learning. Fuzzy logic is an innovative way of representing qualitative or subjective information in numerical form, very useful for technologic and engineering applications where expert criteria are required. Referring to risk evaluation, Carreño, Cardona, and Barbat (2004) and Cardona (2005), applied soft computing techniques to make evaluations of urban seismic risk before and after earthquakes (*ex-ante* and *ex-post* evaluations) and to measure the disaster risk management performance and effectiveness at national, subnational and local level). Considering these features and applications of the computational intelligence techniques and the decision-making needed to determine the habitability and reparability of affected buildings after a seismic disaster, an expert system for post-earthquake building damage and safety evaluation, using a nonsupervised learning Kohonen neuro-fuzzy algorithm, was designed to avoid the mistakes usually made by nonexpert building inspectors when handling subjective and incomplete information. This model considers the possibility of damage in structural and architectural elements and the potential site seismic effects. It also takes into account the preexisting conditions that increase the building vulnerability, such as the bad quality of the construction materials.

Postearthquake Building Damage Evaluation

Seismic Damage of Buildings

After a strong earthquake strikes a vulnerable urban centre, too much damage may occur on the exposed elements like buildings, facilities, and infrastructure lifelines. The damaged buildings could be many and scattered in the city and the damage degrees could be several. The population and governmental officials usually become very concerned about the security of their lives and they need to know if their buildings are safe or not. This question only can be answered by engineers and architects experts in structural and soil mechanics, damage evaluation and building rehabilitation. The decision-making on the habitability and reparability of buildings is urgent; a bad decision could jeopardize human lives.

The damage evaluation is a difficult task and its results depend on the experience of the inspector. Sometimes, a building is obviously unsafe due to the observed damage (see Figures 1-4), but the most cases can generate doubts (see Figures 5-7). In the diagnostic of a building, it is necessary to take into account not only the different damage levels of the elements, but also the overall structural stability. Affected structural and non structural elements can endanger the human life in different ways. The damage of a building can be isolated or generalised and, in both cases, can put in danger the structural stability, depending on the structural configuration or redundancy and on the adverse ground conditions.

The state-of-the-art of earthquakes and seismic damage of buildings have allowed the development of appropriate earthquake resistant design and construction techniques. These techniques include technical and economical criteria to obtain less strong but more ductile structures permitting to control damage without collapse by dissipating a part of the absorbed seismic energy. In general, the building seismic codes accept heavy damage without collapse of the building in case of severe earthquakes; nonstructural effects without or with minor structural damage in case of moderate earthquakes; and slight or no damage when moderate earthquakes occur (Cardona, 2001).

Dowrick (1987) says that a structure will have the maximum chance of surviving an earthquake if the following are true:

- The load-bearing members are uniformly distributed.
- The columns and walls are continuous and without offsets from roof to foundation.
- All beams are free form offsets.
- Columns and beams are co-axial.
- Reinforced concrete columns and beams have nearly the same width.
- Nonprincipal members does not change sections suddenly.
- The structure is a continuous, redundant, and as monolithic as possible.

The ductility and structural redundancy have been the criteria more effective to assure the security against the structural collapse (García, 1998). In reinforced concrete buildings,

Figure 1. Damaged school in Colombia (1999), obviously unsafe

Figure 2. Damaged school in Colombia (1983), clearly unsafe

the typical damage due to strong earthquakes can be observed by the presence in columns of diagonal cracks due to shear or torsion, vertical cracks, reinforcement concrete cover spalling and concrete crush or buckling of the longitudinal reinforcing bars due to excessive bending and axil stresses (see Figure 6). In beams, the most typical effects are the diagonal cracks (see Figures 5 and 7) and the stirrup failure due to shear or torsion; vertical cracks, longitudinal reinforcement failure and concrete crush due to the bending demand to alternating loads. The connections between the structural members are also critical points. The beam-column joints usually present diagonal cracks as a result of shear stresses and it is common the failure because of the lack of anchorage of the longitudinal reinforcement of the beams into the joint or due to excessive bending stresses. The slabs can present punching shear cracks around the columns and longitudinal cracks due to the excessive bending demand (Cardona & Hurtado, 1992).

Figure 3. Damaged residential building in Turkey (1999), obviously unsafe

Figure 4. Damaged office building in Turkey (1999), clearly unsafe

The seismic performance of a building can be improved if some technical criteria are considered, such as avoiding soft story effect, excessive slenderness of columns, fragile infill panels in framed buildings. It is also recommended to use columns stiffer than beams, thicker slabs and adequate gaps between buildings, among others.

The damage in nonstructural elements represents a high proportion of the total damage caused by an earthquake; the most part of it is unrelated to the structure of the building (Coburn & Spence, 2002). Usually, this damage occurs due to inappropriate connection between infill panels, installations or other nonstructural components and the structure, or due to the lack of stiffness of the structure, which result in excessive inter story-drift and lateral deformation. In masonry partitions and façades are common the diagonal cracks, as it is shown in Figure 8.

Figure 5. Affected bus terminal in Colombia (1983)

Figure 6. Damaged column in a residential building in Colombia (1999)

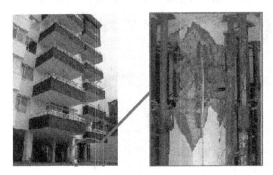

Figure 7. Cracked beam in a office building in Colombia (1999)

Figure 8. Cracks in fragile masonry walls

Existing Seismic Damage Evaluation Methodologies and Guidelines

When the seismic damage is extended and widespread, the number of the required professionals with the experience to tackle the damage assessment is always insufficient. All massive evaluation processes depart from the need to involve voluntary professionals without expertise and experience to determine the habitability and reparability of many buildings; decision-making is really difficult for the nonexpert professionals. The need of avoiding unnecessary demolitions and of helping to define accurate and cost-effective measures of repairing or reinforcing the damaged buildings are the main objectives of the existing damage evaluation methodologies and guidelines.

The development of damage evaluation guidelines has been necessary in countries with high seismic activity. These guidelines have the aim of defining, as soon as possible, whether the buildings may continue being used or not, and identifying safe buildings which can be used as temporary shelters for the evacuated persons. In addition, damage evaluations are essential to make decisions not only about the structural safety, but also to improve the effective earthquake-resistant construction codes by identifying the types of failures of the different structural systems. Using the data of building damage, it is also possible to develop empirical vulnerability functions which are useful to classify and study the affected buildings.

Some countries have developed systematic guidelines and procedures to evaluate the building damage. The main efforts, described by Carreño, Cardona, and Barbat (2005), are the following:

After the Skopje's earthquake of 1963, and particularly after the Montenegro earthquake of 1979, the Institute of Earthquake Engineering and Engineering Seismology (IZIIS) of the University "Kiril and Metodij" of the Former Yugoslav Republic of Macedonia, developed a methodology for the damage evaluation after an earthquake (Instituto de Ingeniería Sísmica y Sismología [IZIIS], 1984). Its main objectives were the reduction of human losses in buildings with low resistance or in damaged buildings which could be destroyed by aftershocks; data acquisition regarding the magnitude of a disaster in terms of available housing, destroyed buildings and unsafe buildings; data acquisition for civil protection and rescue planning and organization after earthquakes; and the improvement of the design specifications of the earthquake resistant construction codes. This methodology and procedure was translated to Spanish and used after the Mexico earthquake of 1985.

Applied Technology Council (ATC; 1989) proposed for California and, in general, for United States the "Procedures for post-earthquake safety evaluation of buildings" (ATC, 1989, p. 20) with three steps. The first one is the rapid evaluation, in which it is decided if a building is obviously unsafe or apparently habitable. These evaluations are often cursory, because there is no sufficient personnel available to perform more thorough inspections. The second step corresponds to the detailed evaluation, in which the buildings obviously unsafe are visually evaluated by a structural engineer. The third step is the engineering evaluation, which is performed for questionable and severely damaged buildings that have to be rehabilitated by to the owner's engineer. A second version of the procedure was published in 1995 with the title *Addendum to the ATC 20 Postearthquake Building Safety Evaluation Procedures* (ATC, 1989, pp. 20-22). In 2003, a mobile postearthquake building safety evaluation data acquisition system was developed—ATC 20i (ATC, 2003), and in 2005, the second edition of the ATC 20-1 *Field Manual: Postearthquake Safety Evaluation of Buildings*, (ATC, 2005) was published.

After the Miyagiken-Oki earthquake, in 1978, the *Guides for Damage Evaluation After an Earthquake and Restore Techniques* were published and tested after the Nihonkai-Chumbu earthquake in 1983 and after the Mexico earthquake in 1985. The methodology was reviewed in 1989 and was published by the Japanese Association for the Disaster Prevention in Buildings (CENAPRED, 1996). Accordingly, the buildings which have to be evaluated are selected by a general inspection after the earthquake. The evaluation is performed in two steps: an immediate visual evaluation of risk level or habitability and a visual evaluation of the degree of structural damage. The first step establishes if the damaged structure or a part of it puts in danger the human life by overturning, failure or collapse. In the second step, the evaluation is based on the level of damage of the building and its components. As a result of this process, a suggestion is made to the owner regarding the necessity of structural rehabilitation (repair, reinforcement, or demolition).

In Mexico, the Institute of Engineering of the National University (UNAM) developed the *Guideline for Post-Earthquake Evaluation of the Structural Safety of Buildings* (Rodríguez & Castrillón, 1995). This method was reviewed and published by the Mexican Society for Earthquake Engineering (Sociedad Mexicana de Ingeniería Sísmica; SMIS) and the government of Mexico City in 1998 (SMIS, 1998) and, like the ATC 20, has three steps: a rapid evaluation, a detailed evaluation and a specialized engineering evaluation..

After the earthquake of Friuli in 1976, in Italy was developed a procedure for estimating the economic losses. More recently, a proposal was published by Goretti (2001) based on a research programme started in 1995. Guidelines and forms were published in 2000 after the earthquakes of Umbría-Marche in 1997 and Pollino in 1998, where the major part of the published decrees for evacuation or limited use of buildings were in agreement with the suggestions made by the inspectors. Another important initiative has been the development of the self-training multimedia tool called MEDEA (Manuale di Esercitazioni sul Danno Ed Agibilità) promoted by the Servizio Sismico Nazionale (Papa & Zuccaro, 2003; Zuccaro & Papa, 2002), which was proposed as a handbook for a consistent classification of the structural elements of masonry and reinforced concrete buildings and of their relevant damage typologies.

In Colombia, evaluation methodologies for some important cities have been developed and later have been reviewed using actual damage data. After the coffee-growing-area earthquake of 1999, several studies have been made in Colombia on seismic hazard and vulnerability to promote seismic risk reduction of buildings and infrastructure (Campos, 1999). In one of the most important projects, lead by the Colombian Association for Earthquake Engineering (Asociación Colombiana de Ingeniería Sísmica; AIS), a methodology for habitability and reparability evaluation of buildings in case of earthquakes has been developed. This method was adopted officially by cities like Bogotá (AIS, 2002) and Manizales (AIS, 2003) and includes an evaluation form, a field manual for the evaluation of the damaged buildings and a neuro-fuzzy system used in the habitability and reparability evaluation (AIS, 2004), which is described in this chapter.

Other works are related to the detection of damaged building in disaster areas using satellite images. One of them is the method proposed by Matsuoka and Yamazaki (2004), which uses satellite synthetic aperture radar (SAR) to identify the distribution of the damaged buildings in the area after a disaster by comparing the pre- and postevent images. This tool helps to detect the extension and magnitude of disasters and is useful for disaster management activities.

Common Problems of Damage and Habitability Evaluations

Taking into account the experiences acquired during different earthquakes and, particularly, the lessons learned in California in 1989 and 1994 about the application of the ATC 20, and after the earthquakes of Colombia occurred in 1995 and 1999, it is possible to say that the damage evaluation processes presents similar problems and difficulties in different countries. The most important shortcomings are the following:

- **The lack of training and qualification of inspectors:** According to the findings of risk perception researchers, the tendency of nonexpert inspectors is to aggravate or to underestimate the damage level of buildings. The information obtained during the damage evaluation process is highly subjective and depends on the inspector's heuristic and biases. For this reason, it is desirable the previous identification and training of inspectors and the anticipated organization of the evaluation activities.

- **Subjectivity of the evaluation:** The damage levels in all methodologies and evaluation guidelines are defined using linguistic qualifications like light, moderate or severe; which may have different meanings according to the judgement of each person. Moreover, the limit between these assessments is not clear. This is the reason why it is necessary to improve the standardisation of the meaning of damage levels before any evaluation.

- **Building location problems:** Other difficulty is the lack of standardisation of the postal addresses in certain cities; this makes difficult the visits of the professional teams, the organisation of the obtained data and the development of maps of damage. One suggestion made after several disasters, was to cover the total number of affected buildings to avoid confusion and misunderstanding of people. In addition, inspectors should have clear guidelines to provide appropriate information about damage to the owners due to the fact that, in some cases, bad communications can exacerbate legal problems and government liabilities.

- **The lack of data organization and systematization:** It is important to have official records of every action and decision; bad quality data and lack of systematization contribute to the confusion and to delay relevant decisions from the perspective of disaster management. On the other hand, it is necessary to have in advance a contingency plan in which the damage evaluation process is one of the needed tasks in order to avoid coordination difficulties among all the governmental emergency agencies, from the local and regional to the national level.

Application of Neural Networks and Fuzzy Sets to Damage Assessment

As mention above, the lack of experience of the inspectors and the unavoidable need of involving in the process nonexpert professionals, make very difficult the decision making regarding safety of the affected buildings. Techniques of computational intelligence facilitate the massive and correct evaluation of damage, risk, habitability and reparability of the state of the affected buildings after an earthquake. Taking into account the different perspectives in using these techniques and tools, some basic definitions are useful to understand their applications:

- **Neural networks:** They are systems that make use of the known or expected organizing principles of the human brain. They consist of a number of independent and simple processors: the neurons. These neurons communicate with each other via weighted connections, the synaptic weights (Nauck, Klawonn, & Kruse, 1997).

- **Linguistic variable:** This concept was introduced by Zadeh (1975) to provide a basis for approximated reasoning, as follows: "By a linguistic variable we mean a variable whose values are words or sentences in a natural or artificial language. The motivation for the use of words or sentences rather than numbers is that linguistic characterizations are, in general, less specific than numerical ones" (Rutkowska, 2002).

- **Fuzzy set:** It is a set without a crisp boundary. The transition from "belong to a set" to "not belong to a set" is gradual and this smooth transition is characterised by members.

The imprecisely defined sets play an important role in human thinking, particularly in the domains of pattern recognition, communication of information and abstraction (Zadeh, 1965). Some authors proposed extend the learning methods to fuzzy training application using fuzzy partial matching indices (Cross & Sudkamp, 1991; Dubois & Prade, 1982) to determine the degree of match. Another approach to learning fuzzy rules has been proposed by Kosko (1992). Kosko envisages the construction of a rule base as a search through the space of all fuzzy rule bases over a fixed topology.

- **Soft computing:** It is an emerging approach to computing which parallels the remarkable ability of the human mind to reason and learn in an environment of uncertainty and imprecision (Zadeh, 1992). The principal constituents of soft computing are fuzzy logic, neurocomputing and genetic algorithms. According to Jang, Sun and Mizutani (1997), human expertise, biologically inspired computing models, new optimization techniques, numerical computation, new application domains, model-free learning, intensive computation, fault tolerance, goal driven characteristics, real-world applications, are the main characteristics of soft computing.

- **Approximate reasoning:** Reasoning with fuzzy logic is not exact but rather is approximate. Based on fuzzy premises and fuzzy implications, fuzzy conclusions are inferred (Rutkowska, 2002).

- **Neuro-fuzzy system:** It is a combination of neural networks and fuzzy systems in such a way that neural networks, or neural networks learning algorithms, are used to determine parameters of fuzzy systems (Nauck et al., 1997).

In the framework of the damage evaluation Chou and Ghaboussi (2001) studied the application of genetic algorithms to damage detection. Static measures of displacements are used to identify the changes of the properties of structural members, such as Young's modulus and the cross-sectional area. For this implementation, bridges have been instrumented and remotely monitorized. The earthquake damage evolution was studied by Song, Hao, Murakami, and Sadohara (1996) using fuzzy theory. Zhao and Chen (2002) proposed a fuzzy system for concrete bridge damage diagnosis. They built the membership functions of the input variables with a fuzzy partitioning algorithm and induced the fuzzy rules from the numerical data. The diagnosis is based on three kinds of factors: design factors, like structural type, span length, deck width, number of spans, etc.; environmental factors, like humidity and precipitation, climate of the region, traffic volume, temperature variations, and so forth; and other factors, like the structure age, function class and location of damages. Lagaros, Papadrakakis, Fragiadakis, Stefanou, and Tsompanakis (2005) proposed the application of artificial neural networks for the probabilistic safety analysis of structural systems under seismic loading. Ahlawat and Ramaswamy (2001) proposed a system for structural vibration control using the fuzzy sets theory.

The soft computing tools have been used in solving problems in many other areas in the field of earthquake engineering. Other two works related to structural behaviour are the application of neural networks in the stochastic mechanics (Hurtado, 2001) or the estimation of the service life of reinforced concrete structural members using fuzzy sets (Anoop, Rao, & Rao, 2002).

Proposed Soft Computing Model

Expert System for Damage Evaluation

In spite of the benefits of the damage evaluation methodologies and guidelines for buildings discussed in the previous section, decision mistakes like demolition of noncritical buildings or unnecessary building evacuation are still possible due to the lack of experience and qualification of inspectors,. This represents serious burdens, especially in the case of key buildings. On the other hand, it is possible that building damages that put at risk the structural stability could be ignored, jeopardizing the life of the occupants. This is the reason why a neuro-fuzzy expert system and a computational model have been proposed and designed to be used in the emergency response phase in case of strong earthquakes (Carreño, Cardona, & Barbat, 2003).

This section describes the use of the soft computing as support to the building habitability evaluation. The expert system for the building damage evaluation process is based on artificial neural networks and fuzzy sets. The authors have been working in this model since 2000 and, although this tool has not been tested yet in a real earthquake emergency, recently it has been adopted officially by the administrations of the cities of Bogotá and Manizales, in Colombia, to face future earthquakes and complement its calibration, once the calibration performed, the system will be ready for use in a real case of earthquake emergency. The model uses a fuzzy logic approach, required by the subjective available information which can be based on linguistic qualifications for the damage levels and can be incomplete. This enables the use of computational intelligence for the damage evaluation by nonexperts.

For the building evaluation four groups of elements were identified: structural elements, non structural elements, ground conditions and preexisting conditions. The first three indicate the damage condition of the building that can jeopardize the life of the occupants. Preexistent conditions are related to the quality of the construction materials, plane and vertical shape irregularities of the building, and the structural configuration and help to identify the building reparability condition.

The proposed model uses an artificial neural network (ANN). Its structure consists of three layers. The neurons in the input layer are grouped in four sets, namely structural elements (SE), nonstructural elements (NE), ground conditions (GC) and preexistent conditions (PC). Each one contributes with information to the neurons in the intermediate layer. They only affect the intermediate neuron in the group to which they belong. The number of input neurons or variables in the model is not constant; this number varies depending on the structural system and on the importance of the groups of variables for the evaluation. In some cases, it is not necessary to evaluate the ground conditions or the preexistent conditions if the damage is important. The number of neurons of the input layer used to analyse the state of the structural elements group changes according to the class of building. Table 1 shows the structural elements or variables considered according to the structural system. A qualification is assigned, depending on the observed damage using five possible damage levels that are represented by means of fuzzy sets. For structural and nonstructural elements, the following linguistics damage state qualifications are used: none (N), light (L), moderate (M), heavy (H) and severe (S). Figure 9 illustrates the membership functions for these qualifications.

Table 1. Structural elements according to structural system

Structural system	Structural elements
RC frames or (with) shear walls	Columns/walls, beams, joints and floors
Steel or wood frames	Columns, beams, connections and floors
Unreinforced/Reinforced/Confined masonry	Bearing walls and floors
Bahareque or tapial walls	Bearing walls and floors

Figure 9. Membership functions for linguistic qualifications

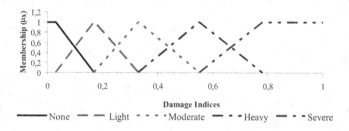

The damage levels in different structural and nonstructural elements of buildings having different typologies can be seen in Figures 10 and 11. The membership functions of the fuzzy sets reach their maximum membership point for the values of the damage indices whose selection will be explained later and is given in Table 3.

Damage in of nonstructural elements does not affect the overall stability of the buildings, but may put at risk the security of the occupants. The nonstructural elements are classified in two groups: elements whose evaluation is compulsory and elements whose evaluation is optional (see Table 2).

The ground and preexistent conditions variables are valued through the qualification of their state in the evaluation moment. The used linguistic qualifications are: very good (VG), medium or poor (M), and very bad (VB). Ground conditions consist of variables that can affect the stability of the building, such as landslides and soil liquefaction; examples of these situations can be observed in Figure 12. Preexistent conditions are illustrated in Figure 13 and can increase the seismic vulnerability of a building.

In the intermediate layer, an index is obtained by defuzzification for each group of variables. Taking into account the four available indices, it is possible to define in the output layer the building damage using fuzzy rules with the structural and nonstructural evaluations. The concept of linguistic variable was a stepping-stone to the concept of a fuzzy IF-THEN rule. Fuzzy rules and their manipulation refer to the so-called calculation of fuzzy rules, the largely self-contained part of fuzzy logic often used in practical applications (Rutkowska, 2002; Zadeh, 1975, 1996). The concept of fuzzy rule is important when the dependencies described by these rules are imprecise or a high degree of precision is not required (Rut-

Figure 10. Damage in structural elements: (a) Severe damage in a reinforced concrete joint, (b) moderate damage in a reinforced concrete beam, (c) heavy damage in a masonry wall, (d) heavy damage in a bahareque wall

(a)

(b)

(c)

(d)

Figure 11. Damage in nonstructural elements: (a) Severe damage in masonry partitions, (b) heavy damage in stairs

(a)

(b)

Table 2. Nonstructural elements

Compulsory evaluation elements	Partitions
	Elements of façade
	Stairs
Optional evaluation elements	Ceiling and lights
	Installations
	Roof
	Elevated tanks

Figure 12. Ground conditions: (a) Soil settlement and liquefaction, (b) landslides and ground failure

(a)   (b)

kowska). The fuzzy rule base is the knowledge base which consists of a collection of fuzzy if-then rules.

Consequently, following the proposed fuzzy rules, the building habitability is obtained by means of structural and nonstructural evaluations but also assessing the ground conditions. Finally, using the preexistent conditions, the system defines the required level of reparation. Thus, habitability and reparability recommendations can be made after an earthquake by using this tool. Remarks as: "habitable after minor adequateness" or "restricted: usable after reparation" or "unsafe: usable after structural strengthening or reinforcement" or "dangerous: possible demolition or total building rehabilitation," are decisions obtained from this expert system. Figure 14 shows the structure of the neural network used in the proposed model.

Figure 13. Preexistent conditions: (a) Bad construction quality, (b) vertical shape irregularities, soft floor, (c) plane shape irregularities, (d) bad structural configuration: some elements are out of the main frames

(a)   (b)

(c)   (d)

Figure 14. Structure of the neural network

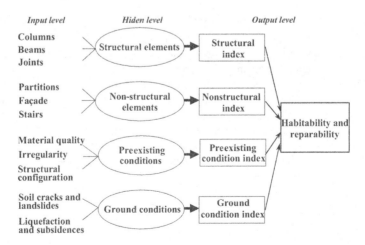

Description of the ANN

Input Layer of the Artificial Neural Network

The fuzzy sets for each element or variable i (for instance columns, walls or beams) of the input layer are obtained from the linguistic qualifications made by the inspectors after a visual inspection of the building, which provide the damage D_j at each level j and its extension or weight w_j. The damage extension, or percentage of each damage level in each element, varies from 0 to 100 and it is normalised:

$$w_j = \frac{D_j}{\sum_N D_j}, \quad \sum_N w_j = 1 \tag{1}$$

The accumulated qualification of damage D_i for each variable is obtained as the union of the scaled fuzzy sets, taking into account the damage membership functions $\mu_{D_j}(D_j)$ and its extensions or weights assigned by the inspector:

$$D_i = (D_N \cup D_L \cup D_M \cup D_H \cup D_S) \tag{2}$$

$$\mu_{D_i}(D) = \max(w_{N,i} \times \mu_{D_N}(D_{N,i}), \ldots, w_{S,i} \times \mu_{D_S}(D_{S,i})) \tag{3}$$

The union in the theory of fuzzy sets is represented by the maximum membership or dependency. By means of defuzzification, using the centroid of area method (COA), a qualification index C_i is obtained for each variable of each group of neurons

$$C_i = [\max(w_{N,i} \times \mu_{D_N}(D_{N,i}), \ldots, w_{S,i} \times \mu_{D_S}(D_{S,i}))]_{centroid} \tag{4}$$

Each variable has predefined the basic membership functions for the fuzzy sets corresponding to the five possible levels of damage. The linguistic qualifications change in each case.

Intermediate or Hidden Layer of the ANN

This layer has four neurons corresponding to each group of variables: structural elements, nonstructural elements, ground conditions and preexistent conditions. Figure 15 shows a more detailed scheme of the evaluation process. In this neural network model, the inputs of the four neurons are the qualifications C_i obtained for each variable from each group of neurons and its weight W_i or degree of importance on the corresponding intermediate neuron. These weights have to be defined with the participation of experts in earthquake damage evaluation. The weights considered for some structural systems are shown in Table 3, while tables 4, 5, and 6 show the weights for the nonstructural elements, ground conditions and

Table 3. Weights for structural elements according to the building type

Structural system	Beams	Columns	Joints or connections	Walls	Bearing walls	Floors
Reinforced concrete frame	19	46	25	-	-	10
Reinforced concrete structural wall	15	-	20	57	-	8
Confined masonry	-	-	-	-	73	27
Reinforced masonry					73	27
Unreinforced masonry	-	-	-	-	70	30
Bahareque walls	-	-	-	-	77	23
Steel frame	18	39	35	-	-	8
Wood frames	23	45	21	-	-	11

preexisting conditions. Using these qualifications and weights for each variable i, a global index is obtained, for each group k, from the defuzzification of the union or maximum membership of the scaled fuzzy sets:

$$I_{SE} = [\max (W_{SE1} \times \mu_{C_{SE1}} (C_{SE1}),...,W_{SEi} \times \mu_{C_{SEi}} (C_{SEi}))]_{centroid} \tag{5}$$

$$\mu_{CSE}(C) = \max (W_{SE1} \times \mu_{C_{SE1}} (C_{SE1}),...,W_{SEi} \times \mu_{C_{SEi}} (C_{SEi})) \tag{6}$$

Table 4. Weights for nonstructural elements

Element	Weight
Partitions	35
Façade	35
Stairs	30

Table 5. Weights for ground conditions variables

Element	Weight
Soil cracks and land slides	50
Liquefaction and subsidences	50

Table 6. Weights for preexisting conditions variables

Element	Weight
Materials quality	25
Plane shape irregularities	25
Vertical shape irregularities	25
Structural configuration	25

The membership functions $\mu_{C_{ki}}(C_{ki})$ and their weights W_{ki} show the notation for the group of structural elements. The groups of variables related to ground and preexisting conditions are optional; thus they can be or can be not considered within the evaluation. In this last case the habitability and reparability of the buildings is assessed only with the structural and nonstructural information.

Output Layer of the ANN

In this layer, the global indices obtained for structural elements, nonstructural elements, ground and preexistent conditions correspond to one final linguistic qualification in each case. The damage level (qualitative) is obtained according to the "proximity" of the value obtained to a global damage function of reference. In this layer, the training process of the neural network is performed. The indices that identify each qualitative level (centre of cluster) are changed in agreement with the indices calculated in each evaluation and with a learning rate. Once the final qualifications are made, it is possible to determine the global building damage, the habitability and reparability of the building using a set of fuzzy rule bases.

Training Process of the ANN

The neural network is calibrated in the output layer where the damage functions are defined in relation to the damage matrix indices. In order to start the calibration, a starting point is defined, that corresponds to the initial indices of each level of damage. The indices proposed by the ATC-13 (ATC, 1985), Park, Ang, and Wen (1984), the fragility curves used by HAZUS-99 promoted by the Federal Emergency Management Agency (1999) and the indices used by Sanchez-Silva and García (2001) have been considered. The values of these indices correspond to the centre of area for every membership function related to each damage level. Table 7 shows the indices proposed in this work together with the indices proposed by Park et al. (1984) and Sanchez-Silva and Garcia (2001), which have been included with

Figure 15. Structure of the proposed artificial neural network

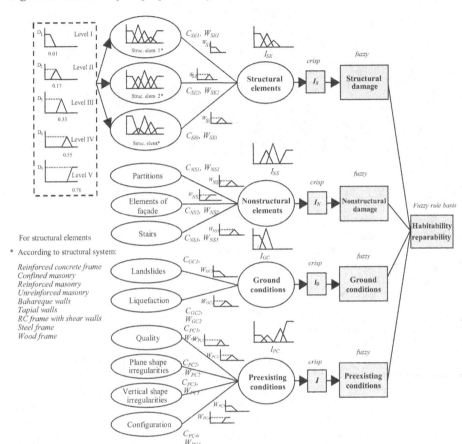

the aim of comparison. The selection of the initial indices is based on the indices of Park; this choice is justified on the basis that they have been calibrated with information of several experimental and numerical studies. Some authors consider that collapse occurs for a value equal to 0.8, although Stone and Taylor (1993) proposed a collapse threshold of 0.77. According to this opinion, a value of 0.76 has been selected to describe the collapse level index. The authors decided to be conservative when selecting the damage index, since the indices corresponding to severe and moderate damage have been highly discussed, and doubts exist on whether they should be smaller.

The calibration is performed for each damage level and only the indices corresponding to the groups of variables considered in each case are calibrated. The network learning is made using a Kohonen network:

Table 7. Comparative table for damage indices

Damage Level	Park, Ang, and Wen	Sanchez-Silva and García	Proposed
Very light	< 0.10 0.07	0.10	0.07
Light	0.10 – 0.25 0.175	0.20	0.17
Moderate	0.25 – 0.40 0.325	0.35	0.33
Severe	0.40 – 0.80 0.6	0.60	0.55
Destruction	>0.80 0.8	0.90	0.76

$$I_{kj}(t+1) = I_{kj}(t) + \alpha(t)[I_{kj}(t) - I_{kj}] \tag{7}$$

where I_{kj} is the value of the index of the variables group k recalculated considering a learning rate α, a function with exponential decay, and the difference between the resulting index of the present evaluation and the previous indices in each damage level j. The learning rate is defined by:

$$\alpha(t) = 0.1 \times \exp(-0.1 \times t) \tag{8}$$

where t is the number of times that has been used the index which is calibrated. For training, the damage evaluations made after the Quindío earthquake in Colombia in 1999 were used. However, more information is necessary to complete the network training for all structural classes, especially for wood and steel framed structures, because these building classes are not common in that area. Reinforced concrete frames with shear walls are also only a few and, therefore, the number of building evaluations to calibrate this structural system were insufficient. The Figure 16 shows a summary of the computational process performed by the proposed model.

Fuzzy Rule Bases For Decision Making

The building habitability and reparability are assessed based on previous results of the damage level of the structural and nonstructural elements, the state of the ground and of the preexistent conditions. Figure 17 displays the used fuzzy rule bases. The global level of the building damage is estimated starting from the structural and nonstructural damage results. The global

Figure 16. Flow chart for the evaluation process

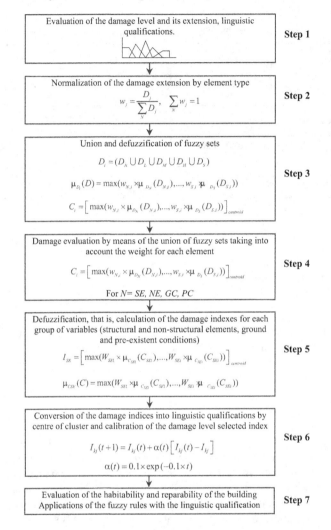

building state is determined taking into account the rule bases of the ground conditions and, by this way, the habitability of the building is obtained. The linguistic qualification for the building habitability has four possibilities: usable, restricted use, dangerous and prohibited. They mean habitable immediately, usable after reparation, usable after structural reinforcement, and not usable at all. Besides, the building reparability depends on other fuzzy rule bases: the preexistent conditions. The building reparability has four possibilities: not any or minor treatment, reparation, reinforcement, and possible demolition.

Figure 17. Method for building habitability and reparability

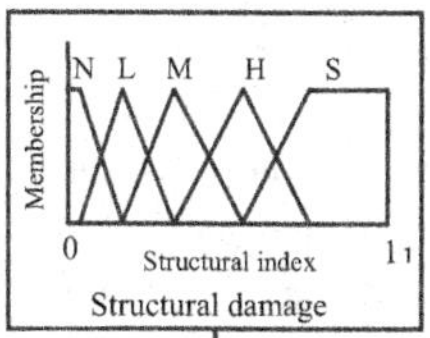

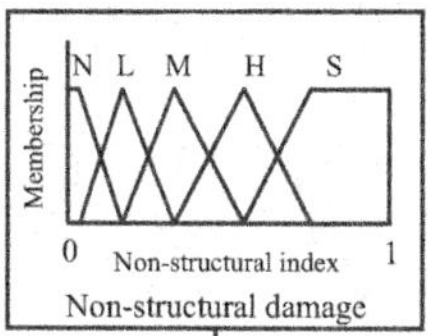

Structural damage — Membership: N L M H S, Structural index 0 to 1

Non-structural damage — Membership: N L M H S, Non-structural index 0 to 1

Rule base for building damage
Structural damage *vs.* non-structural damage

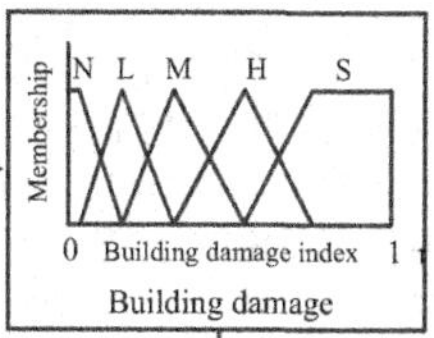

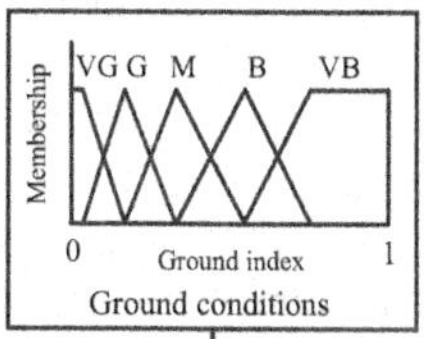

Non-structural \ Structural	None	Light	Moderate	Heavy	Severe
None	N	L	L	M	M
Light	L	L	M	M	M
Moderate	M	M	M	H	H
Heavy	H	H	H	S	S
Severe	S	S	S	S	S

Building damage — Membership: N L M H S, Building damage index 0 to 1

Ground conditions — Membership: VG G M B VB, Ground index 0 to 1

Rule base for building condition
Building damage *vs.* ground conditions

Ground conditions \ Building damage	Very good	Good	Medium	Bad	Very bad
None	N	L	M	H	S
Light	L	L	M	H	S
Moderate	M	M	H	S	S
Heavy	H	H	S	S	S
Severe	S	S	S	S	S

Rule base for building habitability
Building damage *vs.* ground conditions

Ground conditions \ Building damage	Very good	Good	Medium	Bad	Very bad
None	U	U	R	P	D
Light	U	U	R	P	D
Moderate	R	R	P	D	D
Heavy	D	D	D	D	D
Severe	D	D	D	D	D

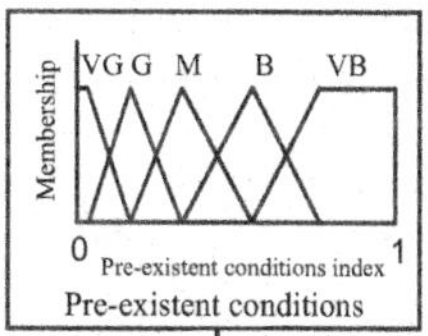

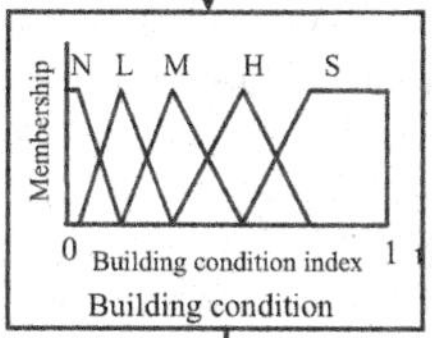

Pre-existent conditions — Membership: VG G M B VB, Pre-existent conditions index 0 to 1

Building condition — Membership: N L M H S, Building condition index 0 to 1

Rule base for building reparability
Building condition *vs.* pre-existent conditions

Pre-existent conditions \ Building condition	Very good	Good	Medium	Bad	Very bad
None	M	M	M	R	S
Light	M	M	R	S	S
Moderate	R	R	S	S	S
Heavy	S	S	S	D	D
Severe	D	D	D	D	D

NOTATIONS:

Structural and non-structural damage:
N: None
L: Light
M: Moderate
H: Heavy
S: Severe

Habitability:
U: Usable
R: Restricted
P: Prohibited
D: Dangerous

Ground and pre-existent conditions:
VG: Very good
minor
G. Good
M: Medium
B: Bad

Reparability:
M: Not any or
R: Reparation
S: Strengthening
D: Possible

Examples of Evaluation
Using the Proposed Computational Model

In this section are given three examples which illustrate the application of the proposed model to the evaluation of the seismic damage and safety of three buildings having different typologies: reinforced concrete, unreinforced masonry and confined masonry. These buildings have been damaged by the January 1999 earthquake occurred in the coffee-growing-area of Colombia. The most important characteristics of each building are given, together with the damage level and extension corresponding to each type of structural and nonstructural element. Photos of the most characteristic damage observed after the earthquake are shown. The results obtained for each example correspond to four aspects of the problem: damage, risk, habitability and reparability, and are given in four sections of a table. The section of damage provides as results numerical and linguistic qualifications for each group of elements. The section of risk gives a qualification of the risk level corresponding to the structure, the nonstructural elements and to the ground and also evaluates the overall state of the building. The habitability section provides a decision about the building habitability and suggests security measures to be undertaken urgently. The reparability section suggests certain measures which have to be applied, but without a detailed description. Obviously, the detailed reparation measures require the intervention of a structural engineer. All the numerical and linguistic results, comments and descriptions are given by a computer program in which the proposed model has been implemented. At present, this program, Earthquake Damage Evaluation of Buildings (EDE), is used as an official tool by the disaster risk management offices of the cities of Bogotá and Manizales, in Colombia. At the end of the table which describes the result for each example, a flow chart is included, which describes the application of the fuzzy rule basis in order to obtain the qualification of the building damage state, habitability and reparability.

Example 1.

General information		
Building inspection:	Outside and inside	
Construction year:	1950 to 1984	
Number of stories above ground:	5	
Bellow ground:	0	
Structural system:	Reinforced concrete frame	
Kind of diaphragms	Solid slabs	
Localization in block:	Corner	

General conditions of the building

Collapse:	Not
Building or any story tilting:	Not
Failure or settlement in the foundation:	Not
Story with the most damage:	3

Damage in structural elements

Beams:	None:70	Light: 30	Moderate:0	Heavy: 0	Severe: 0
Columns:	None: 45	Light: 50	Moderate: 5	Heavy: 0	Severe: 0
Joints:	None: 80	Light: 20	Moderate: 0	Heavy: 0	Severe: 0
Floors (slabs):	None: 40	Light: 60	Moderate: 0	Heavy: 0	Severe: 0

Column with moderate damage   Beam and slab with light damage

Damage in non structural elements

Partitions: Moderate   *Façade:* Light   *Stairs:* None

Partitions with moderate damage

Ground conditions

Crack, slope instability and landslide:	None (very good)
Ground settlement and liquefaction:	None (very good)

Preexistent conditions

Material and construction quality:	Good
Irregular configuration in plan:	Medium
Vertical irregularities or discontinuities:	Medium
Structural configuration:	Poor (bad)

Results Example 1:

Damage	
Structural damage:	Light 0.1519
Non structural damage:	Light 0.2395
Ground conditions:	Very good 0.0541
Preexistent conditions:	Medium 0.33

Risk	
Structural risk:	Low
Non structural risk:	Low
Ground risk:	Low
Building damage:	Light
	The building has structural and nonstructural slight damage. The earthquake resistance has been not reduced
Building condition:	Good
	The state of the building and the ground conditions are good

Habitability	
	Usable
	The building can be normally inhabited
Security measures:	It is not necessary to take security measures in particular

Reparability	
	Reparation needed
	The building needs some reparation possibly due to minor damages and preexistent conditions. Since the building was constructed between 1950 and 1984, it may need a structural reinforcement to be adapted to the current regulations for the earthquake resistant construction. It is recommended to undertake a study of seismic vulnerability

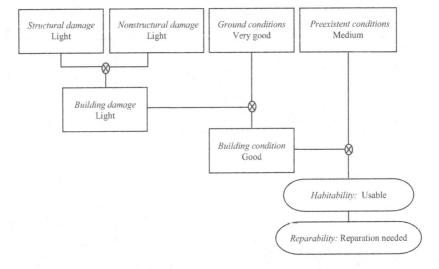

Example 2.

General information	
Building inspection:	Outside and inside
Construction year:	1950 to 1984
Number of stories above ground:	3
Bellow ground:	0
Structural system:	Unreinforced masonry
Type of diaphragms	Solid slabs
Localization in block:	Intermediate

General conditions of the building	
Collapse:	Partial
Building or any story tilting	Not
Failure or settlement in the foundation:	Not
Story with the most damage:	1

Damage in structural elements

Bearing walls:	None: 0	Light: 0	Moderate: 30	Heavy: 20	Severe: 50
Floors (slabs):	None: 0	Light: 0	Moderate: 30	Heavy: 60	Severe: 10

Severe damage in bearing walls of unreinforced masonry (partial collapse)

Damage in non structural elements

Partitions: Heavy	*Façade:* Heavy	*Stairs:* Heavy

Stairs with heavy damage (below)

Ground conditions	
Crack, slope instability and landslide:	Widespread (very bad)
Ground settlement and liquefaction:	Widespread (very bad)

Preexistent conditions	
Material and construction quality:	Poor (very bad)
Irregular configuration in plan:	Very bad
Vertical irregularities or discontinuities:	Very bad
Structural configuration:	Very bad

Results Example 2:

Damage	
Structural damage:	Severe 0.76
Non structural damage:	Heavy 0.5533
Ground conditions:	Very bad 0.76
Preexistent conditions:	Bad 0.6537
Risk	
Structural risk:	Very high
Non structural risk:	High
Ground risk:	Very high
Building damage:	Severe
	The building suffered severe structural damage and strong nonstructural damage. The building partially collapsed or suffered damage that puts the building in danger of collapse.
Building condition:	Very bad
	Severe damage of the building and the very bad ground conditions; due to this situation the building condition is very bad.
Habitability	
	Dangerous
	The building habitability should be prohibited due to the danger of collapse by very bad building and ground conditions.
Security measures:	Exterior barriers should be installed for preventing the traffic of cars and pedestrians close to the building.
Reparability	
	Possible demolition
	The building possibly needs to be demolished due to the very bad conditions of the building and the ground.

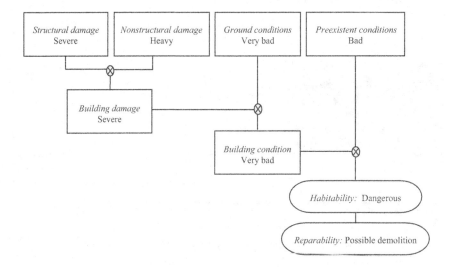

Example 3.

General information	
Building inspection: Outside and inside *Construction year:* 1985 to 1997 *Number of stories above ground:* 3 *Bellow ground:* 0 *Structural system:* Confined masonry *Type of diaphragms* Solid slabs *Localization in block:* Free by all sides	

General conditions of the building

Collapse:	Not
Building or any story tilting	Not
Failure or settlement in the foundation:	Not
Story with the most damage:	1

Damage in structural elements

Bearing walls:	None: 20	Light: 30	Moderate: 50	Heavy: 0	Severe: 0
Floors (slabs):	None: 10	Light: 40	Moderate: 50	Heavy: 0	Severe: 0

Moderate damage in confined masonry bearing walls

Damage in non structural elements

Partitions: Heavy	*Façade:* Heavy	*Stairs:* Heavy

Ground conditions

Crack, slope instability and landslide:	*Not important (very good)*
Ground settlement and liquefaction:	*Not important (very good)*

Preexistent conditions

Material and construction quality:	*Poor (very bad)*
Irregular configuration in plan:	*Significant (Medium)*
Vertical irregularities or discontinuities:	*Significant (Medium)*
Structural configuration:	*Poor (very bad)*

Results example 3

Damage	
Structural damage:	Moderate 0.35
Non structural damage:	Heavy 0.5533
Ground conditions:	Not important information
Preexistent conditions:	Medium 0.35

Risk	
Structural risk:	Low, after some measures
Non structural risk:	High
Ground risk:	Low
Building damage:	Heavy
	The building suffered moderate structural damage and heavy nonstructural damage that can affect the structural stability in case of an aftershock. The earthquake resistance has been reduced.
Building condition:	Bad
	The building was seriously damaged although the ground conditions are good.
Habitability	
	Prohibited
	The building occupancy is dangerous due to the bad conditions of the building
Security measures:	It is necessary to remove and to anchor the elements that can collapse, to demolish some nonstructural elements and to install internal barriers to mark the high risk zones.
Reparability	
	Strengthening
	The building needs reinforcing due to its general bad conditions and the medium preexistent conditions. In this case, an expert structural engineer must be consulted to make a decision.

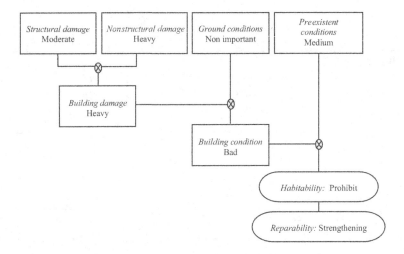

Conclusion and Future Trends

As a support to the complex task of building damage evaluation after an earthquake, an innovative expert system has been proposed, based on computational intelligence techniques such as artificial neural networks and fuzzy logic approach. This computational model improves the existing conventional existing methodologies, making possible a more accurate seismic building damage evaluation by nonexpert professionals. This type of tool is suitable in the practice because building damage evaluation deals with subjective and incomplete information which requires the use of linguistic qualifications that are appropriately handled by fuzzy sets. In addition, an artificial neural network has been used to calibrate the system using the judgment of specialists. The training of the neural network was performed by using a database of real evaluations made by expert engineers.

The proposed neuro-fuzzy expert system enabled the development of a user-friendly computer program called *Earthquake Damage Evaluation of Buildings, EDE*. At present, this program is used as an official tool by the disaster risk management offices of the cities of Bogotá and Manizales, in Colombia, and it is a component of a "National Program on Building Evaluations," in which new inspection guidelines and forms have been also developed.

The possibility of calibrating the expert system to be used directly in different seismic areas depends on the use of accurate and reliable databases of building damage obtained by means of inspections and qualifications made by experts. These databases are essential because it is necessary to have available detailed and relevant damage information for the learning process of the artificial neural network. Unfortunately, these databases are not available at present in the desired amount. Therefore, in order to refine the quality of the proposed expert system, it is necessary, throughout the coordination of professional associations of earthquake engineering and universities, to develop appropriate databases of evaluations after future earthquakes, taking into account all the building types existing in the seismic areas, to complete the learning process for these constructions. Any way, more information on the structural types for which evaluation databases are already available is also desirable in order to improve the knowledge on the seismic behaviour of buildings having the same type but which are located in different areas and, therefore, have different details.

Although it would be also possible to perform future developments in the field of damage evaluation algorithms, first of all it is necessary to improve the data collection of building damages and the organization of procedures and of the coordination of the emergency response after earthquakes. The current conventional methodologies are well designed but their implementation is still not adequate. The possibility of avoiding mistakes made by nonexperts is far to be assured if a tool, like the expert system here described, is not available in the time of a strong earthquake.

Lastly, the use of soft computing is relevant in many activities of civil engineering, in which computational intelligence tools can be successfully applied; however, they had not an appropriate diffusion until present. Therefore, it is recommended to promote their use to provide suitable and versatile solutions to several problems in this field of knowledge.

Acknowledgments

This work has been partially sponsored by the program Applied Research Grants for Disaster Reduction of ProVention Consortium of the World Bank, the Colombian Association for Earthquake Engineering, AIS, and the Spanish Ministry of Science and Technology, projects EVASIS, reference REN2002-03365/RIES and SIGRIS, reference CGL2005-04721/BTE.

References

Ahlawat, A. S., & Ramaswamy, A. (2001). Multiobjective optimal structural vibration control using a fuzzy logic control system. *Journal of Structural Engineering, 127*(11), 1330-1337.

Anoop, M. B., Rao, K. B., & Rao, T. V. S. R. A. (2002). Application of fuzzy sets for estimating service life of reinforced concrete structural members in corrosive environments. *Engineering structures, 24*, 1229-1242.

Applied Technology Council. (1985). *Earthquake damage evaluation data for California, ATC-13*. Redwood City, CA: Author.

Applied Technology Council. (1989). *Procedures for postearthquake safety evaluation of buildings, ATC 20*. Redwood City, CA: Author.

Applied Technology Council. (2003). *Users manual: Mobile postearthquake building safety evaluation data acquisition system, ATC 20i* (Version 1.0). Redwood City, CA: Author.

Applied Technology Council. (2005). *Field manual: Postearthquake safety evaluation of buildings, ATC 20-1* (2nd ed.). Redwood City, CA: Author.

Asociación Colombiana de Ingeniería Sísmica. (2002). *Guía técnica para inspección de edificaciones después de un sismo. Manual de Campo* [Technical guideline for building evaluations after an earthquake. Field manual]. Bogotá, DC, Colombia: Author.

Asociación Colombiana de Ingeniería Sísmica. (2003). *Manual de Campo para inspección de edificios después de un sismo* [Field manual for building evaluations after an earthquake]. Manizales, Colombia: Author.

Asociación Colombiana de Ingeniería Sísmica. (2004). *Sistema experto para la evaluación post-sísmica del daño en edificaciones—EDE Bogotá* [Expert system for the post-seismic damage evaluation of buildings; CD-ROM]. Bogotá, Colombia: Author.

Campos, A. (1999). *Memoria técnica del censo de inmuebles afectados por el sismo del 25 de enero de 1999 en el Eje Cafetero* [The technical memory of the buildings affected by the January 25th, 1999, earthquake in the coffee growing-area]. Bogotá DC:Colombia: Ministerio de Desarrollo Económico.

Cardona, O. D. (2001). *Estimación holística del riesgo sísmico utilizando sistemas dinámicos complejos* [Holistic assessment of the seismic risk using complex dinamic systems].

Doctoral thesis, Technical University of Catalonia, Barcelona, Spain. Retrieved from http://www.desenredando.org/public/varios/2001/ehrisusd/index.html

Cardona, O. D. (2005). *Sistema de indicadores para la gestión del riesgo de desastre: Informe técnico principal* [Indicador system for disaster risk management of: Main technical report]. IDB/IDEA Program of indicators for disaster risk management, National University of Colombia, Manizales. Retrieved from http://idea.unalmzl.edu.co

Cardona, O. D. & Hurtado, J. E. (1992). Análisis de Vulnerabilidad Sísmica de Estructuras de Concreto Reforzado [Analysis of seismic vulnerability for reinforced concrete structures]. *III Reunión del Concreto*, Cartagena, Colombia.

Carreño, M. L., Cardona, O. D., & Barbat, A. H. (2003). Expert system for post-earthquake building damage evaluation. In *Proceedings of the Ninth International Conference on Civil and Structural Engineering Computing.*

Carreño, M. L., Cardona, O. D., & Barbat, A. H. (2004). *Metodología para la evaluación del desempeño de la gestión del riesgo* [Evaluation methodology for the performance of risk management]. Monographs on earthquake engineering, Barcelona, Spain: International Centre of Numerical Methods in Engineering.

Carreño, M. L., Cardona, O. D., & Barbat, A. H. (2005). *Evaluación "ex-post" del estado de daño en los edificios afectados por un terremoto* ["Ex post" evaluation of the damage state in buildings affected by an earthquake]. Monographs on earthquake engineering, Barcelona, Spain: International Centre of Numerical Methods in Engineering.

CENAPRED. (1996). *Norma para la evaluación del nivel de daño por sismo en estructuras y guía técnica de rehabilitación (estructuras de concreto reforzado)* [Guidelines for the evaluation of the seismic damage level in structures and technical guide for rehabilitation (reinforced concrete structures)]. Mexico: Author.

Coburn, A., & Spence, R. (2002). *Earthquake protection* (2nd ed.). Chichester, UK: Wiley.

Chou, J.-H., & Ghaboussi, J. (2001). Genetic algorithm in structural damage detection. *Computers & Structures, 79,* 1335-1353.

Cross V., & Sudkamp, T. (1991). Compatibility measures and aggregation operators for fuzzy evidential reasoning. *Proceedings of the North American Fuzzy Information Processing Society, 13-17.*

Dowrick, D. J. (1987). *Earthquake resistant design for engineers and architects* (2nd ed.). Chichester, UK: Wiley.

Dubois, D., & Prade, H. (1982). On several representations of an uncertain body of evidence. In M. Gupta & E. Sanchez (Eds.), *Fuzzy information and decision processes* (pp. 167-181). North Holland.

Federal Emergency Management Agency. (1999). *Earthquake loss estimation methodology HAZUS* [Tech. Man., Vol. I, II, III, 1st ed. 1997]. Washington, DC: National Institute of Buildings Sciences of Federal Emergency Management Agency.

Garcia, L. E. (1998). *Dinámica estructural aplicada al diseño sísmico* [Structural dynamics applied to the seismic design]. Bogotá, Colombia: Asociación Colombiana de Ingeniería Sísmica.

Goretti, A. (2001). *Post-earthquake building usability: An assessment* [Tech. Rep. No. SSN/RT/01/03]. Italy.

Hurtado, J. E. (2001). Neural networks in stochastic mechanics. *Archives of computational methods in engineering, State of the art reviews, 8*(3), 303-342.

Instituto de Ingeniería Sísmica y Sismología. (1984). *Metodología y procedimiento para la evaluación de daños producidos por terremotos* [Methodology and procedure for the evaluation of seismic damage]. Skopje, Yugoslavia: Universidad "Kiril y Metodij."

Jang, J.-S. R., Sun C.-T., & Mizutani, E. (1997). *Neuro-Fuzzy and soft computing. A computational approach to learning and machine intelligence.* London: Prentice Hall.

Kosko, B. (1992). *Neural networks and fuzzy systems: A dynamical systems approach to machine intelligence.* Englewood Cliffs, NJ: Prentice Hall.

Lagaros, N. D., Papadrakakis, M., Fragiadakis, M., Stefanou, G., & Tsompanakis, Y. (2005). *Recent developments in neural network aided stochastic computations and earthquake engineering.* Paper presented at the International Symposium on Neural Networks and Soft Computing in Structural Engineering (NNSC-2005), Cracow, Poland.

Matsuoka, M., & Yamazaki, F. (2004). Use of satellite SAR intensity imagery for detecting building areas damaged due to earthquakes. EERI, *Earthquake Spectra, 20*(3), 975-994.

Nauck D., Klawonn F., & Kruse, R. (1997). *Foundations of neuro-fuzzy systems.* Chichester, UK: Wiley.

Papa, F., & Zuccaro, G. (2003). *Manuale di esercitazioni sul danno ed agibilità* [Training manual for damage and habitability]. 13th intensive course, Local seismic cultures and earthquake vulnerability reduction in traditional masonry buildings, Ravello, Salerno, Italy.

Park, Y. J., Ang, A., & Wen, Y. (1984). Seismic damage analysis and damage-limiting design of R.C. buildings. *Structural Research Series* [Tech. Rep. No. 516]. Urbana: University of Illinois at Urbana-Champaign.

Rodríguez, M., & Castrillón, E. (1995). Manual de evaluación postsísmica de la seguridad estructural de edificaciones [Manual for postseismic evaluation of the structural safety of buildings]. *Series del Instituto de Ingeniería, 569.* UNAM, México: Instituto Nacional de Ingeniería.

Rutkowska, D. (2002). *Neuro-fuzzy architectures and hybrid learning.* Heidelberg, Germany: Physica Verlag.

Sanchez-Silva, M., & García, L. (2001). Earthquake damage assessment based on fuzzy logic and neural networks, EERI, *Earthquake Spectra, 17*(1), 89-112.

Sociedad Mexicana de Ingeniería Sísmica. (1998). *Manual de evaluación postsísmica de la seguridad estructural de edificaciones* [Manual for postseismic evaluation of the structural safety of buildings]. A. C. Secretaria de Obras y Servicios Gobierno del Distrito Federal, México: Author.

Song, B., Hao, S., Murakami, S., & Sadohara, S. (1996). Comprehensive evaluation method on earthquake damage using fuzzy theory. *Journal of Urban Planning and Development, 122*(1), 1-31.

Stone, W. C., & Taylor, A. W. (1993). Seismic performance of circular bridge columns designed in accordance with AASHTO/CALTRANS standards (NIST Building Science Series, 170). Gaithersburg MD: National Institute of Standards and Technology.

Zadeh, L. A. (1965). Fuzzy sets. *Information and Control, 8*, 338-353.

Zadeh, L. A. (1975). The concept of a linguistic variable and its application to approximate reasoning. *Information science* (Part I, Vol. 8, pp.199-249; Part II, Vol.8, pp. 301-357; Part III, Vol. 9, pp. 43-80).

Zadeh, L. A. (1992). *Fuzzy logic, neural networks and soft computing* (One-page course announcement of CS 294-4, Spring 1993). Berkely: University of California at Berkeley.

Zadeh, L. A. (1996). *Fuzzy logic and the calculi of fuzzy rules and fuzzy graphs: A precise, multiple valued logic* (Vol. 1, pp. 1-38).

Zhao, Z., & Chen, C. (2002). A fuzzy system for concrete bridge damage diagnosis. *Computers & Structures, 80*, 629-641.

Zuccaro, G., & Papa, F. (2002). Multimedia handbook for seismic damage evaluation and post event macroseismic assessment [CD-ROM]. In *Proceedings of the XXIII General Assembly of the European Seismological Commission.*

Chapter VIII

Learning Machines for Structural Damage Detection

Miguel R. Hernandez-Garcia, University of Southern California, USA

Mauricio Sanchez-Silva, Universidad de Los Andes, Colombia

Abstract

The complexity of civil infrastructure systems and the need to keep essential systems operating after unexpected events such as earthquakes demands the development of reliable health monitoring techniques to detect the existence, location, and severity of any damage in real time. This chapter presents an overview of structural health monitoring techniques. Furthermore, it describes a methodology for damage detection based on extracting the dynamic features of structural systems from measured vibration time histories by using independent component analysis. Based on this analysis, statistical learning theory is used to determine and classify damage. In particular, an illustrative example is presented within which artificial neural networks (ANNs) and support vector machines (SVMs) are compared. ANNs and SVMs are of important value in engineering classification problems. They are two applications of the principles of statistical learning theory, which provides a great variety of pattern recognition tools. The results show that data reduction from acceleration time histories using independent component analysis (ICA), followed by an implementation of learning machines applied to pattern recognition provide a grounded model for structural health monitoring.

Introduction

Civil infrastructure systems are critical assets for any country's socioeconomic development. Designing these systems for a particular service life and maintaining them in operation has been recognized as a critical issue worldwide. Civil infrastructure systems are subject to damage and deterioration during their service lives and the cost of maintenance grows exponentially with time. Then, the search for structural integrity becomes an essential part of the so called structural life-cycle analysis. Although life-cycle analysis integrates both technical and socioeconomic aspects, the technical problem, in particular, focuses around defining the interaction among three main processes (1) structural deterioration (e.g., aging, obsolescence); (2) the occurrence of unexpected extreme events (e.g., earthquakes, floods); and (3) the maintenance/rehabilitation program. The first aspect corresponds to the structure's loss of functionality (also expressed in terms of safety) with time. It is usually a slow time-dependant process controlled by a safety/operation threshold specification. Examples of this case are structures in aggressive environments and subject to, for instance, sulfur or chloride penetration, and/or biodeterioration. Other examples are structures subject to fatigue or creep where offshore platforms and sea ports are two cases in point. The occurrences of unexpected or rare events which may affect the structure's safety dramatically are commonly associated to natural (e.g., earthquakes) or man-made disasters (e.g., terrorism). The third aspect, defining an appropriate maintenance program, is definitive when considering a long-term investment policy. Existing evidence has shown that maintenance is an important part of the integrity management process as a means of monitoring the performance of the infrastructure to ensure their safety and serviceability (Bucher & Frangopol, 2006).

It is widely accepted that maintenance can be classified in (a) preventive maintenance, which if it is not done it will cost more at a later stage to keep the structure in a safe condition; and (b) essential maintenance, which is required to keep the structure safe (Yang, Frangopol, Kawakami, & Neves, 2006). Maintenance measures require inspection planning, which is commonly based on general guidelines and engineering judgment; in most cases it is prescriptive and does not take into account the structure specific characteristics or make optimum use of the observed performance data. As a result of this strategy, a significant number of inspections are ineffective by not focusing on the most critical areas or by not using the most appropriate techniques. Furthermore, the ever-increasing complexity of civil infrastructure makes the practicality and reliability of such approaches questionable; in particular, when rapid assessments are required, for instance, after natural disasters such as earthquakes. This results in uneven safety levels and wastage of limited maintenance resources (Onoufriou & Frangopol, 2002).

In recent years, there have been significant developments in the area of reliability-based inspection design and planning for complex structures, such as offshore and bridges, motivated by the need to optimize maintenance expenditure and achieve better safety levels at a lower cost. Specifically, the field of *structural health monitoring* has taken on increased importance in civil and mechanical engineering applications over the last several decades. Structural health monitoring, also refereed to as *damage identification*, is an important application of structural system identification. This is the practice of conducting nondestructive tests or inspection of a structure to determine the existence, location, extent of damage and, in some cases, to make estimation as to the remaining life of the system. Among the advantages of this method are the costs associated with inspection, maintenance, and system downtime which have proved to be a significant driving force for improving inspection and damage identification practices. For these reasons, new methods of structural health monitoring are being explored to better determine the functional safety of structures. Its applications cover a wide range of systems such as bridges, dams, and buildings (Doebling, Ferrar, Prime, & Shevitz, 1996).

Basic Concepts and Conceptual Background

Structural Damage Identification Methodologies

The objective of a monitoring system is to collect information about damage for taking appropriate remedial action to restore the structural system to high quality operation level or at least to ensure safety. The *process* of damage identification can be described as a hierarchical structure as follows (Worden & Dulieu-Barton, 2004):

- **Detection:** Qualitative indication that damage might be present in the structure
- **Localization:** Information about probable location of the damage
- **Classification:** Information about the type of damage
- **Assessment:** Estimation of the extent of the damage
- **Prediction:** Enformation about the safety of the structure

The implementation of algorithms to process data and perform damage identification is the crucial task of an intelligent damage identification strategy (Worden & Dulieu-Barton, 2004). However, before choosing the algorithm, it is necessary to select between two complementary approaches to the problem:

- Inverse problem approach
- Pattern recognition approach

The first approach usually adopts a model of the structure and tries to relate changes in the model to changes in the data measured from the structure. Frequently, linear models are used to simplify the analysis. The algorithms used are mainly based on linear algebra or optimization theory (Doebling et al., 1996). The second approach is based on assigning, to a sample of measured data from the system (e.g., vibration time histories, modal parameters), a damage class label related to location, type or extent of damage. These algorithms are based on statistical learning theory (Worden & Dulieu-Barton, 2004).

Regarding the *damage identification process*, there can be found in the literature a great number of methodologies for structural damage identification (Doebling et al., 1996). In this chapter, only those methods based on the structural vibration characteristics are treated in detail. These *vibration-based* health monitoring algorithms can be broken down into the following two categories:

- Nonmodel-based methods
- Model-based methods

Nonmodel-based methods use direct changes in the sensor's output signal to detect and locate the damage in the structure. These can be thought as methods where signal-processing, feature extraction, and pattern processing algorithms are linked to solve the damage identification problem. Model-based techniques utilize changes in response functions or modal parameters such as natural frequencies, mode shapes, or their derivatives, to identify the damage extension and location. This analysis compares the changes of the structural response between sequential tests over time to characterize the damage (Doebling et al., 1996).

Damage Identification Using Pattern Recognition

As it was stated before, vibration-based structural damage detection can be viewed as a pattern recognition problem because a it must be able to distinguish between various "normal" (i.e., undamaged) and damaged states of the structure. Traditional pattern recognition and classification processes can be divided into three stages (see Figure 1):

- Measurements
- Feature extraction/feature reduction
- Pattern processing

The measurements stage consists of collecting raw data from field tests through a set of sensors strategically placed on the structure (e.g., vibration time histories) and data cleansing (i.e., filtering, removing outliers, etc.). Afterwards, feature extraction/reduction is used to remove redundant information from the measured data reducing it to a data set containing just the dominant features of each measurement, and making it distinguishable from other data classes (i.e., damage and undamaged). In general, the components of the signal that distinguish the various damage classes will be hidden by features that characterize the normal operative conditions of the structure, particularly when the damage is not severe (Goumas & Zervakis, 2002). In other words, the purpose of this task is to magnify the characteristics of the various damage classes and suppress the normal background (Worden & Dulieu-Barton, 2004). These features can be typically obtained by analyzing field measurements in the time domain, frequency domain, wavelet domain or modal domain. Existing statistical algorithms (e.g., principal component analysis, independent component analysis) can be used to reduce the dimensions of measured data resulting in a low dimensional data set containing the principal features of the system.

Figure 1. Pattern Recognition damage detection system

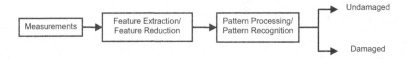

Finally, pattern processing consist on developing algorithms for making decisions about feature vectors (e.g., modal parameters, frequency response functions, auto-regressive coefficients, principal components) to produce a classification or decision rule about the type of damage based on previously accumulated knowledge of at least the normal or undamaged operating condition. In other words, the feature vectors or pattern vectors are classified into one or more classes (nondamaged, i.e., normal, and damaged states) just using the decision rule implemented by the pattern recognition algorithm.

As mentioned before, many modern approaches to damage identification are based on the idea of *pattern recognition* (PR). A pattern recognition algorithm assigns to a sample of features extracted from measured data a class label, usually from a finite set. In the case of damage identification, the features could be modal parameters, frequency response functions, and so forth. The appropriate class labels would encode damage type, location, extent, etc. In order to carry out higher levels of identification using Pattern Recognition (PR), it will almost certainly be necessary to construct examples of data corresponding to each class. That is, the algorithm must have prior knowledge of what measured data correspond to each damage class label (Worden & Dulieu-Barton, 2004). Depending on the type of training process, *learning algorithms* can be classified into:

- Supervised learning
- Unsupervised learning

Supervised learning requires having available data from every conceivable damage situation. Two possible sources for obtaining such data are numerical modeling and experimental test. Numerical modeling maybe cumbersome if the structure or the system of interest is geometrically or materially complex. Furthermore, the damage itself may be difficult to model due to its time-dependant properties and its correlation with the nonlinear dynamic structural performance. On the other hand, experimental work is expensive, time consuming and difficult to validate. In order to accumulate enough training data, it would be necessary to make several specimens (i.e., "copies") of the system and test them to achieve damage levels similar to those that might occur in case of an earthquake; for high value structures this is not possible.

The idea behind *unsupervised learning* is that only training data from the normal operating condition of the structure or system is used to establish the diagnostics. A model of normal condition is created; later, the structure is subjected to ground excitations. Then, data acquired from monitoring the dynamic structural performance are compared with the original undamaged model. Significant deviations in the structural response indicate an unnormal condition which is related to a given damage level in the system. If the training data is generated from a model, only the undamaged condition is required and this simplifies the modeling process considerably. From an experimental point of view, there is no need to damage the structure.

The main requirement of an intelligent damage identification system is that it should return relevant information for making decisions (e.g., maintenance, retrofitting). In other words, as suggested by Zadeh (1965), the analysis should be taken to a level where relevance and precision become mutually exclusive. Data should be measured and processed with the appropriate algorithm. It should take proper account of uncertainty in the data and return a confidence level in its diagnosis. This bold statement raises numerous issues that the following sections will attempt to address.

Within intelligent systems, three types of algorithms can be defined depending on the diagnosis of the damage (detection, localization, classification, etc.) (Doebling et al., 1996; Friswell & Penny, 1997; Hernandez & Sanchez, 2002; Kramers, 1992; Manson, Worden, & Allden, 2003a, 2003b; Masri, Chassiakos, & Caughey, 1993; Masri, Nakamura, Chassiakos, & Caughey, 1996; Masri, Smyth, Chassiados, & Caughey,2000; Worden, 2003; Worden & Dulieu-Barton, 2004; Worden & Lane, 2001; Worden & Manson, 2000; Worden, Manson, & Allman, 2003):

- **Novelty detection:** This algorithm indicates if extracted feature data belongs to an undamaged condition or not. This is a two-class (damaged/undamaged) problem where unsupervised learning can be used. Some methods for novelty detection include: outlier analysis, kernel density methods, auto-associative neural networks, Kohonen networks, and novelty support vector machines.

- **Classification:** This algorithm defines a discrete multi-class label. This is a multiclass classification problem and it is necessary that the damage classification rule (localization, damage type, etc.) could be divided into labeled clusters. This algorithm indicates in which cluster (e.g., story where damage occurred) the feature data can be classified. Examples of algorithms include: neural network classifiers, support vector machines and nearest neighbor classifiers

- **Regression:** The output of the algorithm is one or more continuous variables. In this case the damage could be associated to the assessment of location, intensity, and so forth. The regression problem is often nonlinear and is particularly suited to neural networks and regression support vector machines.

This chapter deals mainly with the novelty detection approach. However, the concepts of pattern recognition have been widely applied within the context of failure and damage detection. For earlier references the reader can consult the Web site of Los Alamos National Laboratory (http://www.lanl.gov/projects/damage_id/index.htm).

Vibration-Based
Structural Damage Detection Model

Based on the discussion presented in the previous sections, this chapter will focus on a model for damage detection based on the structural vibration properties. The structural damage detection problem can be modeled as a pattern recognition problem, where the patterns may be defined by existence of damage. The proposed model is illustrated in Figure 2 and can be divided into the following processes:

- Data acquisition
- Feature extraction and data reduction (e.g., independent component analysis)
- Damage detection (i.e., pattern recognition) by learning machines

First of all, the model requires structure's vibration time histories, from which the dynamic features of system can be obtained. Feature extraction and dimension reduction were performed by using the independent component analysis (ICA) technique (Lee, Yoo, & Lee, 2004). This analysis is carried out based on the work of Zang, Friswell, and Imregun,(2004), who showed that vibration data measured in the time domain (e.g., acceleration records) can be represented as a linear combination of dominant statistical components and a mixing matrix. In the structural dynamic field, those dominant components can be thought as some kind of exciting external forces (e.g., ambient vibrations), the mixing matrix as an accurate representation of the essential dynamic characteristics of the structural system. Finally, due to the fact that most data available for structural health monitoring is usually sampled from the undamaged system, the damage detection process is based on a one-class classification problem. The pattern recognition process (i.e., damage detection process) will be carried out by using (a) artificial neural networks; and (b) support vector machines; that are implemented by novel detection algorithms using the extracted dynamic features (i.e., mixing matrix) from acceleration time histories (see Figure 2).

In the following two sections the theory behind independent component analysis (ICA) and statistical learning theory (SLT), from which ANN and SVM theory is derived is explained in detail.

Figure 2. Proposed damage detection process

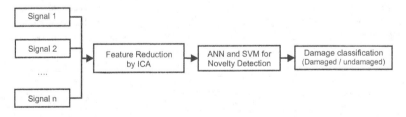

Independent Component Analysis

Independent component analysis is a statistical signal processing technique for revealing hidden factors in random variables, measurements, or signals; as well as feature extraction, noise reduction and blind source separation. ICA was originally proposed to solve the blind source separation problem, which involves recovering independent source signals (e.g., different voice, music, or noise sources) after they have been linearly mixed by an unknown matrix (Hyvärinen & Oja, 2000). The following ICA algorithm is based on the formalism presented in the survey article of Hyvärinen and Oja (2000).

Let's consider a vector $\mathbf{X}$ of n random variables with zero mean, which can be expressed as linear combination of m unknown statistical independent components described by a vector $\mathbf{S}$, where s_j is a known independent component with mean zero and variance one. Then, the relationship between $\mathbf{X}$ and $\mathbf{S}$ is given by:

$$\mathbf{X} = \mathbf{AS} + \mathbf{E}, \tag{1}$$

where $\mathbf{X} = \{\mathbf{x}(1), \mathbf{x}(2),..., \mathbf{x}(k)\} \in \Re^{n \times k}$ is the data matrix obtained from n measured signals; $\mathbf{A} = \{\mathbf{a}_1, \mathbf{a}_2,..., \mathbf{a}_m\} \in \Re^{n \times m}$ is the so called *mixing matrix*, which is unknown; $\mathbf{S} = \{\mathbf{s}(1), \mathbf{s}(2),..., \mathbf{s}(k)\} \in \Re^{m \times k}$ is the independent component matrix, $\mathbf{E} \in \Re^{n \times k}$ is the residual matrix and k is the number of time samples. The main goal of ICA consists of estimating the mixing matrix $\mathbf{A}$, and the independent components $\mathbf{S}$ using only observed data (i.e., $\mathbf{X}$). Using a dynamic systems analogy, the ICA can be summarized as recovering the exciting external forces as well as the system's dynamic characteristics just from the vibration time histories (e.g., accelerations) measured in the system. For that purpose, it is necessary to calculate a separating matrix $\mathbf{W} \in \Re^{m \times n}$ so that components of the reconstructed data matrix, $\hat{\mathbf{S}}$, are computed as:

$$\hat{\mathbf{S}} = \mathbf{WX}. \tag{2}$$

The initial step in ICA is *whitening*, also known as *sphering*, which eliminates all the cross-correlation between random variables. This transformation can be done by traditional methodologies for uncorrelating random data. The uncorrelated data $\mathbf{z}(k)$ can be expressed in terms of a new matrix $\mathbf{B}$ as:

$$\mathbf{z}(k) = \mathbf{Qx}(k) = \mathbf{QAs}(k) = \mathbf{Bs}(k), \tag{3}$$

where $\mathbf{B} = \mathbf{QA}$. At this point, the problem has been reduced from finding an arbitrary full-rank matrix $\mathbf{A}$ to the problem of finding an orthogonal matrix, which much easier. Then, from equation (3), the independent components $\mathbf{s}(k)$ of a sample k, can be estimated as follows:

$$\hat{\mathbf{s}}(k) = \mathbf{B}^T \mathbf{z}(k) = \mathbf{B}^T \mathbf{Qx}(k). \tag{4}$$

Since the main goal of ICA is to express random vectors in terms of linear combination of *independent vectors* (i.e., independent components) it is necessary to guarantee that all these components are effectively independent. The independence is then quantified by measuring their "non-Gaussianity" using *negentropy*, which is a information-theoretic quantity of differential entropy. To estimate the efficiency of the negentropy, Hyvärinen and Oja (2000) suggested a simpler approximation:

$$J(y) \approx (E[G(y)] - E[G(v)])^2, \tag{5}$$

where y is assumed to be a random variable with zero mean and variance one, v is a Gaussian variable with zero mean and variance one also, and G is any nonquadratic function. Based on this approximated form, Hyvärinen (1999) introduced a very simple and highly efficient fixed point algorithm for ICA calculated over sphered zero-mean vectors $\mathbf{z}$. The algorithm can be summarized as follows:

1. Choose the number of independent components to be estimated.

2. Select a random initial unit vector $\mathbf{b}_j$.

3. By using the negentropy approximation (see equation (5)), estimate a new vector $\mathbf{b}_i$ as: $\mathbf{b}_i = E[\mathbf{z}\dot{G}(\mathbf{b}_i^T\mathbf{z})] - E[\ddot{G}(\mathbf{b}_i^T\mathbf{z})]\mathbf{b}_i^T$.

4. Repeat the process until the vector $\mathbf{b}$ has converged (i.e., the number of the desired independent components has been obtained).

5. After obtaining the whole matrix $\mathbf{B}$, the corresponding independent components $\hat{s}(k)$ and demixing matrix $\mathbf{W}$ can then be found using equations (2-4).

For more details on the FastICA algorithm, see Hyvärinen & Oja (2000) and Hyvärinen, Karhunen, and Oja (2001).

Basic Concepts of Statistical Learning Theory

A learning method is an algorithm that estimates an unknown mapping (dependency) between a system's inputs and outputs from available data (i.e., known input-output samples). Once such a dependency has been estimated, it could be used for classification and regression. In the case of damage detection, the learning method estimates the mapping (i.e., decision rule) between extracted features from measured data (i.e., inputs) and the labels associated to undamaged and damaged scenarios (e.g., outputs). The underlying assumption is that observed data are representative of the unknown probabilistic phenomenon. Basic SLT methods focus on finding a optimal mapping function that minimizes some measure of error (i.e., generalization error) between the assumed mapping model and the unknown real mapping function (Hernandez, Sánchez-Silva, & Caicedo, 2005).

The general SLT model consists of learning from examples and the following three elements play a relevant part (Vapnik, 1995):

- **Input random vector,** $x \in \mathfrak{R}^n$: is the input data (e.g., extracted dynamic features) to the system and a has fixed but unknown probability distribution function $f(\mathbf{x})$.

- **Output value,** $y \in \mathfrak{R}^m$: is associated to every input vector $\mathbf{x}$ (i.e., real damage/undamaged condition), with a fixed but unknown conditional probability function $f(\mathbf{y} \mid \mathbf{x})$.

- **Response value,** $\hat{y} \in \mathfrak{R}^m$: is the result of a implemented set of mapping functions $g(\mathbf{x}, \alpha)$, where α belongs to a set of free parameters (e.g., estimated damage/undamaged condition).

The learning procedure consists on selecting the best indicator function (e.g., damage/undamaged decision rule) that minimizes the classification error, in the case of pattern classification; or the best approximating function (e.g., damage extent quantification), in the case of regression estimation, from a set of mapping functions $g(\mathbf{x}, \alpha)$. The selection of this function is based on a training set of l independent input couples $\{(\mathbf{x}_1, \mathbf{y}_1),...,(\mathbf{x}_l, \mathbf{y}_l)\}$ with probability $f(\mathbf{x}_i, \mathbf{y}_i) = f(\mathbf{x}_i) f(\mathbf{y}_i \mid \mathbf{x}_i)$. In damage detection, those input couples would consist of extracted dynamic features and their respective damage or undamaged labels. The selection of the optimal function requires defining the so called *actual risk function*. This function is the expected value of the *loss* and can be expressed as:

$$R(\alpha) = \int L(\mathbf{y}, g(\mathbf{x}, \alpha)) f(\mathbf{x}, \mathbf{y}) d\mathbf{x} d\mathbf{y}, \tag{6}$$

where the *loss function*, $L(\mathbf{y}, g(\mathbf{x}, \alpha))$, is defined depending upon the type of problem (classification or regression) and taking into account the output value, $\mathbf{y}$, and the actual response value, $\hat{y}$ obtained from the learning model. The main problem of this approach is that $R(\alpha)$ is unknown because $f(\mathbf{y} \mid \mathbf{x})$ is unknown, and the only available information is the training data set $\{(\mathbf{x}_1, \mathbf{y}_1),...,(\mathbf{x}_l, \mathbf{y}_l)\}$. Therefore, the actual risk function, $R(\alpha)$, is approximated by an empirical risk function, $R_{emp}(\alpha)$, as follows:

$$R(\alpha) \approx R_{emp}(\alpha) = \frac{1}{l} \sum_{i=1}^{l} L(\mathbf{y}_i, g(\mathbf{x}_i, \alpha)) = \frac{1}{l} \sum_{i=1}^{l} Q(\mathbf{z}_i, \alpha). \tag{7}$$

The loss function, $L(\mathbf{y}, g(\mathbf{x}, \alpha))$ is written, without loss of generality, as $Q(\mathbf{z}_i, \alpha)$. Then, the optimal function that minimizes the actual risk is approximated by the function $Q(\mathbf{z}_i, \alpha_0)$ obtained by minimizing the empirical risk. This approach is known as the *empirical risk minimization* (ERM) principle. One drawback of ERM is that it cannot guarantee a small actual risk by minimizing only the empirical risk. In other words, minimizing the function $R_{emp}(\alpha)$, which lead to the smallest values of error of training set, does not necessarily imply a higher generalization ability (Cristiani & Shawe-Taylor, 2000). To solve this problem, Vapnik and Chervonenkis (1971) developed the theory of uniform convergence in probability, which defines bounds on the actual risk for controlling the generalization ability of learning machines (Vapnik, 1995).

The bounded actual risk is expressed as:

$$R(\alpha) = R_{emp}(\alpha) + \Phi(l, h), \tag{8}$$

where Φ is a confidence interval and it is function of the dimension of the training set (i.e., l), and the so called VC (Vapnik-Chervonenkis) dimension (i.e., h). The VC dimension measures the richness of flexibility of a function class. In other words, it describes the ability of a set of functions, implemented in the learning machine, to generalize the data set (e.g., ability for detecting damage). Then, it is expected that learning machines with many parameters would have higher VC dimension, while those with few parameters have lower VC dimension.

Another disadvantage of ERM is that this principle was developed for dealing with large sizes of training data sets. In this case, the confidence interval, $\Phi(l, h)$, becomes small and the actual risk $R(\alpha)$ is approximately equal to the empirical risk. However, if the dimension of the training set is small (i.e., $l / h < 20$), a minimum actual risk cannot be achieved by minimizing $R_{emp}(\alpha)$ only, therefore, a simultaneous minimization over $R_{emp}(\alpha)$ and $\Phi(l, h)$ are required. In order to do this, Vapnik & Chervonenkis (1974) developed the so called structural risk minimization (SRM) principle, which minimizes the empirical risk and the confidence interval at same time using the VC dimension as a control variable (see Figure 3). This is carried out by decomposing the set of functions $S = \{Q(\mathbf{z}, \alpha)\}$ into a structured subset of functions S_k such that:

$$S_1 \subset S_2 \subset \cdots \subset S_k \subset \cdots \subset S_n. \tag{9}$$

Figure 3. Graphical interpretation of the structural risk minimization principle

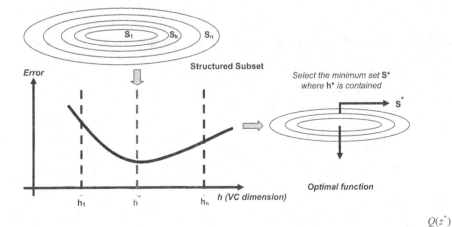

With increasing capacity (VC dimension, h_k) of each subset in such way that it fulfills the following criteria:

$$h_1 \leq h_2 \leq \cdots \leq h_k \leq \cdots \leq h_n. \tag{10}$$

Then, by selecting a subset S_k (with low capacity h_k) that minimizes the bounded risk, a class function $Q(z, \alpha_l^k) \in S_k$ with a small training error is automatically chosen (Scholkopf & Smola, 2002). Then, the bounded actual risk minimization equation can be simplified as:

$$R(\alpha_l^k) = R_{emp}(\alpha_l^k) + \Phi(l, h_k). \tag{11}$$

This solution guarantees a high level of generalization of the learning process (Vapnik, 1995). Two approaches to minimize the bounded risk have been developed and implemented by different learning machines (Hernandez et al., 2005):

1.   The first approach is used by artificial neural networks and it fixes the confidence interval by choosing an appropriate set of admissible functions (i.e., design of the network structure) from which it finds the function that minimizes the empirical risk (i.e., error in training data).

2.   The second approach is used in support vector machines and it fixes the empirical risk and minimizes the confidence interval.

Both artificial neural networks and support vector machines will be applied in this chapter to detect damage in structures by implementing novelty detection algorithms. These two approaches are described in detail in the following sections.

Learning Machines for Novelty Detection

The objective of learning machines for novelty detection is to decide if a given input vector is similar or not to a set of training vectors, assumed to be "normal" or "nonnovel." Relating this to damage detection process, the objective is to determine if a given set of dynamic features matches the date from the undamaged structure. In the case of mismatch, the data set is assumed to belong to a damage scenario in the structure.

Suppose that a set of training vectors $S = \{x_1, ..., x_l\}$ is available in an input space, denoted by X and each vector x_i is supposed to be a "normal" vector. In order to know if a given a vector x belongs to a *normal* or *abnormal* group (e.g., undamaged or damaged state), it is necessary to derive a decision function $g(x, \alpha)$, such that:

$g(\mathbf{x}, \alpha) > 0$ if $\mathbf{x}$ is similar to S (i.e., $\mathbf{x}$ is *normal*),

$g(\mathbf{x}, \alpha) < 0$ if $\mathbf{x}$ is *not* similar to S (i.e., $\mathbf{x}$ is *abnormal*).   (12)

To implement this decision rule, it is required to define how to assess the linguistic variables "similar" and "not similar" (e.g., undamaged or damaged); this will be the main function of support vector machines and artificial neural networks. Assuming that samples or training vectors $S = \{\mathbf{x}_1,\ldots, \mathbf{x}_l\}$ are distributed according to some probability distribution $f(\mathbf{x})$, the decision rule will be reduced to know if the new vector has the same probability distribution or not. In the first case, if the decision rule determines that the new vector $\mathbf{x}$ has the same probability distribution $f(\mathbf{x})$ of the training vectors; it will be considered as normal vector. On the other hand if the decision rule decides that the probability distribution of the new vector is different to $f(\mathbf{x})$, then this vector will belong to the abnormal case (Scholkopf & Smola, 2002).

Artificial Neural Networks

In the literature, the most common learning machines used for pattern recognition are artificial neural networks. As it was explained above, ANNs minimize the bounded risk by a process in which the empirical risk is minimized while keeping the confidence interval fixed. The training process of this learning machines requires, first of all, fixing the confidence interval by defining a structured set of functions (i.e., defined by the ANN architecture). Usually, the set of functions in ANN layers can be defined as sigmoid, linear or radial basis. The next step consists on estimating the optimal weights of all connections between neurons by modifying them iteratively using the gradient of the empirical risk function (Hernandez et al., 2005).

Let us consider a training set $\{(\mathbf{x}_1, \mathbf{y}_1),\ldots,(\mathbf{x}_l, \mathbf{y}_l)\}$, where $x \in \mathfrak{R}^n$, $y \in \mathfrak{R}^n$ and a neural network with $m + 1$ layers (with n_k neurons in the k-th layer, $k = 1,\ldots, m$). The n-dimensional input vector at the k-th layer can be expressed as: $\mathbf{x}_i(0) = [\mathbf{x}_i^1(0),\ldots,\mathbf{x}_i^n(0)]$ where $i = 1,\ldots, l$; and the image $\mathbf{x}_i(k) = [\mathbf{x}_i^1(k),\ldots,\mathbf{x}_i^{n_k}(k)]$ can be computed by a forward pass as:

$$\mathbf{x}_i(k) = S_i\{(\mathbf{w}(k)\,\mathbf{x}_i(k-1)\}. \tag{13}$$

The neuron transfer functions are represented by $S_i\{\}$ and the weighting matrix connecting layer k-1 to layer k by $\mathbf{w}(k)$. By using previous definitions, the backward pass in the learning algorithm can be computed by:

$$\mathbf{b}_i(k) = \mathbf{w}^T(k+1)\nabla S_i\{\mathbf{w}(k+1)\mathbf{x}_i(k)\}\mathbf{b}_i(k+1), \tag{14}$$

where $\mathbf{b}_i(k)$ corresponds to the error propagated back from the output nodes to neurons in hidden layers. Finally, the weight matrix is update by:

$$\mathbf{w}(k) = \mathbf{w}(k) - \eta \sum_{i=1}^{l} \mathbf{b}_i(k) \nabla S_i \{\mathbf{w}(k)\mathbf{x}_i(k-1)\} \mathbf{w}(k)\mathbf{x}_i^T(k-1). \tag{15}$$

This back-propagation algorithm leads to a final matrix of weights for which the empirical risk is minimized for a given confidence interval. At this point, the input to output mapping has been determined and the learning machine is able to classify a new input vector when it is presented to the network.

Support Vector Machines

Support vector machines are a learning process based on the structural risk minimization (SRM) principle (Vapnik, 1995). Currently, SVM are gaining popularity due to their many attractive features and promising empirical performance. SVMs have shown to have a superior performance and a grater ability to generalize due to its robust mathematical formulation. In general, it can be asserted that it is more efficient that other learning machines (Osuna, Freund, & Girosi, 1997; Schölkopf & Smola, 2002). Although the theory of SVM was developed firstly for the case of pattern classification, SVM can be used also to solve regression problems (estimation of real-valued functions) keeping the main feature that characterize SVM for classification; this is, representing the decision boundary in terms of a typically small subset of training examples (Schölkopf & Smola, 2002).

The solution to novelty detection problem consists on determining a region R of the space $\mathbf{X}$ (i.e., input vectors) that captures most of their probability mass. In other words, this is the most probable region where input vectors (e.g., undamaged feature data set) are laying in the feature space. Then R is such that $\int_R f(\mathbf{x})d\mathbf{x} = 1$. The decision function is such that $g(\mathbf{x}, \alpha) > 0$ if $x \in R$, and $g(\mathbf{x}, \alpha) < 0$ otherwise. Estimating the region R, or its bounded hyper-surface, is made by fitting a kernel function $K(\mathbf{x}_i, \mathbf{x})$ to the training vectors $\mathbf{x}_i$. Then, a hyper-surface can be defined by the weighted addition of the kernel functions:

$$\sum_{i=1}^{l} \alpha_i K(\mathbf{x}_i, \mathbf{x}), \tag{16}$$

and the region R is such that:

$$\mathbf{x} \in R \leftrightarrow \sum_{i=1}^{l} \alpha_i K(\mathbf{x}_i, \mathbf{x}) - \rho > 0. \tag{17}$$

The solution of this problem requires optimizing the kernel function K, the weights α_i, and the threshold ρ. Many different kernels have been proposed, but for one-class problems the radial basis function kernel (Gaussian kernel) has shown excellent performance; it is defined as (Unnthorsson et al., 2003):

$$K(\mathbf{x}_i, \mathbf{x}) = \exp(-\gamma \|\mathbf{x} - \mathbf{x}_i\|^2) = \exp\left(-\frac{\|\mathbf{x} - \mathbf{x}_i\|^2}{\sigma^2}\right), \tag{18}$$

where σ is related to the kernel spread.

Once the kernel has been selected, it is necessary to determine the parameters α_i and ρ in such way that the region R captures most of the probability mass. By mapping the input vectors into a high-dimensional feature space (i.e., $\varphi : X \rightarrow F$) and solving the optimization problem, the bounding hyper-surface of R will be limited. In the feature space the training vectors will be distributed by an underlying distribution $f^*(\mathbf{x})$ (related to $f(\mathbf{x})$ by the mapping function φ), and the problem is to determine a region R^* in the F(feature) space (Schölkopf & Smola, 2002).

Since $\|\mathbf{x}_i\|^2 = K(\mathbf{x}_i, \mathbf{x}_i) = 1$ for all $i = 1,..., l$, the training vectors $\mathbf{x}_i$ in the feature space F are located on a hyper-sphere centered at the origin of F with radius equal to 1. In the feature space the training data are located on a portion of the surface of the hyper-sphere, as is shown in Figure 4, and they can be separated from the rest of the feature space by a hyper-plane $H(\mathbf{w}, \rho)$ defined by the vector $\mathbf{w}$ and the distance to the origin $\rho / \|\mathbf{w}\|$. The parameters $\mathbf{w}$ and ρ are computed by maximizing the distance $\rho / \|\mathbf{w}\|$ under the constraint that all the training vectors must be located above the hyper-plane. In many situations, the training set may contain a small number of abnormal vectors or outliers, and the optimal hyper-plane should be positioned such that those outliers be located between the origin and $H(\mathbf{w}, \rho)$. This process is known as soft margin novelty detection, and it is necessary in the case of

Figure 4. Sketch of one-class support vector machine feature space

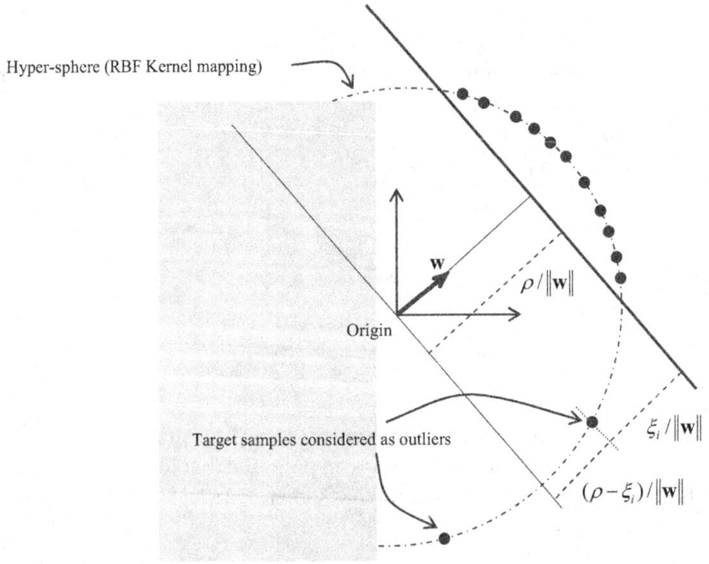

problems that are not linearly separable in the feature space. The novelty detection process can be expressed in term of an optimization problem as (Schölkopf & Smola, 2002):

$$minimize_{w,b,\xi} \quad \frac{1}{2}\langle \mathbf{w} \cdot \mathbf{w} \rangle - \rho + \frac{1}{\upsilon l}\sum_{i=1}^{l}\xi_i$$

$$subject \ to \quad \langle \mathbf{w} \cdot \phi(\mathbf{x}_i) \rangle \geq \rho - \xi_i$$
$$\xi_i \geq 0$$
$$i = 1, 2, \ldots, l$$

$$(19)$$

where ξ_i is the slack variable associated with each data example (vectors with $\xi_i > 0$) are classified as outliers) and $\upsilon \in [0, 1]$ is a regularization parameter tuning the number of possible abnormal training vectors. This parameter υ will be an upper bound for the fraction of outliers and a lower bound for the fraction of support vectors. The region R^* is then defined as the part of the hyper-sphere surface bounded by $H(\mathbf{w}, \rho)$ containing the training vectors with $\xi_i = 0$ (normal vectors). The optimization problem can be expressed and easily solved using dual problem:

$$minimize \quad W(\alpha) = \frac{1}{2}\sum_{i,j=1}^{l} \alpha_i \alpha_j K(\mathbf{x}_i, \mathbf{x}_j)$$

$$subject \ to \quad \sum_{i=1}^{l}\alpha_i = 1$$
$$1/\upsilon l \geq \alpha_i \geq 0, \quad i = 1, 2, \ldots, l$$

$$(20)$$

Now, the decision function for pattern classification/recognition will be defined mathematically as a function of the support vectors (vectors $\mathbf{x}_i$ with $\alpha_i \neq 0$) as:

$$g(\mathbf{x}, \alpha) = sign\left(\sum_{i \in sv} \alpha_i K(\mathbf{x}_i, \mathbf{x}) - \rho\right).$$

$$(21)$$

Similarly the slack values can be recovered by computing for all bounded support vectors $(\alpha_i = 1 / \upsilon l)$ by:

$$\xi_i = \rho - \langle \mathbf{w} \cdot \phi(\mathbf{x}_i) \rangle = \rho - \sum_{j \in sv} \alpha_j K(\mathbf{x}_j, \mathbf{x}_i)$$

$$(22)$$

In the case of support vector novelty detection, there are two types of misclassifications or errors that have to be minimized:

- Rejection of a normal data or false alarm
- Acceptance of an outlier (abnormal) data

In general the acceptance of an outlier (e.g., damaged state classify as undamaged) can be considered worse than a false alarm (undamaged state classify as damaged). Because of this and since the only training data set is normal data, the minimization of just false alarms may lead to acceptance of all data. Therefore, the classifier should be allowed to have some classification error on the training data (Unnthorsson, Runarsson, & Jonsson, 2003).

This learning algorithm leads to a final set of weights $\alpha_i \neq 0$ that characterize the decision rule of learning machine that is able to classify a new input vector into a normal or abnormal condition.

Damage Estimation:
A Case Study of the IASC-ASCE Benchmark Problem

IASC-ASCE Benchmark Problem

In order to illustrate the theory described in previous sections, the IASC-ASCE Benchmark structure was used for testing the proposed model for structural damage detection. This structure has been used by several researchers to identify the capabilities and limitations of various damage detection techniques (Caicedo & Dyke, 2002; Johnson et al., 2000). The SHM Benchmark structure is a 4-story, 2-bay by 2-bay steel-frame scale-model structure (Figure 5) located in the Earthquake Engineering Research Laboratory at the University of British Columbia (UBC). It has a 2.5 m × 2.5 m plan and is 3.6 m tall. The members are

Figure 5. SHM Benchmark Structure (Source: Picture taken from http://wusceel.cive.wustl.edu/asce.shm/)

Table 1. Damaged patterns used in this study

Case	Description	
	Analytical Model	**Experimental Model**
Case 1	Undamaged structure. All braces present.	Undamaged structure. All braces present.
Case 2	All braces on the first floor are removed.	All braces on the east side are removed.
Case 3	All braces on the first and third floor are removed.	Braces on the south half of the east side are removed.
Case 4	One brace on the first floor is removed.	Braces on the first and fourth floor of the south half of the east side are removed.
Case 5	One brace on the first floor and one on the third floor are removed.	Braces on the first floor of the south half of the east side are removed.
Case 6	Case 5 + loosed beam-column connection.	Braces on the second floor of the north side are removed.
Case 7	One brace on first floor with partial reduction in axial stiffness.	

hot rolled grade 300W steel (nominal yield stress 300 MPa (42.6 kpsi)). Slabs are placed at each floor level to simulate the mass of a structure, four 800 kg slabs at the first level, four 600 kg slabs at second and third levels, and, on the fourth floor, either four 400 kg slabs or three 400 kg and one 550 kg to create some asymmetry mass distribution (Johnson, Lam, Katafygiotis, & Beck, 2000, 2004).

Two finite element models for modeling the dynamic behavior of benchmark structure are available. The first model was developed assuming perfectly rigid floors and allowing just translations in parallel planes and rotations about vertical axis (i.e., shearing building model). This model has a total of 12 degrees of freedom (DOF), two translational DOFs along the longitudinal and transverse axes and one rotational DOF per floor. The second model has a total of 120 DOFs. Out-of-plane motion and rotation have been allowed by constraining the displacements and rotations of all nodes in each floor to be the same.

A total of six damage patterns are defined in the analytical part of benchmark problem (see Table 1). For Damage Pattern 1, all the braces in the first floor are removed. All of the braces at the first and third floors are removed for Damage Pattern 2. In the case of Damage Pattern 3 just one brace in the first floor was removed. Damage Pattern 4 was generated by removing one brace at the first floor and one at the third floor. Damage Pattern 5 consists on same scenario that Damage Pattern 4 plus a partially loose beam-column connection in first floor. Damage Pattern 6 is associated with a partial in the axial stiffness of one of the braces of the first floor. The analytical models, cases and patterns of damage are described in greater detail in Johnson et al. (2000, 2004); they will not be discussed further herein.

For the experimental phase, different damaged configurations were tested, as it is described in detail by Dyke et al. (2003). Six configurations (one undamaged and five damaged cases) are based on the braced structure. In these cases the structure was damaged by removing braces at specific locations of the structure. All undamaged and damaged structural configurations were excited by three different loading conditions: ambient vibration, hammer impacts, and broadband forced vibration produced by an electro-dynamic shaker. For the ambient vibration case, the excitation was induced by several factors such as wind, ground excitations, traffic, and working machinery. Hammer impacts were all located in southwest corner at the first floor (between first and second stories), and responses were recorded for both, impacts directed in the north-south and east-west directions. Finally, a random input was introduced with an electro-dynamic shaker placed at the top floor of the structure. A total of 15 uniaxial accelerometers were placed on the structure to record its response under all load cases and damage scenarios. Each floor (including the base) was equipped with three sensors, two of which measured accelerations in the north-south direction at opposite sides of the structure, and the third measured east-west accelerations. A detailed description of the experimental phase of benchmark problem can be found in Dyke, Bernal, Beck, and Ventura (2003). Data records and more related information can be found at http://wusceel. cive.wustl.edu/asce.shm/.

Data Considered in This Study

For this study, the analysis focuses on the acceleration measurements for undamaged and damaged configurations using (a) the analytical model (i.e., 120 DOFs model), where records were obtained for the case of a full set of sensors (i.e., accelerations measured at 4 DOFs per floor); and (b) the experimental model where ambient vibration excitation was considered recording accelerations at 3 DOFs per floor. Table 1 summarizes the damage cases investigated.

Simulated data used in this study were obtained from the analytical model. The finite element model was excited with random forces applied at each floor in the y-direction to simulate ambient vibration in the structure. Gaussian white noise was used to excite the model and accelerations were recorded on the center columns of each side of the structure (i.e., 16 accelerometers in total). For providing a more realistic example, the acceleration time histories were contaminated with 0, 5, 10, 20 and 30% of noise simulated by adding Gaussian pulse processes to the records. In total, 560 noise-corrupted accelerations were obtained for the undamaged structure and 480 sample data sets for damaged model (i.e., 80 records for each damage case).

In the experimental case, just the accelerations obtained from the ambient vibration loading case by the 12 accelerometers mounted in the structure (i.e., three sensors per floor) were considered. Each acceleration record was divided in windows of 10 seconds with a 25% of overlap, obtaining 35 time history records per sensor. Thus the total number of data sets generated for the undamaged case was 420 and for the damaged structure was 2100 (i.e., 420 records for each damage case).

Independent Component Analysis for Data Reduction

The 560 time histories corresponding to the analytical healthy state of the structure were used to form a 560×2001 matrix, where 2001 is the number of points in each time history. Clearly, such a large data set cannot be easily managed by either ANNs or SVMs directly;

Figure 6. Original and reconstructed time histories for sensor 14 in analytical model (undamaged state)

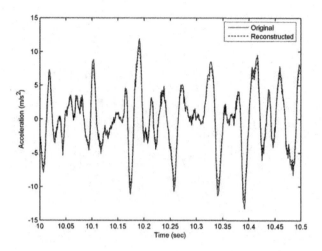

Figure 7. Original and reconstructed time histories for sensor 14 in analytical model (damage case 3)

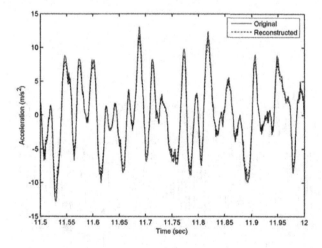

Figure 8. Original and reconstructed time histories for sensor 13 in experimental model (undamaged case)

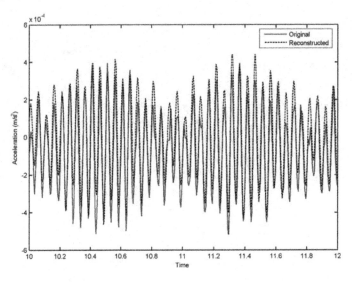

Figure 9. Original and reconstructed time histories for sensor 13 in experimental model (damage case 2)

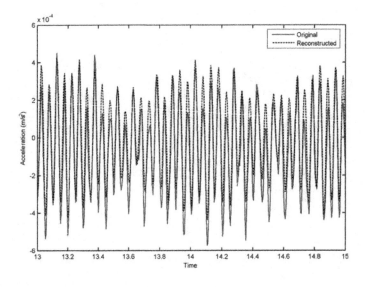

therefore a data reduction method, ICA was used. During the sphering (whitening) stage the first 20 dominant components containing 98.39% of the original information were retained. Then 20 independent components and the corresponding mixing matrix A with dimension of 560×20 were estimated by fastICA algorithm (Hyvärinen, 1999). To estimate the extraction quality, the time histories were reconstructed using the 20 independent components and the corresponding mixing matrix. The same procedure was applied to 480 samples belonging to damaged states. The reconstructed signals are compared with the original time histories, for a limited time range, in Figures 6 and 7. As it can be seen, very good matches between the original and the reconstructed responses were achieved, and the independent components clearly characterize both dynamics of the healthy structure and the effects of damage in the analytical model.

On the other hand, for the experimental model, independent component analysis were performed to 420 acceleration records obtained from the undamaged structure and the 2100 from the different damage cases. Again in this case the number of points in each acceleration time history was 2001. After sphering the data matrices, just the first 15 dominant components were retained for characterize the dynamics of both damaged and healthy experimental structure. These 15 components contain the 95.03% of the original data. Comparing again the original and reconstructed accelerations (Figures 8, 9) we can conclude again that mixing matrix A can represent the essential dynamic characteristics of the damage and undamaged systems.

Neural Networks and Support Vector Machines for Damage Detection

Using ICA all acceleration records obtained from the analytical and experimental model for the undamaged and all damaged scenarios were transformed into a linear combination of the mixing matrix A and the independent components containing a high percentage of the original data. The mixing matrix A, which represents the dynamic characteristics of the healthy structure and the damaged system, is used as input data to the artificial neural network model and the support vector machine for novelty detection (i.e., damage detection). The available 560 and 420 data sets, for the analytical and experimental case respectively, obtained from the healthy or reference structure were divided into training sets of 420 and 315 records (i.e., 75% of available data) and a validation set consisting in 140 and 105 vectors for the analytical and experimental system.

For the current specific application, the neural network used for novelty detection in each model, were auto-associative neural networks trained by a Levenberg-Marquardt version of back propagation. These neural networks are conventional feedforward networks, where the observed data (i.e., mixing matrix) is both the output target and the input of the neural network, mapping the underlying dependency of the extracted features and operational variations (Sohn, Worden, & Ferrar, 2001). Therefore, when the auto-associative network is fed with the inputs obtained from a damage state of the system, the novelty index, which is defined as the Euclidean distance between the target outputs and the outputs of the neural network, that is, $NI(\mathbf{y}) = \|\mathbf{y} - \hat{\mathbf{y}}\|$, will increase (Sohn, Allen, Worden, & Ferrar, 2005). If the learning has been successful, the index will be $NI(\mathbf{y}) \approx 0$ for data obtained from undamaged

Figure 10. Damage classification using artificial neural network based on undamaged and damaged cases for analytical model (validation and testing phase)

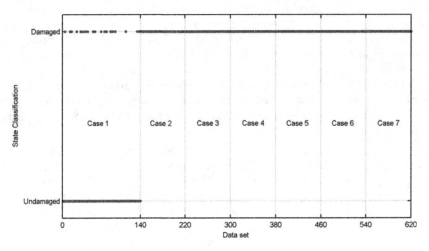

Figure 11. Damage classification using artificial neural network based on undamaged and damaged cases for experimental model (validation and testing phase)

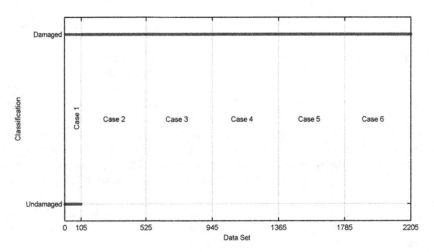

state. However, if data is obtained from the damaged system, the novelty index will show a value different to zero providing an indication of an abnormal condition.

The neural network used in the analytical case had a structure of 20 neurons in the input layer; 15, 10 and 15 neurons in three hidden layer respectively; and 20 neurons in the output

Table 2. Confusion matrix obtained for damage detection using artificial neural network

True class	Prediction (Analytical model)		Prediction (Experimental model)	
	Undamaged	Damaged	Undamaged	Damaged
Undamaged	80.7%	19.3%	81.9%	18.1%
Damaged	0%	100%	0%	100%

layer. For the experimental case, the network had a structure of 15:10:5:10:15 neurons in each layer. In both neural networks, a sigmoid activation function was used in each neuron. The number input and output neurons are determined by the number of independent components obtained by ICA. After a successful training of the network, the validation set corresponding to data from undamaged state and testing sets (i.e., data sets from all damaged states) were sequentially introduced to both networks (one for analytical model and other one for experimental model) and then the outputs compared using the novelty index. In Figures 10 and 11, the results of the damage detection process using neural networks are shown. These figures show the classifying decision (i.e., undamaged or damaged) obtained for analytical and experimental model by the neural networks for each one of the all new testing feature vectors from undamaged and damaged states. For the damage case 1 (i.e., undamaged structure), it can be observed that neural network classifies some vectors from healthy states into a damaged state in both analytical and experimental models. In other words, the learning machine decided that damage has occurred but the structure is still healthy. This is known as a false alarm. For other damage cases, the neural network detected correctly all damaged occurrences. These results can be summarized using a confusion matrix containing infor-

Figure 12. Characteristic curves used for optimal SVM selection for $\upsilon = 0.05$

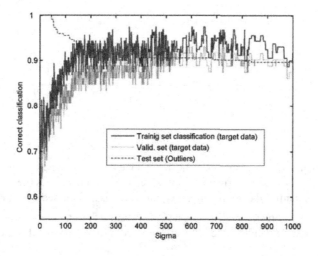

Figure 13. Damage classification using support vector machines based on undamaged and damaged cases for analytical model (validation and testing phase)

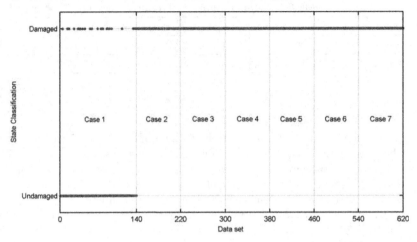

Figure 14. Damage classification using support vector machines based on undamaged and damaged cases for experimental model (validation and testing phase)

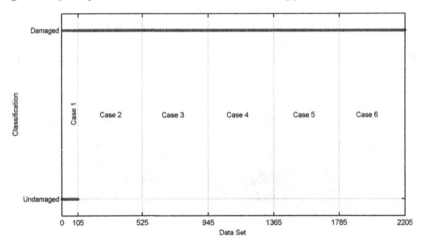

mation about actual and predicted classifications done by a classification system. In other words, it shows the percentages of correct classification for both classes (i.e., damaged and undamaged), and confused classification (i.e., false alarms and missed detection).

The corresponding confusion matrix obtained after testing the artificial neural network for damage detection is shown in Table 2. It can be stated from the results that the ANN used

Table 3. Confusion matrix obtained for damage detection using support vector machines

True class	Prediction (Analytical model)		Prediction (Experimental model)	
	Undamaged	Damaged	Undamaged	Damaged
Undamaged	93.6%	6.4%	92.8%	7.2%
Damaged	0%	100%	0%	100%

for novelty detection was able to detect all damage cases in both analytical and experimental models. On the other hand, analyzing the percentage of false alarms that correspond to misclassification of undamaged states, it was similar in both analytical and experimental models and was around the 18.5%. Based on this, it could be stated that type of signal used for damage detection (experimental or analytical data) did not affect the performance of neural network. However, it might be possible that the percentage of false alarms in novelty detection will be an inherent characteristic of the type of learning machine used for implementing the damage detection process.

The second learning machine used in this study was the one-class SVM trained for novelty detection by using the algorithm proposed by Unnthorsson et al. (2003) were the RBF kernel is tuned finding the parameter σ that produce a fraction of errors (on the undamaged data set) less or equal to parameter υ. As it was explained before, υ is an upper bound of the fraction of outliers (proportion of errors on undamaged data set). This procedure for selecting σ had the highest classification accuracy of outliers for a given value of υ compared to other tuning methods like the minimization of misclassified healthy data or minimizing the number of support vectors, which divided by the number of training examples gives a leave-one out bound on the test error of training data (Unnthorsson et al.). The criteria defined for selecting σ starts with a small σ, which is then increased until the rate of training data misclassification first reaches υ. This results in the highest classification accuracy of outliers for given value of υ.

In the case of support vector machines, it is not necessary to previously define any type of architecture or structure. For implementing the damage detection algorithm using novelty detection, two different support vector machines will be design for the analytical and experimental case respectively. The first step in the model selection process consists of defining the maximum classification error on the training data υ, which in this study was fixed to values of 0.05, 0.1, 0.2 and 0.3 (%). The training procedure on each learning machine was carried out for different values of σ (from 1 to 1000) using the independent component vectors obtained by applying ICA to acceleration time histories from undamaged analytical and experimental models respectively. After training, the optimal σ value (i.e., the optimal one-class SVM) that maximize the percentage of correct damaged/undamaged classification is selected. The optimal σ values were 210 for the SVM using analytical data and 160 for the SVM implemented for the experimental model. Figure 12 shows the performance for novelty detection of SVM used in experimental model (i.e., correct classification) obtained during the training, validation and testing using undamaged and damaged data for different values of σ. In Figures 13 and 14, the results of the damage detection process using support vector machines for novelty detection in both models (i.e., analytical and experimental) are shown.

The corresponding confusion matrix obtained after testing the SVM for damage detection is shown in Table 3. The one-class support vector machine for novelty detection identifies all damage cases in both analytical and experimental models. Analyzing the percentage of false alarms that correspond to misclassification of undamaged states, it can be seen that it was fewer than 10% in both SVM implemented for the analytical and experimental model. This percentage was much less, in both models, than misclassification obtained by neural network which was around the 18%. Looking at results obtained by support vector machines and neural networks, it could be stated that performance of SVM for damage detection is much better than neural networks in terms of the percentage of false alarms in the novelty detection process using experimental and analytical dynamic features extracted from each model by applying an independent component analysis to measured acceleration time histories. These results agreed with Vapnik (1995), who expressed and shown that support vector machines are much robust than neural network in pattern recognition problems.

Conclusion

1. The basic concepts of statistical learning theory and the main principles of learning machines were reviewed. Additionally, it was shown that learning machines (ANN and SVM) used for novelty detection are attractive alternatives for dealing with pattern classification when only data from the normal operation of the structural system is available.

2. The results demonstrate that data reduction from acceleration time histories using independent component analysis (ICA), followed by an implementation of learning machines applied to pattern recognition are a suitable methodology for structural health monitoring. Independent component analysis is a powerful tool to decompose signals into uncorrelated independent components, to compress the dynamic characteristics of the undamaged and damaged structure into the corresponding mixing matrix, reduce the size of measured data sets and provide robustness to noise contamination.

3. The results show that the SVM classifier is at least competitive with other methods previously investigated, when applied to real engineering data sets. In all classification problems, there is always a decision to be made on the relative costs of false alarms and the acceptance of abnormal data; SVM provides the required flexibility.

4. The critical task in damage identification using learning machines is the correct classification of undamaged vectors and vectors with a low level of damage. This is caused due to the feature extraction from vibration data, which is similar and could share in some cases the same region (cluster) in the feature space. For that reason, for structural damage detection it has been preferred to use supervised learning for training learning machines when data from damaged states is available. However, the drawback of this approach is that most of information available for health monitoring is obtained from undamaged systems.

5. ANN and SVM for novelty detection were trained using only information from the healthy system and tested trying to classify damaged and undamaged feature vectors correctly. The results show that in this case the feature vectors obtained from all damaged scenarios in both analytical and experimental models were classified correctly (100% of precision) but the accuracy to classify the undamaged patterns was around 80% for neural networks and 93% for support vector machines. The optimal selection of a learning machine for novelty detection depends on the level of false alarms and the level of acceptance of outliers.

6. ANNs and SVMs are of important value in engineering classification problems. ANN and SVM are just some applications of the principles of SLT, and it may well be that the ideas of SLT will bring to engineers a great variety of pattern recognition tools to be used in structural health monitoring approaches.

References

Bucher, C., & Frangopol, D. M. (2006). Optimization of life maintenance strategies for deteriorating structures considering probabilities of violating safety, condition, and cost thresholds. *Probabilistic Engineering Mechanics, 21*(1), 1-8.

Caicedo, J. M., & Dyke, S. J. (2002). Solution for the second phase of the analytical SHM benchmark problem. In *Proceedings of the 15ᵗʰ ASCE Engineering Mechanics Conference*, New York.

Cristianini, N., & Shawe-Tylor, J. (2000). *An introduction to support vector machines and other kernel-based learning methods*. Cambridge University Press.

Doebling, S. W., Farrar, C. R., Prime, M. B., & Shevitz, D. (1996). *Damage identification and health monitoring of structural and mechanical systems from changes in their vibration characteristics* (Tech. Rep. No. LA13070). Los Alamos National Laboratory.

Dyke, S. J., Bernal, D., Beck, J., & Ventura, C. (2003). Experimental phase of the structural health monitoring benchmark problem. In *Proceedings of the 16ᵗʰ ASCE Engineering Mechanics Conference*, Seattle, WA.

Friswell, M. I., & Penny, J. E. T. (1997). Is damage location using vibration measurements practical? In *Proceedings of the Second International Conference on Structural Damage Assessment using Advanced Signal Processing Procedures* (pp. 351-362).

Goumas, S. K., & Zervakis, M. K. (2002). Classification of washing machines vibration signals using discrete wavelet analysis for feature extraction. *IEEE Transactions on Instrumentation and Measurement, 51*(3), 497-508.

Hernández, M. R., & Sánchez, M. (2002). Structural damage detection based on changes in the vibration properties. In *Proceedings of Third World Conference on Structural Control*, Como, Italy.

Hernández, M. R. Sánchez-Silva, M., & Caicedo, B. (2005). Estimation of the mechanical properties of existing multi-layer pavement structures. In *Proceedings of the Inter-*

national Conference on Structural Safety and Reliability (ICOSSAR-2005), Rome, Italy.

Hyvärinen, A. (1999). Fast and robust fixed-point algorithms for independent component analysis. *IEEE Transactions on Neural Networks, 10*, 626-634.

Hyvärinen, A., Karhunen, J., & Oja, E. (2001). *Independent component analysis.* New York: Wiley.

Hyvärinen, A., & Oja, E. (2000). Independent component analysis: algorithms and applications. *Neural Networks, 13*(5), 411-430.

Johnson, E. A., Lam, H. F., Katafygiotis, L., & Beck, J. (2000). A benchmark problem for structural health monitoring and damage detection. In *Proceedings of the 14th ASCE Engineering Mechanics Conference*, Austin, TX.

Johnson, E. A., Lam, H. F., Katafygiotis, L., & Beck, J. (2004). The phase I IASC-ASCE structural health monitoring benchmark problem using simulated data. *Journal of Engineering Mechanics, 130*(1), 3-15.

Kramers, M. A. (1992). Autoassociative neural networks. *Computers in Chemical Engineering, 16*, 233-243.

Lee, J. M., Yoo, C. K., & Lee, I. B. (2004). Statistical process monitoring with independent component analysis. *Journal of Process Control, 14*, 467-485.

Manson, G., Worden, K., & Allman, D. J. (2003a). Experimental validation of structural health monitoring methodology II: novelty detection on an aircraft wing. *Journal of Sound and Vibration, 259*(2), 345-363.

Manson, G., Worden, K., & Allman, D.J. (2003b). Experimental validation of structural health monitoring methodology III: damage location on an aircraft wing. *Journal of Sound and Vibration, 259*(2), 365-385.

Masri, S. F., Chassiakos, A. G., & Caughey, T. K. (1993). Identification of nonlinear dynamic systems using neural networks. *Journal of Applied Mechanics, 60*, 123-133.

Masri, S. F., Nakamura, M., Chassiakos, A. G., & Caughey, T. K. (1996). Neural network approach to detection of changes in structural parameters. *Journal of Engineering Mechanics, 122*(4), 350-360.

Masri, S. F., Smyth, A. W., Chassiakos, A. G., Caughey, T. K., & Hunter, N. F. (2000). Application of neural networks for detection of changes in nonlinear systems. *Journal of Engineering Mechanics, 126*(7), 666-676.

Onoufriou, T., & Frangopol, D. M. (2002). Reliability based inspection optimization of complex structures: a brief retrospective. *Computers and Structures, 80*, 1133-1144.

Osuna, E., Freund, R., & Girosi, F. (1997). An improved training algorithm for support vector machines. In *Proceedings of the VII IEEE Workshop on Neural Networks for Signal Processing*, New York.

Pouliezos, A. D., & Stavrakakis, G. S. (1994). *Real time fault monitoring of industrial processes.* Norwell, MA: Kluwer.

Schölkopf, B., & Smola, A. J. (2002). *Learning with kernels: support vector machines, regularization, optimization, and beyond.* Cambridge, MA: MIT Press.

Sohn, H., Worden, K., & Farrar, C. R. (2001). Novelty detection under changing environmental conditions. In *Proceedings of SPIE 8th Annual International Symposium on Smart Structures and Materials*, Newport Beach, CA.

Sohn, H., Allen, D. W., Worden, K., & Farrar, C. R. (2005). Structural damage classification using extreme value statistics. *Journal of Dynamic Systems, Measurement, and Control, 127,* 125-132.

Unnthorsson, R., Runarsson, T. P., & Jonsson, M. T. (2003). Model selection in one-class v- SVMs using RBF kernels. In *Proceedings of the 16th International Congress and Exhibition on Condition Monitoring and Diagnostic Engineering Management*, Växjö, Sweden.

Vapnik, V. (1995). *The nature of statistical learning theory*. New York: Springer.

Vapnik, V., & Chervonenkis, A. (1971). On the uniform convergence of relative frequencies of events to their probabilities. *Theory of Probability and its Applications, 16*(2), 264-280.

Vapnik, V., & Chervonenkis, A. (1974). Ordered risk minimization. *Automation and Remote Control, 35,* 1226-1235, 1403-1412.

Worden, K. (2003). Cost action F3 on structural dynamics: benchmarks for working group 2—structural health monitoring. *Mechanical Systems and Signal Processing, 17,* 73-75.

Worden, K., & Dulieu-Barton. (2004). An overview of intelligent fault detection in systems and structures. *Structural Health Monitoring, 3*(1), 85-98.

Worden, K., & Lane, A. J. (2001). Damage identification using support vector machines. *Smart Materials and Structures, 10,* 540–547.

Worden, K., & Manson, G. (2000). Damage identification using multivariate statistics: Kernel discriminant analysis. *Inverse Problems in Engineering, 8,* 25-46.

Worden, K., Manson, G., & Allman, D. J. (2003). Experimental validation of structural health monitoring methodology I: novelty detection on a laboratory structure. *Journal of Sound and Vibration, 259*(2), 323-343.

Yang, S. I., Frangopol, D. M., Kawakami, Y., & Neves, L. C. (2006). The use of lifetime functions in the optimization of interventions on existing bridges considering maintenance and failure costs. *Reliability Engineering and Systems Safety, 91*(6), 698-705.

Zang, C., Friswell M. I., & Imregun, M. (2004). Structural damage detection using independent component analysis. *Structural Health Monitoring, 3*(1), 69-83.

Zadeh, L. A. (1965). Fuzzy sets. *Information and Control, 8,* 338-353.

Chapter IX

Structural Assessment of RC Constructions and Fuzzy Expert Systems

Mauro Mezzina, Technical University of Bari, Italy

Giuseppina Uva, Technical University of Bari, Italy

Rita Greco, Technical University of Bari, Italy

Giuseppe Acciani, Technical University of Bari, Italy

Giuseppe Leonardo Cascella, Technical University of Bari, Italy

Girolamo Fornarelli, Technical University of Bari, Italy

Abstract

The chapter deals with the structural assessment of existing constructions, with a particular attention to seismic risk mitigation. Two aspects are involved: the appraisal of the actual conditions of the structure (material deterioration, preexisting damages) and the evaluation of the structural "vulnerability," that is, the propensity to suffer damage because of the intrinsic geometric and structural arrangement, boundary conditions, specific structural details. Attention is first focused on the investigation protocol, which is organized through a multilevel, hierarchical scheme: the procedure includes visual inspections, surveys, experimental testing on site and in laboratory, and gradually proceeds into the details of the problem, progressively refining and verifying hypotheses and preliminary judgments. In a second part, the definition of effective tools for uncertainty management and decision making is performed, by presenting a genetic-fuzzy expert system which handles the procedure of the assessment properly accounting for uncertainty and errors, and is able to tune the parameters involved on the basis of experts' knowledge, "training" the system. Finally, a case study is presented, applying the whole assessment procedure and the fuzzy genetic algorithm.

Introduction and Definition of the Problem

History, even the most recent, has shown how devastating can be the effects of natural catastrophes—in particular earthquakes—over unprepared territories and communities. Actually, "disaster prevention" is the buzzword of the last few years at a scientific, administrative and political level, and risk mitigation strategies have become a field of great political and economical engagement, with a consistent financial and technical effort. Of course, the scientific community plays a crucial role in these processes, and is charged with the task of defining—at a theoretical level—proper methods for the analyses and the interventions, and to transfer this knowledge to the community in the form of practical decisional tools.

The question should be framed in a definition of *risk*—by now widely acknowledged in the scientific world—as the "average expected losses" from a given disastrous event over a specified future time period, whether expressed in terms of numbers of lives lost or in terms of expected economic loss, physical damage to property, structures, and activities. An element of a society is considered "at risk" or "vulnerable" when it has a high propensity to suffer damage under the occurrence of a given disaster (hazard). There are three essential components in the determination of risk, each of which should be separately quantified:

1. **The hazard occurrence probability:** The likelihood of experiencing any natural or technological hazard at a specific location
2. **The elements at risk:** Elements which would be affected by the hazard
3. **The vulnerability of the elements at risk:** The propensity to suffer damage under a given hazard

The elements at risk of a society can be (besides human lives, buildings, and properties) facilities; transportation networks; strategic, economical, and directional activities; and even the same structure and social cohesion. With specific regard to the *urban environment,* eight components at risk can be singled out: population, space, functions, activities, government, regulation, identity, and image.

For instance, transportation facilities represent a strategic element, fundamental in order to guarantee the organizational functionality in the emergency phase and to preserve the continuity of the economical and productive activities. Risk is symbolically defined through a simple mathematical formula (see Figure 1), which expresses the dynamic relationship existing among the concepts of vulnerability, hazard, and risk: The greater the potential

Figure 1. The fundamental relationship for risk assessment: Factors involved in the definition of risk

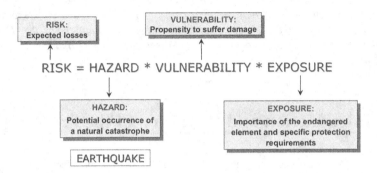

occurrence of a hazard, the more vulnerable the elements and the greater their importance and implications, the higher is the likelihood that the occurrence of a natural catastrophe will result in a disaster.

Hence, the assessment of the risk related to a particular disaster involves the quantification of the expected losses in a specific location during a fixed time frame. Such an evaluation includes, on the one hand, the availability and analysis of data about the probability occurrence of a particular catastrophic event (hazard evaluation) and on the other one the definition of the response of the element at risk (vulnerability assessment). Since most natural hazard events can not be controlled but only reasonably predicted (e.g., an earthquake cannot be reduced but could be forecast through proper modelling), the only way of reducing disaster risk is to decrease vulnerability (see Figure 2). Among the different components of elements at risk in the urban environment, the present chapter is focused on RC structures—and more specifically bridges—for which methodologies for structural vulnerability assessment organized according to a multilevel scheme will be presented.

Figure 2. Strategies for seismic risk mitigation:Iidentification of the elements at risk and vulnerability assessment vs. loss reduction

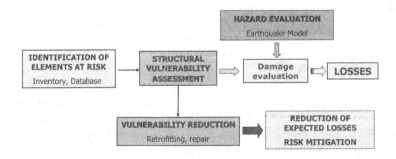

Procedures for Structural Assessment
of Existing Structures

Actually, structural vulnerability involves two distinct aspects (see Figure 3): (a) the survey of the actual conditions of the structure, in terms of preexisting damages, from earthquakes or other external events or degradation of materials and structural elements; and (b) the appraisal of the potential damageability that results from general deficiencies, inappropriate geometrical configuration, defects in design, execution, and detailing. For the assessment of existing constructions, especially RC constructions, the first point is particularly relevant. In fact, a severe-damage condition detected for the structure has a great impact on a global vulnerability judgement. In this case, the prosecution of the analysis on with more extensive investigation is generally recommended, and in the most critical cases, immediate interventions are necessary. Moreover, RC building and facilities stock include a large number of structures, belonging to the second half of the 19th century, which exhibit severe degradation due to a poor durability performance.

In these cases, the availability of an effective and feasible methodology for damage and safety assessment is widely needed, especially in seismic areas, in order to plan systematic maintenance and inspection programs. Indeed, the great attention devoted in the last few years to the development of reliable diagnostics techniques on the one hand, and to the proposal of tools for data processing, uncertainty management, and decision making on the other hand, clearly points out the relevance of the problem. In particular, crucial questions are:

- The identification of relevant data on which structural analyses should be based
- The systematization of the investigation protocol
- The implementation, management, and automatic processing of available data on an expert basis
- The formulation of a concise judgment on the damage state and safety level (in which, besides the different data and variables of the problem, the uncertainty and errors involved are properly taken into account)

In order to obtain the information for the structural assessment, it is necessary to perform a diagnostic investigation, following a systematic procedure that will progressively proceed into the details of the problem. This will include preliminary visual inspections, surveys, experimental testing on site and in laboratory (both destructive and nondestructive) aimed at the verification of the hypotheses formulated in the preliminary phases. Presently, from a technical point of view, a lot of appropriate and specific technologies are available in this field. Since 1983, the subject has been extensively studied producing a number of technical documents and guidelines, such as the FIP CEB proposals (CEB-FIP, 1998, 2002).

Actually, all the most advanced technical codes have introduced specific sections devoted to the assessment of existing structures, pointing out the different steps that should be followed for the analyses and adopting a hierarchical approach, characterized by increasingly accurate levels of knowledge. In the European context, the reference is represented by the

Eurocode 8 (2003), whose contents have been largely reproposed by the Italian Ordinance 3274 (OPCM, 2003). Recently, a very important European research project, involving many different countries, has dealt with the question of the Assessment, specifically applied to Bridge management: Project BRIME—Bridge Management in Europe (2001). The project was aimed at developing a framework for a bridge management system (BMS) for the European road network and represents a fundamental reference in this field.

Besides the problems related to the methodological approach for the diagnostic inspections and the related protocols (that is particularly interesting and stimulating both for the professionals and the scholars), a separate and quite complex question is represented by the interpretation of the data and consequent formulation of the final diagnosis about the "health" of the structure. In fact, such a matter heavily involves the uncertainty factors characterizing all the phases of the procedure (e.g., shortage/incompleteness of available data; instrumental or human errors during tests and surveys; reliability and intrinsic errors of the chosen structural model or in the calculations). These aspects can be conveniently managed through soft computing (SC) methods, which are just aimed at emulating human mind ability in terms of storing and processing information which is by nature imprecise, vague, uncertain and difficult to be categorized. Fuzzy logic (FL) and genetic algorithms (GA), which are two of the main subjects of SC, particularly lend themselves to this work.

Over the last 20 years, fuzzy logic has been proved to be an effective solution for classification, modelling or control in a large number of applications. FL is a language that uses syntax and local semantics to imprint any qualitative knowledge about the problem to be solved. Its main attribute is the robustness of the interpolative reasoning mechanism. In most cases, the key for success was the ability of fuzzy systems to incorporate human expert knowledge. Indeed, the spread of FL-based applications highlights the shortage of learning capabilities in this field.

In order to overcome this problem, many approaches aimed at the integration of learning capabilities have been proposed, and one of the most powerful and successful is represented by the hybridisation attempt carried out in the field of soft computing. A very promising direction, in this sense, is the hybridization by genetic algorithms. GAs are search algorithms based on the mechanics of natural selection and genetics. GAs proved to be a robust optimization method that effectively copes with multiobjective problems involving a large number of parameters.

Preliminary Remarks

The theoretical framework to which we will refer from now on, is that traced by CEB–FIB (1998, 2002) proposals. Structural assessment is a complex and uncertain process, relying on an integrated path of preliminary knowledge, analysis, decision making, and action. Among the aspects involved we can point out are:

- Structural, environmental and service data
- Data from existing documents
- Data from visual inspections

- Test data from in-situ and laboratory investigations
- Consideration of potential remedial actions

Moreover, for large stocks of structures (i.e., roads, bridges, hospitals, schools, strategic buildings, etc.) it is generally required a systematic and periodic assessment in order to safeguard safety and functionality and accordingly planning the necessary maintenance and rehabilitation measures. The scopes and extension of the assessment activities will depend, of course, on the severity of damage, importance of the structure, and hazard level.

The acquisition of the data required for an exhaustive and complete assessment, in most cases, is only achievable by applying time-consuming and expensive procedures. On the other side, the dimension and extension of the problem (considering the great number of endangered structures) would instead require faster and noninvasive diagnostic methods, even at the cost of reducing the accuracy level of the assessment.

Actually, considering that a prioritization is necessary to identify imminent dangers (or to prescribe further investigations), even a coarse appraisal of the safety level of an existing building is often a significant outcome, above all if this is achieved in a short time and causing the minimum damage. More detailed and precise analyses will be performed later, if appropriate, when the first coarse screening has pointed out that the structural safety level is lower than a minimum threshold. In this perspective, it seems reasonable to define proper "diagnostic protocols"; such a definition is intended to include not only the choice of the survey and testing methods but also the methodology to adopt for the operational procedures and for the data processing) according to a *hierarchical scheme*, including both quick observational models (able to provide satisfactory results as quickly as possible and with moderate destructive actions) and more accurate, mechanically based approaches.

Multilevel Assessment Scheme

The assessment procedure is based on a decisional tree in which, on the basis of the information acquired and the indication provided by their interpretation the future step to be taken is established.

- **Preliminary "quick" knowledge (I LEVEL ASSESSMENT):** In this phase, that is essentially based on visual inspections and on site basic testing, it is possible to point out situations of severe and imminent danger needing a prompt provision and a judgment about the opportunity of extending the investigation is expressed. It could also happen that vulnerability, damage and deterioration levels and are very low, and it can immediately recognized that no further investigation or intervention is needed.

- **Extensive and detailed investigation (II LEVEL ASSESSMENT):** In most cases, the preliminary investigation has to be supplemented and completed with more exhaustive data in order to assess the seismic performances, mechanical consistency of the materials, actual extent of the damage, and deteriorating phenomena in progress.

Figure 3. Multilevel strategy of the structural assessment process

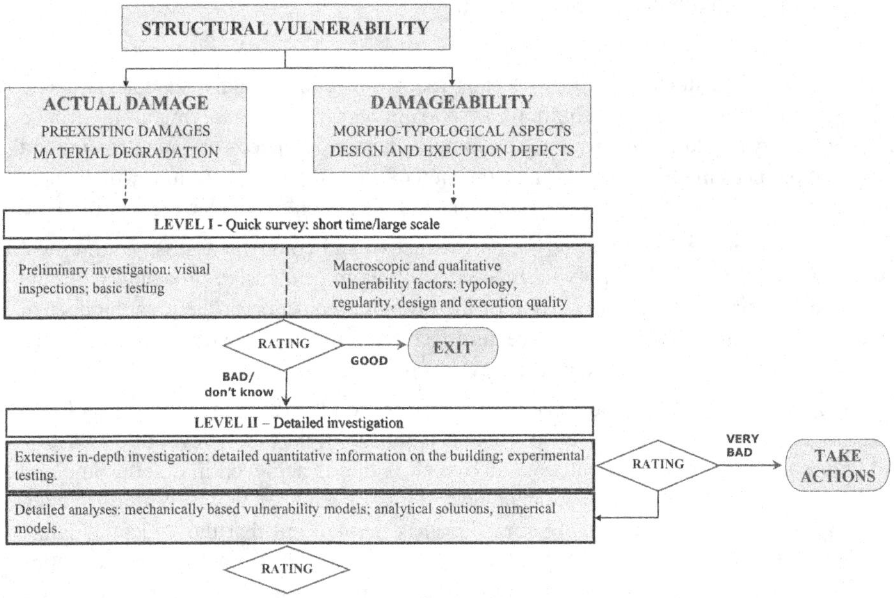

This phase will obviously require more extensive surveys and will be supported by laboratory and on site experimental testing, and by numerical analyses and modeling as well.

- **Data processing and interpretation:** Once the on-site operations have been accomplished, the collected data must be inventoried and processed, in order to provide the final diagnosis—that is to say, a judgment about the safety level of the building and the damage extent. If appropriate, the rehabilitation measures aimed at guaranteeing the prescribed safety conditions can be outlined.

Such a hierarchic, multilevel organization is also strictly functional to the administrative and political management of the built environment, with regard to the budgetary planning, assessment, and allocation of economical resources and prioritization of interventions.

In Figure 3, a concise logical flow chart of the multilevel procedure for the structural assessment is shown.

Diagnostics Protocols

The assessment of an existing concrete structure usually includes the following activities:

- Gathering information about the "*history*" of the structure
- Preliminary routine inspections (visual inspection, basic testing); preliminary reporting and evaluation about damage and macroscopic vulnerability factors; planning a detailed investigation (if necessary)
- Detailed investigation: special testing on materials and deterioration phenomena (concrete durability; prediction of corrosion progress); structure response testing; categorization of degraded areas on the structures; residual prestressing forces

Intelligent Systems in Structural Assessment

The treatment of natural phenomena is usually based on the derivation of a relationship between measurements of variables that relate to the phenomena. Theoretically, this is appropriately done in terms of abstract models actually translated into mathematical laws. For example, the probability theory represents a tool frequently used in the structural design in order to assess reliability, thanks to the possibility of treating the uncertainty and randomness of the data involved in the problem. Unfortunately, it is not able to manage those kind of uncertainties and ambiguities that have not a random nature: these are the so-called subjective uncertainties that derive from the judgment and experience of the technicians in charge of the structural safety assessment for existing buildings. The difficulties involved in such an evaluation, in fact, not only arise from the imprecision of the numerical data utilized in the analyses but also on the involvement of subjective opinions that definitely have a non-random nature but belong to the class of the so-called linguistic data, qualitative judgments, or inaccurate information. But, above all, standard probability theory is not able to process perception-based information and has no mechanism for (a) representing the meaning of perceptions, and (b) computing and reasoning with representations of meaning.

Dealing with Uncertainty

If we reason on uncertainty, we discover that uncertainty is linked to information through the concept of granular structure, a concept which plays a key role in human interaction with the real word.

In essence, a granule is a clump of physical or mental objects (points) drawn together by indistinguishability, similarity, proximity or functionality. A granule may be crisp or fuzzy, depending on whether its boundaries are or are not sharply defined.

Granulation is pervasive in human cognition. For example, the granules of "damage" are fuzzy sets labelled "light," "medium," and "severe." The concept of granularity underlies the concept of a linguistic variable. Fuzziness of granules, their attributes, and their values is characteristic of ways in which human concepts are formed, organized, and manipulated. In effect, fuzzy information granulation (fuzzy IG) may be viewed as a human way of employing data compression for reasoning and, more particularly, making rational decisions in an environment of imprecision, uncertainty, and partial truth. More specifically, perceptions are f-granular; that is:

Figure 4. Definitions of damage: Crisp; fuzzy, fuzzy granular; precisiated natural language and granulation in human cognition (Source: Zadeh, 2002)

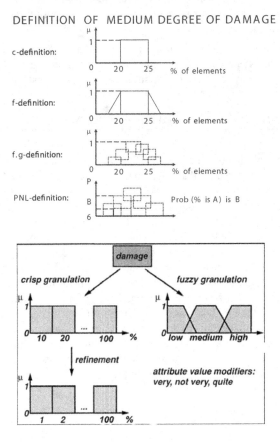

1. Perceptions are fuzzy in the sense that perceived values of variables are not sharply defined.

2. Perceptions are granular in the sense that perceived values of variables are grouped into granules.

According to Zadeh (2002), there are four principal rationales for the use of fuzzy logic–based systems:

1. **The "don't know" rationale:** In this case, the values of variables and/or parameters are not known with sufficient precision to justify the use of conventional methods of numerical computing. An example is decision making with poorly defined probabilities and utilities.

2.   **The "don't need" rationale:** In this case, there is a tolerance for imprecision which can be exploited to achieve tractability, robustness, low solution cost and better rapport with reality. An example in the structural assessment field is the problem of the determination of chloride penetration profile.

3.   **The "can't solve" rationale:** In this case, the problem cannot be solved through the use of numerical computing. An example is the problem of giving a concise judgement based on visual inspection.

4.   **The "can't define" rationale:** In this case, a concept that we wish to define is too complex to admit a definition in terms of a set of numerical criteria. A case in point is concept of causality, like that between damage assessment and structural vulnerability.

Why Evolutionary Algorithms?

A growing spread of FL-based applications have highlighted the lack of learning capabilities characterising most of the works in the field. To overcome this problem, one of the most successful approaches consists of integration of learning capabilities have been the hybridisation attempts made in the framework of soft computing. A very promising direction is the hybridization by evolutionary algorithms (EAs) (Herrera & Verdegay, 1996). EAs are search algorithms based on the mechanics of natural selection and genetics. EAs provide learning capabilities to the FL system because they optimize the latter according to training examples. In this way, the proposed FL system can integrates the experience of experts in assessment evaluations.

Mitigation of Seismic Risk for Bridges

The primary objective of risk assessment performed on a population of structures is to identify the elements most likely to undergo severe losses during an earthquake. The results of such studies are particular important for the elaboration of mitigation programs, disaster management, and the definition of strengthening intervention. In this section, attention will be focused on a particular class of structures: bridges. There are many reasons that make this structural type particularly relevant with respect to assessment and management issues.

Generally speaking, bridges structures are key elements of the transportation network. Particularly in seismic areas, their efficiency is crucial during the immediate post–earthquake phase, in order to guarantee the effectiveness of the emergency actions. Indeed, bridges are included in the list of essential facilities, whose structural integrity and functionality must be preserved during disasters.

The majority of catastrophic bridge failures around the world have shown that these structural systems are strongly vulnerable. In fact, bridges are often exposed to adverse environmental condition, are subjected to intense use and severe degradation, and—in seismic areas—could have been severely damaged by past earthquakes. Moreover, above all for existing bridges, whose structural organization and design does not comply with recent antiseismic require-ments, it is necessary to appraise the potential vulnerability deriving from design and morpho-typological deficiencies. The high damage levels observed on bridges after strong earthquakes have raised the efforts directed to improve their stability and reliability.

The assessment and prediction of seismic risk (i.e., the estimation of expected losses) for transportation systems provide valuable information both for pre–earthquake planning and risk mitigation, and for post–earthquake recovery purposes. In the case of bridges, risk assessment is generally oriented to retrofitting decisions, disaster response planning, estimation of direct monetary loss, and evaluation of loss of functionality. Indeed, a careful planning of routine inspections, maintenance programs, and rehabilitation interventions is particularly crucial, especially considering the importance of a proper allotment of economic resources.

Objectives

The primary objective of the vulnerability assessment—for bridges as well as for any kind of civil structure—is to guarantee the safety of the users. In order to do this, it will be necessary to detect possible damage caused by previous earthquakes and by deterioration processes, monitor the conditions of each structural element, and perform the vulnerability analysis. The main motivation, in this sense, is the capability of providing "right decisions in a reasonable time," since every judgment should be followed by a proper action aimed at preserving, if possible, the safety and serviceability of the structure, or alternatively at repairing it in order to reestablish such conditions.

Figure 5. Vulnerability assessment for bridges

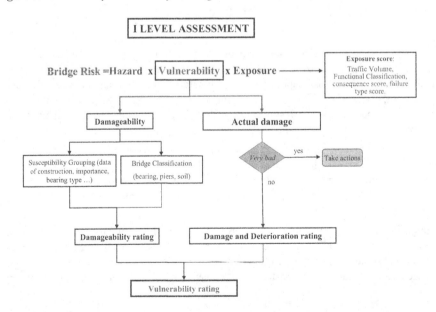

As previously stated, there are three essential components in the determination of seismic risk of structures: the seismic *hazard* occurrence probability, the *vulnerability,* and the *exposure* of the structure. For bridges, exposure is a measure of the impact that a failure would have on the bridge users and on the whole highway network and is determined on the basis of the traffic volume and the functional classification of the highway carried by the bridge. This parameter also takes into account the likelihood of failure, the expected consequences, and the failure type (see Figure 5).

I Level Assessment for Bridges and Priority Lists

Seismic risk assessment for bridges involves a number of complex issues (e.g., inventory, evaluation, ranking, definition of retrofitting/strengthening measures), and requires considerable professional expertise. The vulnerability assessment of a large stock of bridges is intended to identify the one characterized by an unsatisfactory safety level, because of design/morpho-typological deficiencies, damages caused by past earthquakes, and material deterioration. The final objective of the assessment is to establish a priority list of urgency with regard to intervention or provisional measures, that is, retrofitting and replacement.

Prioritization is one of the most difficult aspects of management. The simplest methods are generally founded on the coupling of a condition index and a strategic index. This approach allows, at least, to obtain a set of priorities: individual bridges or groups of bridges urgently needing seismic rehabilitation and maintenance can be selected and are and are included, by means of a proper ranking score, in a priority list for the development of planning actions (see Figure 6). This policy is especially applied in developing countries, which often have a vulnerable transportation network. However, it is also widely used by those developed countries that have neglected maintenance programs over many years and, because of the occurrence of seismic events, are now in charge of extensive repair and rehabilitation on a

Figure 6. Prioritization of bridges based on condition rating (Source: BRIME, 2001)

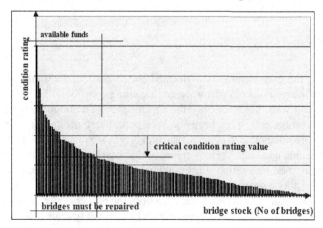

significant part of their stock. In countries where maintenance has been regularly applied and where the condition of the stock is generally satisfactory, it is instead possible to establish priorities using the optimisation of resources based on a true socioeconomic approach.

Damage and Vulnerability Rating for Bridges

Procedures for vulnerability assessment of bridges aimed at monitoring their conditions and preserving their reliability under seismic events have been developed in many countries. The various approaches developed are different in several aspects and also in the way of expressing the final result. The first step is represented by the definition of a programme of inspections, aimed at collecting data on which decision about seismic rehabilitation, repair or strengthening will be based. Generally, criteria used for establishing the priority in pre–earthquakes planning actions are based on a vulnerability rating, which is obtained through progressive steps by considering, on one side, the damage and deterioration rating, and on the other side, a "deficiency rating" derived by design and morpho-typological features (see Figure 5).

For bridges, a structural deficiencies rating is developed by means of screening and classification procedures, which result in a final rating for each bridge. More in detail, in a first phase the bridge is assigned to a susceptibility group, depending by a number of factors:

- Date of construction
- Importance factor (strategic role; number and importance of the facilities carried; bypass length, function)
- Single or multiple spans
- Simple or continuous girders
- Bearing type
- Number of girders per span (girder redundancy)
- Skew
- Pier type
- Footing type

In a second phase, the classification is performed on the basis of additional information, which strongly influences seismic performance, including connection bearings and seats, piers, abutments, and soils. In fact, even if the performance of a bridge is based on the interaction of all its components, it has been observed in past earthquakes that certain bridge components are more vulnerable to damage than others. These are: connection bearings and seats, piers, abutments, and soils. Among these, bridge bearings seem to be a critical element and, moreover, their retrofitting/substitution is easier and more convenient than other structural parts. For this reason, the classification is usually developed by examining the connections, bearings, and seat details separately from the remainder of the structure. The connections distinguish the parts where the superstructure is continuous by those separated by joints. A separate score is calculated for these components. The final judgement is determined from the sum of partial scores for each of the component (piers, abutments, and soils) susceptible to failure.

In order do complete the bridge vulnerability assessment, it is then necessary to detect all damage and deterioration occurred in a structural element. Observed damage caused by previous seismic actions consist in crack patterns (flexural and shear cracks), piers and abutments movements, joints and bearings fractures, and displacements, reinforcements yield, and buckling. A more detailed description will be given in Table 3.

In bridges, the most common form of concrete deterioration is represented by reinforcement corrosion, caused by the penetration of carbon dioxide or chloride ions. Concrete structures can also deteriorate because of cyclic freezing and thawing during the winter. Chemical reactions within concrete caused by sulfate attack or alkali silica reaction can also cause severe degradation of the concrete surface or structural element.

It is evident that the damage and deterioration level detected on the structure represent a crucial point in the emanation of the final judgement to be used in the prioritization. In fact, on the basis of this partial score in specific cases, especially when damage and deterioration level is marked, it is possible to plan retrofit and rehabilitation measures without performing further vulnerability analyses, which require additional more detailed investigations. The next part of this chapter will be focused, therefore, on the damage and deterioration assessment, since this is the critical point to be faced in order to plan the first actions aimed at immediate rehabilitation interventions.

Damage and Deterioration

Among the different methods proposed for the structural assessment, condition rating is a particular technique developed for "quantifying" the damage and deterioration level of viaducts and bridges, in order to identify the most damaged elements, scheduling the maintenance program and prioritize retrofitting or repair interventions.

The term *condition rating* (CEB-FIB, 1998, 2002) specifically indicates a numerical index able to express a concise judgment about the safety/serviceability level of a structure on the basis of a miscellaneous set of data, deriving from the visual inspection alone. In order to perform the above-mentioned evaluations, the main structural components have to be first identified and examined (in the case of bridges: piles, deck, girders, bearing). Then, the numerical index is computed from a set of partial contributions. Especially in the case of very large and complex constructions, it is generally convenient to perform a preliminary division into substructures (e.g., in the case of multispan bridges, the typical substructure could be represented by a single bay), so that the final assessment of the whole structure is derived from a simple summation of the individual performance conditions, provided that is at all time possible to carry out a detailed condition rating of each substructure or structural element. Different methods have been developed for evaluating bridge inspection data to give the bridge a condition rating. Among these, it is here presented one of the most used procedures, based on the work developed in Slovenia by the Slovenian National Building and Civil Engineering Institute (Lubiana: Department of Transportation, 1994).

Basic Procedure

The numerical computation of the condition rating simply consists in a score, obtained by assigning a set of penalty factors on the base of the expert's judgment (according to a chosen reference scale for the damage/deterioration level). In particular, for the numerical evaluation of each type of damage, the following elements are considered:

- Potential effect of the damage type on the safety and/or durability of the observed structural element or on the whole construction

- Potential effect of the specific damage observed within a single structural element on the safety of the whole structure

- Damage intensity

- Extension and potential propagation of the damage

According to this scheme, the global condition rating R is determined by adding the final scores associated to each damage type detected on the structure (after examining the elements that constitute the sub-structures). If this criterion is used, that is to say the condition rating is calculated by adding all the penalty scores assigned to the different damage classes, the following expression is used:

$$Condition\ rating = \sum^{32} G_i\ H_{1i}\ H_{2i}\ H_{3i}\ H_{4i} \tag{1}$$

where:

- G_i – **Damage type:** Basic value associated with the damage type i. The value of G_i expresses the potential effect of the damage type on the safety and/or durability of the observed structural element. For bridge located in adverse environmental conditions, for instance, 32 damage classes have been defined, also including damages to secondary elements or finishing. The numerical score ranges between 1 and 5, according to the severity level. For each damage type, a detailed description of the extension degree and repair needs is provided.

- H_{1i} – **Damage extension:** This is expressed from a number in the range *0-1* either referred to single components of the bridge or to the whole structure. It could be given a linguistic definition, through expressions like "small," "frequent," "very frequent," or "extended." Damage extension, as a rule, is not expressed by a direct measure (length, amplitude).

- H_{2i} – **Damage intensity:** This parameter is described by a numerical value comprised in the range 0-1 or by linguistic expressions like "negligible," "little," "medium," "severe," or "very severe." The description of intensity is usually related to the type of damage (e.g., width of the cracks, thickness of the delamination, etc.).

Table 1. Damage classes for the final condition rating

Damage Class	Description	Condition rating
1	Not present or negligible	0-3
2	Light	2-8
3	Medium – Severe	6-13
4	Severe	10-25
5	Very severe	20-70
6	Total	>50

- H_{3i} – **Importance of the element:** It is a factor which describes the importance of the structural component or member for the safety of the entire structure: structural components are classified in primary, secondary, others. The values range between 0 and 1.

- H_{4i} – **Urgency of intervention:** The values range between 0 and 10. The chosen value depends of the type of structure, and seriousness and risk of collapse of the affected structure or its part.

Following this approach, the condition rating R is expressed by a number ranging from 0 to 70. This interval is then subdivided into 6 intervals, corresponding to damage classes with decreasing grades of severity, described in Table 1, into which the bridge structure is classified.

On this ground, the main bridge inspector should be able to take decision about the global condition class of the whole bridge structure: perform a first ranking of the bridges in a management system; screen those bridges that should be commissioned for in-depth inspection, and undergo a more thorough maintenance, repair, or rehabilitation intervention. However, the method, in this specific formulation, has a minor drawback. In fact, different structural types (e.g., slab superstructure, superstructure made of girders, transverse beams and deck slab) may actually have different condition rating values, even if the apparent damage types, intensities and extent are estimated as being the same. Above all for this reason, the method has been modified and improved, expressing the condition rating as the ratio between the sum of the actual damage and the sum of the potential ones (i.e., evaluated in the hypothesis of a maximum extension and severity level).

Modified Method for Condition Rating

According to this approach, the condition rating of the structure is not expressed by the simple sum of damage values, but by the ratio between:

- the actual sum of the damage values obtained by the damage types detected during the inspection (choosing them from a pre-defined list of potential damage types); and

- the sum of the values of the potential damages (choosing them from the pre-defined list of potential damage types) that could realistically occur, considered at the maximum intensity and extent factors.

Thus, the condition rating of a construction (or a single part/component) is defined as the fraction (or percentage) of a "reference" damage rating value. It should be noticed that such a reference sum/condition level is not a fixed value, but depends from the particular structural type, from the arrangement of the inspected structure from the possible damages which can affect the structure. Therefore, the reference sum should only take into account those damages that have a real probability to appear, and shall ignore those that can never occur on it. For instance, in order to calculate the condition rating of an ordinary reinforced structure, the reference sum shall not consider damage values related to prestressed concrete. Hence, the condition rating is expressed in the following way:

$$R_C = \frac{\sum V_D}{\sum V_{D,ref}} 100 \tag{2}$$

The rating value is now expressed as a percentage of a conventional reference deterioration level, and is much less affected by the number of damage types, and by the number of members composing the observed structure, or by number of spans.

In equation (2), $R = \sum V_D$ is the sum of all the damage values V_D actually detected on the structure (or a substructure), selected from the list of all the potential damage types:

$$R = \sum V_D = \sum_1^{32} B_i\, K_{1i}\, K_{2i}\, K_{3i}\, K_{4i} \tag{3}$$

where:

- B_i – **Reference value associated to the *i-th* damage:** This factor accounts for the potential effect of the *i-th* damage type on the safety and/or durability of the observed structural element and ranges in the field 1-4 (see table 3).

- K_{1i} – **Importance of the element:** It is a factor which describes the importance of the structural component or member for the safety of the entire structure: structural components are classified in primary, secondary, others. The values range between 0 and 1.

In the case of a bridge structure, the K_{1i} of individual members shall be selected in the way that the total value for the standard bridge components remains within the following limits:

Table 2. Factor K_{1i}: general criteria for the importance of an element

Bridge substructure	Structural member	K_{1i}	Σ
Pier	Piles (when inspectable)	0.2	
	Foundation or pile cap	0.3	
	Columns or wall	0.4	1.2
	Pier cap	0.3	
Abutment	Piles (when inspectable)	0.2	
	Foundation or pile cap	0.3	
	Abutment wall	0.1	1.2
	Back wall	0.1	
	Wing walls	0.2	
Connections	Anchorage	0.3	
	Bearings	0.3	
	Shear keys	0.2	1.0
	Isolation and dissipation devices	0.2	
Bridge Super structure	**Structural member**	K_{1i}	Σ
Superstructure, type 1: Girders	Girders	0.6	
	Deck slab	0.4	
	End diaphragms	0.2	1.4
	Diaphragms	0.2	
Superstructure, type 2: Stringers	Stringers	0.6	
	Deck slab	0.4	1.2
	End diaphragms	0.2	
Superstructure, type 3	Solid or voided slab	1.2	1.2
Superstructure, type 4: Box girder arch	Top (deck) slab	0.4	
	Bottom slab	0.3	
	Webs	0.3	1.2
	Diaphragms	0.2	
Bridge deck	Sidewalk	0.1	
	Barrier	0.2	
	Parapet	0	0.5
	Median	0	
	Expansion joints	0.2	

- Substructure: 1.0 ± 0.2
- Superstructure, reinforced 1.2 ± 0.2
- Superstructure, prestressed 1.45 ± 0.2
- Bridge deck 0.4 ± 0.1

For every member in prestressed concrete, the basic K_{1i} value shall be multiplied by 1.2. Some examples for the determination of K_{1i} are shown in Table 3.

- **K_{2i} – Intensity of the i-th damage:** The general criteria for the determination of this parameter, and the corresponding values are provided in Table 4.

- **K_{3i} – Damage extension and propagation in an individual structural element or in a class of members of the same type:** The reference values are given in Table 5.

- **K_{4i} – Urgency of the intervention with regard to the i-th damage:** This factor is directly related to the safety of the individual element that is being analyzed. The general criteria for the evaluation and the corresponding values are provided in Table 6. $\sum V_{D,ref}$ is the reference sum, obtained by taking into account all the damage that can occur on the structure (they are extracted by the list of all potential damages), that is to say, considering the maximum value for each K- factor ($K_{2i} = K_{3i} = 2$, $K_{4i} = 5$).

In the case of multispan bridges and viaducts, where the inspection is usually performed span after span, the condition rating of the whole structure or of its main components is expressed by the average sum of damage values calculated for each individual span. Namely, the intensity and the extent of the detected damages can be more adequately defined and be better balanced when the evaluation is carried out for individual spans. In the practice, equation (2) is applied in the following modified form:

$$R_c = \frac{\sum_1^K K_{1m} M_m}{\sum_1^K K_{1m} M_{ref}} 100 \tag{4}$$

where:

- $M_m = \sum^n B_i K_{2i} K_{3i} K_{4i}$, ($i=1$ to "n") is the "reduced" sum of the damages, by applying the factor K_{1i} computed for the m-th structural member ($M_m = \sum^n_1 V_D / K_{1m}$).
- $M_{m,ref} = \sum^t B_i K_{2i} K_{3i} K_{4i}$, ($i=1$ to "t") is the sum of the reference damages, reduced by the factor K_{1i} referred to the same m-th structural element ($M_{m,ref} = \sum^n_1 V_{D,ref} / K_{1m}$).
- K is the number of structural elements belonging to the class "m" within the considered substructure.

Table 3.(a) Damage types to be evaluated and associated basic values B_i (displacements and deformations of the structure, concrete), (b) damage types to be evaluated and associated basic values B_i (reinforcing and prestressing steel, bearings)

Item	Damage type	B_i
1.0 Displacements and deformations of the structure		
1.1	Vertical deflection deck and girders	2.0
1.2	Span unseating at movement joint	2.0
1.3	Relative movements at spans in the longitudinal and transversal direction	4.0
1.4	Pounding of structures	3.0
1.5	Abutments slumping	3.0
1.6	Unsmooth approach, bump	1.0
1.7	Footing failure (flexural, shear, anchorage)	4.0
1.8	Loss of supporting resulting from large relative transversal and longitudinal movements	4.0
2.0 Concrete		
2.1	Poor workmanship: peeling, stratification, honeycomb, voids	1.0
2.2	Plastic shrinkage and plastic settlement cracks, crazing, cracks caused by inefficient joints	1.0
2.3	Strength lower than required	2.0
2.4	Depth of cover less than required for the ambient condition	2.0
2.5	Carbonation front (pH < 10), with reference to the reinforcement level	2.0
2.6	Chloride penetration, with reference to the reinforcement level	3.0
2.7	Cracking caused by direct loading, imposed deformations and restraint	3.0
2.8	Cracking due to inadequate shear strength	3.0
2.9	Cracking due to inadequate flexural strength or ductility	
2.10	Mechanical damages; erosion, collision	1.0
2.11	Efflorescence, exudation, popouts	1.0
2.12	Leakage through concrete	2.0
2.13	Leakage at cracks, joints, embedded items	2.0
2.14	Wet surface	1.0
2.15	Freezing and thawing	2.0
2.16	Freezing in presence of de-icing salts, scaling	2.0
2.17	Cover defects caused by reinforcement corrosion	2.0
2.18	Spalling caused by corrosion of reinforcement (bars and prestressing tendons or ducts) and/or by buckling of reinforcements	3.0
2.19	Open joint between segments	2.0

(a)

Table 3. continued

3. Reinforcing and prestressing steel		
3.1	Corrosion of stirrups	1.0
3.2	Corrosion of main reinforcing bars, reduction of steel area in the section (if in critical section, than: $K_4 \geq 2$)	3.0
3.3	Buckling of reinforcements	4.0
3.4	Duct deficiencies	2.0
3.5	Corrosion of prestressing tendons, depth (if in critical section: $K_4 \geq 2$)	4.0
3.6	Ejection or exposure of anchorage devices	4.0
4. Bearings		
4.1	Sliding of shear keys	3.0
4.2	Instability and/or overturning of rocker bearings	3.0
4.3	Misaligning and horizontal displacement of roller bearings	3.0
4.4	Wacky and instability of elastomeric bearings due to inadequate fastening	3.0
4.5	Excessive sliding of PTFE and stainless steel palte	3.0
4.6	Damage in the bearing-structure connections (keeper bars and anchor bolts)	3.0

(b)

- n is the number of actual damages of "i" type detected on the *n-th* structural element.
- t is the total number of the potential damages for the *n-th* member.

In Table 3a and 3b, the reference values B_i (ranging from 1 to 4) associated to each damage type are reported. Damage types are classified into:

- Displacements and deformations of the structure
- Damage to concrete
- Damage to reinforcing and prestressing steel

Tables 4, 5, and 6, respectively, report the values for the factor K_{2i}, K_{3i}, K_{4i}, and the general criteria to determine the degree of each damage type.

Finally, in Table 7 the complete description of the damage/degradation classes is provided. It is possible to identify 6 different classes, each corresponding to an assigned range of the *condition rating R_c*. The value obtained by applying the procedure previously described has to be compared with this table, in order to categorize the structure. Table 7a and 7b also includes the description of the serviceability and safety conditions for the structure, and the repair requirements, besides some reference examples.

Table 4. Factor K_{2i} – general criteria for the degree of a damage type

Class	Degree	Criterion	K_{2i}
I	Low	Damage is of small size, generally appearing on single localities of a member.	0.5
II	Medium	Damage is of medium size, confined to single localities, or, damage is of small size appearing on few localities or on a small area of a member (e.g., ≤ 25%).	1.0
III	High	Damage is of large size, appearing on many localities or on greater area of a member (e.g., 25% to 75%).	1.5
IV	Very high	Damage is of very large size, appearing on the major part of a member (e.g., > 50%).	2.0

Table 5. Factor K_{3i} general criteria for the extent of a damage type

Criterion	K_{3i}
Damage is confined to a single unit of the same bridge member.	0.5
Damage is appearing on several units (e.g., less than ¼) of the same bridge member.	1.0
Damage is appearing on the major part of units (e.g., ¼ to ¾).	1.5
Damage is appearing on the great majority of units (e.g., more than ¾) of the same bridge member.	2.0

Table 6. Factor K_{4i} – to stress the urgency of intervention

Criterion	K_{4i}
Intervention is not urgent because the damage does not impair either the overall safety and/or durability (service life) of the bridge structure or the durability of the affected member.	1
Damage must be repaired to prevent further impairment of the overall safety and/or serviceability and/or durability of the bridge structure or solely the durability of the affected member.	2 to 3
Immediate repair is required, as the damage is already jeopardizing the overall safety and/or serviceability and/or durability of the bridge structure, or if there is direct danger to people.	3 to 5
Temporary propping or limitation of traffic loads is required.	5

Table 7. (a) Deterioration classes (I-III), (b) deterioration classes (IV–VI)

Damage and Deterioration class	Damage and Deterioration Degree–Recommended Interventions	Rating values
I	**Absent.** No repair; only regular maintenance is needed.	0 to 5
II	**Low.** If not repaired in a proper time, safety, serviceability or durability of the examined structural component could be affected. Damaged parts can still be repaired with a low cost, within regular maintenance programs.	3 to 10
III	**Medium.** Safety, serviceability and durability of the examined structural component could be affected. No limitation in the use of the structure is required yet. Repair in a reasonably short time is required.	7 to 15

(a)

Damage and Deterioration class	Damage and Deterioration Degree–Recommended Interventions	Rating values
IV	**High** Safety, serviceability, and durability of the examined structural component is reduced. Local limitation in the use of the structure could be required. Immediate repair is required.	12 to 25
V	**Very high** Limitation in the use of the structure is required: restricted vehicle weight on bridges; propping of most critical components or other protective measures. Immediate repair and strengthening of the structure is required.	22 to 35
VI	**Critical** Immediate propping of the structure and total limitation of the use (e.g., closing of the bridge) are required. Immediate and extensive rehabilitation works are needed. It is possible that the restoration of the original safety and serviceability levels can no more be achieved with an acceptable economic cost.	≥ 30

(b)

Hybrid Evolutionary Fuzzy Expert System for Structural Safety Assessment of RC Constructions

The assessment procedure based on condition rating is a typical situation in which the fuzzy approach can be conveniently introduced. Indeed, the appraisal of the damage parameters involved (K_{2i}, K_{3i}, K_{4i}) is usually very difficult and extremely sensitive to expert judgements' swaying, even if the referee is a skilled engineer. The automation of safety analyses by means of a fuzzy system allows, in this sense, a better management of data deriving from those human opinions that cannot be appropriately and comprehensively represented by a crisp value. Hence, in order to perform the condition rating previously described, it is here proposed an expert system based on the hybridization of a fuzzy logic system by hybrid evolutionary algorithms (HEAs), which is characterized by flexibility, robustness, and comprehensibility and can be trained to learn by examples. The motivation for using HEA in combination with the fuzzy system is not only given by the possibility of easily optimizing the accuracy of the procedure but also by the general improvement in the comprehensibility of the whole process. In fact, the transparency features of fuzzy expert systems is particularly convenient for higher order knowledge representation.

But the most interesting aspect of the proposed hybrid evolutionary fuzzy expert system (HEFES) surely is the "acquisition" and "integration" of human expertise about structural assessment, which is then made available for future evaluations. Moreover, these are flexible systems, that can be trained and updated at any time by new expert data, and can be specialized according to the training data sets.

The Fuzzy System:
Membership Functions, Rules, and Defuzzification

The input data of the proposed fuzzy system are: K_{2i}, K_{3i} and K_{4i}, defined as fuzzy variables ranging according to the 4 intervals specified in Table 9. Accordingly, 4 *"membership functions"* $\mu_{kj}(x)$: $[0,1] \rightarrow [0,1]$ (defined as a mapping from the universe X into the unit interval) are assigned for each of the input fuzzy-variable. If the value assumed by the membership function is *1*, then x completely belongs to the fuzzy set. If it is 0, x does not belong to the set. If the membership degree is between 0 and 1, x is a partial member of the fuzzy set:

$$\mu_K(x) = \begin{cases} = 1 & x \text{ is a full member of } X \\ \in (0,1) & x \text{ is a partial member of } X \\ = 0 & x \text{ is a not member of } X \end{cases}$$

In particular, trapezoidal membership functions have been adopted in the algorithm:

$$\mu_K(x;a,b,c,d) = \max\left(0, \min\left(\frac{x-a}{b-a}, 1, \frac{d-x}{d-c} \right) \right)$$

where a, b, c and d define the vertexes of the trapezoid, and their initial values are reported in Table 8.

In this way, the result of the evaluation associated to each K-parameter will belong to the unit interval [0,1]. For instance, if no damage is detected, the value of K_{2i} will be 0, whereas, if the observed damage reaches a maximum severity level degree, the value will be 1. All intermediate conditions are described by internal values of the interval [0,1].

Table 8. Initial values for the parameters a, b, c and d of the trapezoidal membership functions

Classes	Membership, A_r	[a, b, c, d]
I	Low	[-0.28, -0.05, 0.05, 0.28]
II	Medium low	[0.05, 0.28, 0.38, 0.61]
III	Medium high	[0.38, 0.61, 0.71, 0.95]
IV	High	[0.71, 0.95, 1.05, 1.28]

Figure 7. Implementation of the fuzzy system by means of the Matlab Fuzzy Toolbox: the membership functions for the factor K_2

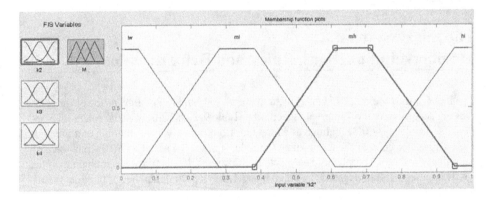

As a second step, the set of the fuzzy rules is built in order to fit the product $K_{2i} K_{3i} K_{4i}$, which is the core of the calculation of M_j, in the following way:

If $(K_{2i}$ is $A_{r2})$ and $(K_{3i}$ is $A_{r3})$ and $(K_{4i}$ is $A_{r4})$ **then**   output is B_s

where $r2, r3$, and $r4$ are subscripts denoting the active membership function, and ranging from 1 to 4; subscript s denotes the singleton resulting by the product of values listed in Tables 4, 5, and 6. In other words, the value of B_s is the result of all the combinations of products of the three parameters K_{2i}, K_{3i}, K_{4i} when the values defined in Table 8 (that is to say, the values defining the four classes of each parameter in the original classification) are used.

Table 9. Values of K_i used to evaluate the output membership functions

K_i	Values
K_{2i}	[0.5, 1.0, 1.5, 2.0]
K_{3i}	[0.5, 1.0, 1.5, 2.0]
K_{4i}	[1.0, 2.5, 4.0, 5.0]

Considering all the possible products, 64 singleton output membership functions are obtained. Each of these products corresponds to the activation of three input membership functions: for example, considering the values $K_{2i} = 0.5$, $K_{3i} = 1.0$ and $K_{4i} = 2.5$, the active input membership functions are: $\mu_{K_2 Low}(x)$, $\mu_{K_3 Medium-Low}(x)$ and $\mu_{K_4 Medium-Low}(x)$. In fact the values displayed in Table 6 correspond to the classes I, II, III and IV of the original classification, as well as the membership functions listed in Table 5.

Now, the fuzzy rules of the system are completely defined. Indeed, a set of 3 input membership functions activates the singleton output function through the value B_s provided by the corresponding values of the K-parameters (associated to the class represented by each of the

Figure 8. Output singleton membership functions

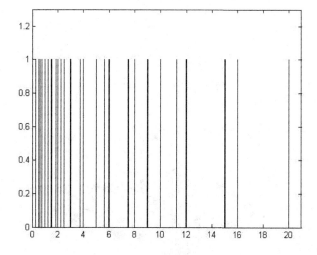

input membership function). Considering the previous example again, it can be seen that the activation of the functions $\mu_{K_2 Low}(x)$, $\mu_{K_3 Medium-Low}(x)$ and $\mu_{K_4 Medium-Low}(x)$ activates the singleton output function defined by $B_s = 1.25$. The output singleton membership functions are shown in Figure 8.

Under these conditions, the output singleton functions actually are a mapping of the universe [0,1], the definition domain of each input parameter, in the interval [0.25, 20].

In fact, when the values of K_{2i}, K_{3i} and K_{4i} are different from those listed in Table 9, the output functions (two or more) are partially activated, and all the possible values of the interval [0.25, 20] are permitted. Figure 9 shows the output surface in the three-dimensional space (K_{2i}, K_{3i}, M_j), that is, the result of M_j as function of the only two parameters K_{2i}, K_{3i}, for a fixed value of $K_{4i} = 0.5$.

At last, the fuzzy system is completed by the defuzzification step. Here, the centre of gravity (COG) method has adopted, which produces the following output:

$$y = \frac{\sum_{r=1}^{n} \beta_r b_r}{\sum_{r=1}^{n} b_r}$$

where β_r is the degree of fulfilment over the activated output singleton rules. The proposed method allows the fuzzy evaluation of M_j. The j-th addendum of the equation defining M_j is obtained by multiplying the output of the fuzzy system by the B_i coefficient. After the visual inspection performed by the expert, all the values of B_i and K-parameters (for each kind of damage, and for each the structural element) are provided, and the coefficient M_j is the automatically calculated. Finally, the *condition rating* of the whole building is obtained as the weighted-sum of the different M_j.

Figure 9. Output surface for a fixed value of K_{4i} (= 0.5)

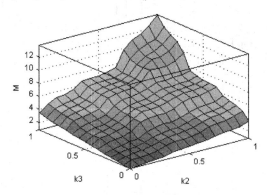

The fuzzy evaluation of M_j is a first step in order to overcome part of the problems previously discussed about condition rating, but there is still a crucial point that remains unaddressed.

Hybrid Evolutionary Algorithm

Let us suppose that an expert operates the condition rating of a given structure, following the standard procedure or adopting a fuzzy method, but that the result provided by the procedure does not match with his or her general "instinctive" perception about the static conditions of the construction. Then the same experiment is repeated with a number of experts who, again, do not completely agree with the condition rating. This leads to two possible conclusions. The former is that all the interviewed experts missed a proper classification of the damages, but this should be very unlikely if the number and the experience of the experts is proportionate. The latter is that the formula on which the condition rating is based on is not so accurate for the given building. In this case, it is very important to integrate the experience of the interviewed experts in the system for the condition rating.

Another similar example is given by a case in which a "I level" condition rating of a given structure is operated, and afterwards a more accurate "II level" analysis is performed, which does not match with the preliminary evaluation, possibly because the procedure on which the I level condition rating is based on is not so accurate for the given building, whereas the II level is able to account for factors and circumstances which are peculiar in the analysis. Also in this case, it is very important to integrate the experience of the II level assessment to the procedure used.

Speaking of a fuzzy system, this means, in other words, that it is necessary to change the membership functions in order to match the expert's II level classifications with their own I level condition rating. In particular, an optimization algorithm has to locate the best set of the vertexes of the four trapezoids for each input. Due to the large number of parameters to be tuned, 12 parameters per input $\Rightarrow$ 36 parameters, EAs can be an attractive choice.

Design for a Standard EA

The choice of representation, search domain, evaluation, and genetic operators is problem oriented. Considering the suggestions of literature on EAs several configurations have been tested and the most satisfactory one will be described in this subsection.

A. Representation

An individual or potential solution to the design problem is encoded as chromosome or string. The floating point representation has been chosen, hence the chromosome is a vector of floating point numbers, known as genes, one for each parameter, as shown in Figure 12.

B. Search Domain

There are two conflicting demands when the search domain is defined. The search domain has to be wide enough to ensure a good result, but the wider the search domain the slower the optimization. For our purpose the search domain is a 36-dimensional hyperrectangle given by the Cartesian product of the intervals which each parameter is limited to. The upper and lower bounds of each interval have been set as [0,1]. As a first step of the EA, N_{ind} individuals are randomly sampled over the search domain to create the first population. It should be noted that the random sampling is properly modified to avoid the overlapping of the top flat of the membership functions of the inputs.

C. Evaluation

The second step consists of the evaluation of how successful each individual is at solving the problem. The performance evaluation of a potential solution is a crucial point for every optimization method. In our case, the evaluation consists of the absolute value of the difference between the result given by the fuzzy system and the II level investigation. This can be considered a function, also known as objective function, to be minimized.

D. Genetic Operators

After the evaluation of all the individuals of a population, the selection process starts. The selection consists of ranking and sampling. In order to avoid preconvergence (Baker, 1985), individuals are initially ranked according to their objective values. Particularly the linear ranking algorithm (Mühlenbein & Schlierkamp-Voosen, 1993) has been adopted. Hence, the individuals fitness values are calculated as follows:

$$Ftn = (2 - SP) + 2\frac{SP - 1}{N_{ind} - 1}(Pos - 1)$$

Figure 10. Fitness as function of the individual position and the selective pressure

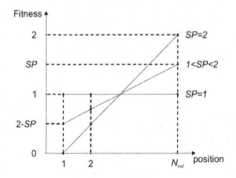

where *Ftn* is the vector of the element fitness, *Pos* is the vector of the ranking position of each element, and $SP \in [1, 2]$ is the selective pressure. Figure 10 shows how the linear ranking, according to the selective pressure, is a diversity diminishing and contracting operation. The population diversity, often listed among the advantages of the evolutionary approach, can defeat the optimization when the difference between some individual and the others is too much. This because the evolution tends to use only the few test individuals to create the new generation. In other words, the whole population tends to replicate those individuals drastically reducing the exploration and causing the pre-convergence. Usually the selective pressure is set between 1.5 and 2 because a too low value causes a too slow convergence rate.

The N_{pa} individuals of the mating pool are also called parents. The non-linear ranking and stochastic universal sampling methods have been chosen to avoid preconvergence (Baker, 1985) and to achieve zero bias and minimum spread (Baker, 1987). The probability of the individual being selected is:

$$SelP = 1 - \frac{1}{\sum\limits_{1}^{Nind} Ftn(i)} Ftn$$

The exchange of genetic information is known as recombination. The recombination of the N_{pa} parents generates the same number of new individuals called offspring. The algorithm adopted is known as intermediate recombination (Mühlenbein & Schlierkamp-Voosen, 1993), and is based on the following formula for offspring calculation:

$$x_{of} = x_{p1} + B(x_{p2} - x_{p1})$$

where x_{p1}, x_{p2}, and x_{of} are the two parents and the offspring, respectively. B is a diagonal matrix with diagonal elements b_i randomly chosen in the interval $[b_{min}, b_{max}]$, where typical

Figure 11. Example of intermediate recombination for a two-dimension problem

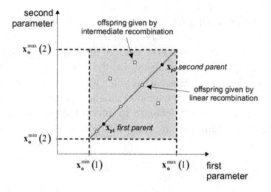

Figure 12. Scheme of the evolutionary algorithm

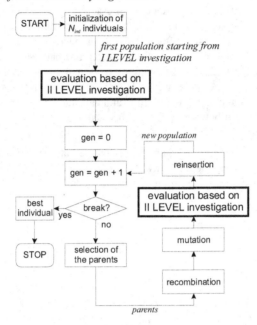

values are b_{min} = -0.25 and b_{max} = 1.25. If all the b_i have the same value, the intermediate recombination is called linear recombination. An example of intermediate recombination for a two-dimension problem is shown in Figure 11.

Then the offspring genes can be low-probabilistically altered during the mutation process (Mühlenbein, Schomisch, & Born, 1991), which operates directly on the floating point encoding. Finally, fitness-based reinsertion has been used to create the new population of N_{ind} individuals. The least fit individuals of the previous population are probabilistically replaced by the most fit offspring.

All the genetic operators can be tuned by means of coefficients such as the mutation probability and the selective pressure. Those values, which are usually constant in GAs, can vary throughout the optimization. We implemented the adaptive technique, proposed by Caponio et al. (2006), to estimate the state of convergence of the algorithm. This technique is based on the estimation a high or low diversity (in terms of fitness) among the individuals of the population. If there is a high diversity, it means that the solutions are not exploited enough, on the contrary, it means that the convergence is going to happen and since this convergence can be premature. In the latter case, for example, a higher search pressure is needed. In this way the adaptive algorithmic parameters aim to inhibit the premature convergence and stagnation and therefore to guarantee a more robust algorithm.

Figure 13. The chromosome contains the 36 parameters defining the input membership function; in particular, the figure shows the first third of the chromosome considering only the input K_2

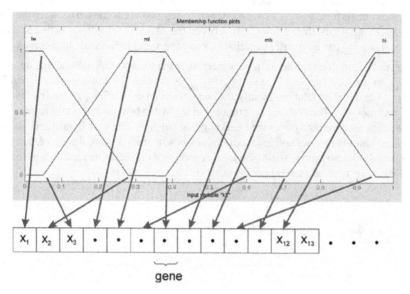

gene

E. Termination

At the end of every iteration, or generation, the EA provides a new population whose individuals can generally solve the problem better than the previous ones. Because of the stochastic nature of the EA, the formulation of convergence criteria is still an open problem. For this research the EA is terminated after a predefined number of generations N_{gen}.

Hybridization of EAs

Unfortunately, EAs are guided stochastic search methods that suffer of a slow convergence rate. A careful configuration only partly solves such a problem that lies in the nondeterministic nature of evolutionary operators that locate the optimal "hill"; that is, the zone where the optimal solution is, but are not able to quickly refine it. On the contrary, classical methods, often classified as hill climbing, can efficiently exploit local information to speed up the optimization but properly work when the function to be optimized is smoothed and unimodal; but they often fail in real-word problems that usually are ill behaved. In order to get the benefits of both the techniques, many hybrid methods were proposed. A hybrid method consists of a combination between different search methods. Although it is a very general definition, it is the only one that can include the huge number of possibilities. These are given not only by the number of search methods, but also by the hybrid architecture, that is, the way in which the different methods are integrated in a framework to cooperate.

As shown in Figure 14, an initial subevolution transforms the initial population, pop_{in} into the intermediate population pop_{int}. This stage consists of running the EA. Consequently, the top-ranking individuals are extracted from pop_{int} creating $pop_{top,in}$. Analogously, the medium-ranking individuals form $pop_{med,in}$. The size of $pop_{top,in}$ top ranking is chosen between the *10%* and *15%* of pop_{int}, whilst that of $pop_{med,in}$ is between the *95%* and *85%*. It has to be noted that $pop_{top,in}$ and $pop_{med,in}$ can be overlapped, that is, can have some individuals in common.

The probabilistic multidirectional simplex method, is the local search method that operates on $pop_{top,in}$ to produce $pop_{top,out}$. Again, the $pop_{med,out}$ is obtained by the aforementioned EA with $pop_{med,in}$ as initial population. Finally, the pop_{out} is obtained by a fitness-based merging of $pop_{top,out}$ and $pop_{med,out}$. The iteration reported in Figure 14 can be repeated until a termination criterion, such as a prefixed number of iterations, is satisfied. The new hybrid architecture better coordinates the global and local search methods in order to save fitness evaluations. The proposed solution rations use of the local search, and for the same reason, the simplex method has been adopted as local search method.

Figure 14. Iteration of the proposed hybrid EA

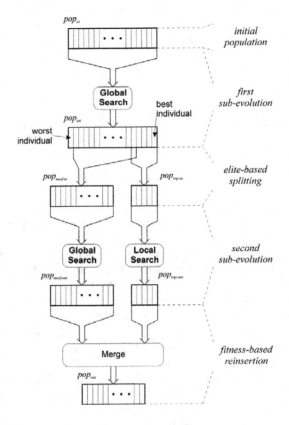

Figure 15. K_2 membership functions after the optimization

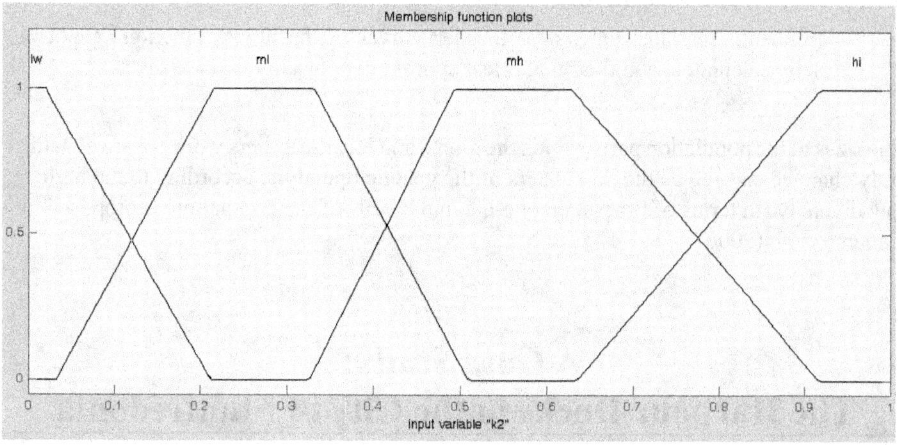

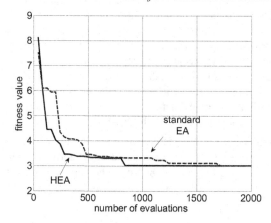

Figure 16. Comparison between the evolution of the standard EA and HEA

The simplex method, popularized by Nelder and Mead (1965) with their effective version, belongs to the class of the direct search methods whose two main properties are:

- No gradient, or any gradient approximation, can be used.
- Only the values of the fitness function can be used.

These properties make the direct search methods an efficient alternative to Newton's, and quasi-Newton methods that are impracticable:

- If the fitness evaluation is very time-consuming and noisy such as when calculated through experimental tests.

- If gradient, Hessian, and first partial derivatives of the fitness function cannot be exactly calculated, and their numerical approximations are too expensive.

The sizes of the population parts, which the global and local searchers work on, are dynamically changed, as well as the parameters of the genetic operators, according to the high or low diversity (in terms of fitness) among the individuals of the population as proposed by Caponio et al. (2006).

A Case Study:
The Harbour Docks in the City of Manfredonia

General Remarks

The case study described is concerned with the damage and deterioration assessment of the structures of an industrial harbour in southern Italy (City of Manfredonia, see Figure 17). Actually, the location is characterized by a medium-high seismicity level, and this would require the application of a complete procedure of vulnerability assessment. Indeed, ever from the first investigations, severe and widespread degradation and damage have been observed on the structural elements, inducing the administration to require, as a priority objective, the extensive analysis of the deterioration state. On the basis of the results obtained, it was decided to take immediate and urgent actions for repairing most of the structural elements and reestablish the original performance of the structure.

The complete assessment of the seismic risk for the harbour will follow the completion of the urgent restoration works. The harbour includes five berths and an approach pier for ships, 2 km long, connecting the land with two of the berths. A shorter pier leads to the Dam, consisting in an additional three-berths wharf and a breakwater barrier. The struc-

Figure 17. General view of the harbour

Figure 18. General plan of the harbour

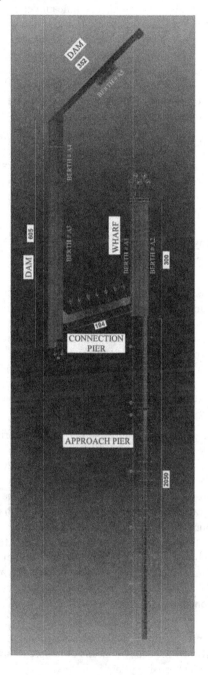

ture, designed by SNAM Progetti – Milano, was built in the 1970s, from 1971 to 1978. Although being a quite recent construction, a significant degradation affects great part of the structural elements either underwater, near or far from the seafront. Damages exhibited by the structure comprise generalized and widespread corrosion of reinforcements, concrete delamination, spalling, cracking, local pulling out of steel bars, and loss of protection of the anchoring device. The actual condition of the structure, so severe that the ship wharf is presently closed, is not only related to the hard environmental conditions but also to an unsatisfactory maintenance program. For instance, the device for the cathodic protection of the piles is out of order from several years.

An additional problem is represented by the very design and execution of the structures: some of the adopted solution results in fact are scarcely fit for a construction plunged in the sea. Moreover, several deficiencies and faults in the execution have been observed. Within the harbour, four main parts can be identified: approach pier, connection pier, wharf, and dam. These are built by differently assembling three basic types of structural elements: tubular piles, fixed into the depth of the sea and filled with RC; prestressed RC pile caps monolithically connected to the piles; and flanked box girders which constitute a ribbed plate, simply supported by pile caps.

A general visual survey of the harbour has immediately pointed out a significant degradation involving most parts of the structural elements, with a particular extension and severity in the members exposed to sprinkles and tide (most of the pile caps in the approach pier; the totality of the pile caps in the connection pier, wharf, and dam). The static deficiency of many structural members has been also confirmed by some preliminary on site diagnostic investigation. In particular, the experimental investigations performed have highlighted an abnormal corrosion process and a serious level of chloride penetration involving the pretensioned cables. With regard to the structural elements not directly exposed to the sea, the main problem is represented by carbonation of concrete, besides some local damage phenomena, such as exposition of anchorage devices, lack of embedding mortar, corrosion of loose reinforcements and tendons, and so forth.

Actually, the situation appears to be critical, above all considering the rapid progression of the degradation phenomena reported in the last few months and the involvements of most part of the main structural parts, indicating a very high level of risk. Unfortunately, the great extension of the work, and the limited funds that were available did not allow a generalized repair or substitution of the ineffective structural members. Hence, the use of a fuzzy based condition rating as been here exploited in order to rank the level of damage within the different groups of elements and prioritize the immediate actions, on the one hand, or the planning of further experimental tests. With regard to this strategy, it is here reported the assessment of the deck of the approach pier, and in particular the evaluation of the two structural member types which compose it: pile caps and box girders. This pier presents a total number of 82 pile caps, regularly spanned 25 m each other. They are in prestressed reinforced concrete, fully supported by the piles, with a trapezoidal section and an approximate length of 20 m.

The deck is realized with prestressed Π-shaped girders (length 24.50 m; width 5.15 m; height 170 cm), simply supported by the caps by means of neoprene supports. Each bay is made of two flanked girders, and expansion joints are placed every 150 m.

Figure 19. An example of pile caps belonging to the macro class a (left) and b (right)

Besides the two ribbed plates, there is also a prestressed RC truss for the accommodation of the pier equipment. Girders are connected by a transversal prestressed beam and, at the end sections, by two cast-in beams. The deck is completed by a cast-in slab having a thickness of 20 cm and stiffened by ribs 30 ÷ 40 cm large and 30 cm thick.

Condition Rating of the Approach Pier

The structural members of the deck, that is, pile caps and girders, were grouped into two macroclasses, with regard to the damage level:

- **(class a) Severe damage:** visual inspection, even supported by basic testing, revealed serious degradation both for the importance of the damage type and its extension, or at least for the damage importance (see Figure 19 – pile caps, and Figure 20 – girders).
- **(class b) Light damage:** only visual inspection has been performed, showing no serious degradation or deficiency (see Figure 19 – pile caps, and Figure 20 – girders).

Figure 20. An example of girders belonging to the macro class a (left) and b (right)

 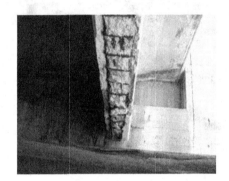

The survey of the structural elements of the type "pile caps" and "girders" including visual inspection and some basic testing, has allowed the evaluation of the K-coefficients involved in the condition rating procedure, as summarized in Table 10. According to the method previously described, first of all the condition rating of the individual structural members constituting the pier is performed. The final overall rating is then obtained introducing the importance factor K_{1i}, which graduates the effect of each element on the safety/durability of the entire construction. The condition rating for the structural components "pile caps" and "girders" is performed on the basis of the classification in the two macroclasses just defined.

The analysis of the individual members, and the corresponding fuzzy condition rating, has allowed the concise judgment about the whole pier, as shown in Table 11.

By considering the importance of the elements belonging to different damage classes, the final rating for the structural parts of the harbour is: high damage and degradation level. As a consequence, the serviceability and durability condition are reduced, but no extended limitations in the use of the facility are prescribed. Urgent repair are instead required.

The application of the procedure for the damage/degradation appraisal on the different structural elements of the harbour has allowed to categorize their condition. Different levels have been singled out, ranging from a *low* to a *very high* damage level. In particular, this last

Table 10. Condition rating for the "Class A" pile caps of the approach pier

Structural component: Deck								
Structural Element: Class a pile caps								
				Factors				
				Intensity	Extension	Urgency	FUZZY PRODUCT	Mm	Mrif
	Damage type	Bi		K_{2i}	K_{3i}	K_{4i}	$B_i \times K_{2i} \times K_{3i} \times K_{4i}$	$S B_i \times K_{2i} \times K_{3i} \times K_{4i}$	$B_i \times 2 \times 2 \times 5$
Displacements and deformations	Vertical deflections	2,0		0,0	0,0	0,0	0,5		40,0
	Unsmooth approach, bump	1,0		0,0	0,0	0,0	0,3		20,0
Concrete	peeling, stratification, honeycomb, voids	1,0		0,6	0,3	0,7	5,0		20,0
	Strength lower than required	2,0		0,0	0,0	0,0	0,5		40,0
	Depth of cover less than required for the ambient condition	2,0		1,0	1,0	0,8	35,6		40,0
	Carbonation front	2,0		0,9	1,0	0,8	34,0		40,0
	Chloride penetration	3,0		1,0	0,9	0,7	48,9		60,0
	Cracking caused by direct loading, imposed deformations and restraint	3,0		0,0	0,0	0,0	0,8		60,0
	Mechanical damages; erosion, collision	1,0		0,3	1,0	0,7	7,4		20,0
	Efflorescence, exudation, popouts	1,0		0,0	0,0	0,0	0,3		20,0
	Wet surface	1,0		0,3	1,0	0,3	3,8		20,0
	Cover defects caused by reinforcement corrosion	2,0		1,0	1,0	0,8	35,6		40,0
	Spalling caused by corrosion of reinforcement	3,0		0,9	0,9	0,7	41,7		60,0
Reinforcements	Corrosion of stirrups	1,0		1,0	1,0	0,8	17,8		20,0
	Corrosion of main reinforcing bars	3,0		0,3	0,7	0,7	17,8		60,0
	Duct deficiencies	2,0		0,4	0,3	0,7	8,5		40,0
	Corrosion of prestressing tendons	4,0		0,6	0,4	0,7	26,1		80,0
	injection defects	4,0		0,3	0,3	0,7	13,6		80,0
	unthreading of pretensioned bars	4,0		0,4	0,3	0,7	16,9		80,0
	exposition/expulsion of anchorage devices	4,0		0,4	0,3	0,8	18,3		80,0
								333,3	920,0
	Condition rating of the element							Rc=Mm/Mrif	36,2

Note: The rating is obtained by applying the procedure sketched in para. 4.1. In the table, the values assigned to K_{2i}, K_{3i} and K_{4i} are "normalized," that is, a mapping of the original domain (see Table 9) into the interval [0,1] is performed by the fuzzy algorithm. The final fuzzy product, instead, is expressed in the original reference system. For example, when $K_{2,3,4\,i}$ are zero in the table, actually the values in the original scale are 0.5, 0.5, 1 respectively, and the fuzzy product is 0.5.

condition is the most widespread. For those structural elements in a severe damage state, it is recommended the immediate repair/retrofitting or substitution; propping of most critical components, or similar provisional measures. With regard to the structural elements belonging to a low/medium damage class, repair or retrofitting could be postponed, but no more than one year, and attention should be paid to use repair techniques and materials suited

Table 11. Condition rating for the approach pier

Structural component	Pile caps		Girders		
Structural element	"Class a" pile caps	"Class b" pile caps	"Class a" girders	"Class b" girders	
Prestressed concrete	1,20	1,20	1,20	1,20	
K1i	0,80	0,80	0,60	0,60	
K1i f	0,96	0,96	0,72	0,72	
Mref	920,00	920,00	760,00	760,00	
K1f x Mref	883,20	883,20	547,20	547,20	
^{S}K1if x Mref					1430,40
Mmi	333,30	143,88	146,78	91,92	
K1if x Mmi	319,97	138,12	105,68	66,18	
W	0,15	0,85	0,10	0,90	
K1if x Mmi x W	48,00	117,40	10,57	59,56	
Weighted sum					235,53
$(R_c)_e = \dfrac{(g)}{(b)} x100$					16,47

Figure 21. Rating obtained after the assessment of the different substructures of the harbour

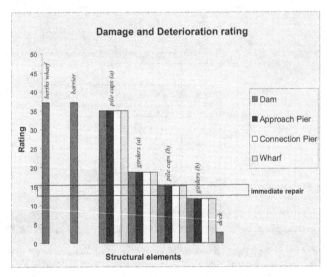

for an aggressive environment. Anyway, for these elements, the executive definition of the intervention should be preceded by in-depth investigations.

The final result of the analysis, that is, the high level of damage and degradation detected, have induced the management to immediately contract the repair interventions necessary in order to guarantee the required safety level, delaying further vulnerability analyses in a future phase.

References

Baker, J. E. (1985). Adaptive selection methods for genetic algorithms. In *Proceedings of the First International Conference on Genetic Algorithms and Their Applications* (pp. 101–111).

Baker, J. E. (1987). Reducing bias and inefficiency in the selection algorithm. In *Proceedings of the Second International Conference on Genetic Algorithms* (pp. 14–21).

BRIME—Bridge Management in Europe. (2001). *Project founded by the European Commission under the Transport RTD Program of the Fourth FrameWork Program* (Final report).

Caponio, A., & Cascella, G. L., Neri, F., Salvatore, N., & Sumner, M. (2006). A fast adaptive memetic algorithm for on-line and off-line control design of PMSM drives. *IEEE Transactions on System Man and Cybernetics—part B—Special issue on memetic algorithms.*

CEB-FIP. (1998). Strategies for testing and assessment of concrete structures. *Bullettin d'Infomation, No. 243.*

CEB-FIP. (2002). Management, maintenance and strengthening of concrete structures. *Bulletin d'Infomation, No. 17.*

Eurocode 8. (2003). *Design of structures for earthquake resistance – Part 3: Strengthening and repair of buildings.*

Herrera, F., & Verdegay, J. L. (Eds.). (1996). *Genetic algorithms and soft computing. Number 8 in studies in fuzziness and soft computing.* Physica Verlag.

Lubiana: Department of Transportation. (1994). *Evaluation of the carrying capacity of existing bridges: Final Report* (USA Project No. JF 026). Ministry of Science and Technology, Republic of Slovenia.

Michalewicz, Z. (1996). *Genetic algorithms + data structures = evolution programs.* Berlin, Germany: Springer-Verlag.

Mühlenbein, H., & Schlierkamp-Voosen, D. (1993). Predictive models for the breeder genetic algorithms: I. Continuous parameter optimization. *Evolutionary Computation, 1,* 25–49.

Mühlenbein, H., Schomisch, M., & Born, J. (1991). The parallel genetic algorithm as a function optimizer. *Parallel Computing, 17,* 619–632.

Nelder, A., & Mead, R. (1965). A simplex method for function optimization. *Computation Journal.* (7), 308–313.

OPCM. (2003). *Ordinanza presidente del consiglio dei ministri n. 3274 del 20/03/2003. GU n. 105.*

Port Authority of NY and NJ, The. (1998). *Guidelines for condition survey of waterfront structures.*

Ross, T. J., Sorensen, H. C., Savage, S. J., & Carson, J. M. (1990). DAPS: Expert system for structural damage assessment. *Journal of Computing in Civil Engineering ASCE, 4*(4), 327–348.

Whitley, D. (1989). The genitor algorithm and selection pressure: Why rank-based allocation of reproductive trials is best. In *Proceedings of the Third International Conference on Genetic Algorithms* (pp. 116–121).

Zadeh, L. A. (1965). Fuzzy sets. *Information Control, 8,* 338–353.

Zadeh, L. A. (2002). Towards a perception-based theory of probabilistic reasoning with imprecise probability. *Journal of Statistical Planning and inference, 105,* 233–264.

Chapter X

Life-Cycle Cost Evaluation of Bridge Structures Considering Seismic Risk

Hitoshi Furuta, Kansai University, Japan

Kazuhiro Koyama, NTT Comware Ltd., Japan

Abstract

This chapter introduces a life-cycle cost (LCC) analysis of bridge structures considering seismic risk. Recently, LCC has been paid attention as a possible and promising method to achieve a rational maintenance program. In general, LCC consists of initial cost, maintenance cost, and renewal cost. However, when considering LCC in the region that often suffers from natural hazards such as typhoons and earthquakes, it is necessary to account for the effects of such natural hazards. Using the probability of damage occurrence, LCC can be calculated for the bridge structures with earthquake excitations. The LCC analysis method proposed in this chapter can be applied to optimal maintenance planning by using genetic algorithms and can be extended to the life-cycle cost analysis of road network.

Introduction

Many existing bridges in Japan are suffering from damage and deterioration of materials due to heavy traffics and aging. In the future, it is evident that serious social problems will arise as the number of damaged bridges increases. Considering the present social and economic situation of Japan, it is urgent and important to establish an optimal maintenance strategy for such existing bridges so as to ensure their safety in satisfactory levels. Life-cycle cost (LCC) has been paid attention as a possible and promising method to achieve a rational maintenance program (Frangopol & Furuta, 2001). LCC of bridges consists of initial construction cost, maintenance cost, and failure cost (renewal cost, user cost, social and environmental costs and so on). In usual, LCC analysis considers the damage and deterioration of materials and structures (Furuta, Nose, Dogaki, & Frangopol, 2002). However, in the region that often suffers from natural hazards such as typhoons and earthquakes, it is necessary to account for the effects of such natural hazards.

In this chapter, based on the seismic risk analysis, LCC is evaluated focusing on the effects of earthquakes that are major natural disasters in Japan (Japan Society of Civil Engineers, 1996). At first, LCC analysis is formulated to consider the social and economical effects due to the collapse of structures occurred by the earthquakes as well as the minimization of maintenance cost (Furuta & Koyama, 2003). The loss by the collapse of structures due to the earthquake can be defined in terms of an expected cost and introduced into the evaluation of LCC. The LCC analysis method proposed can be applied to optimal maintenance planning by using genetic algorithm (GA) and can be extended to LCC analysis of road network (Furuta, Kameda, Fukuda, & Frangopol, 2004; Liu, & Frangopol, 2004a, 2004b).

Life-Cycle Formulation Including Seismic Risk

In general, life-cycle cost is defined in terms of initial construction cost, maintenance cost, and replacement cost. As the initial cost, only pier is considered here, because the sufficient data for the whole bridge is not available. Seismic risk includes both losses due to earthquake and user cost (Furuta, Koyama, Ohi, & Sugimoto, 2005).

$$LCC = C_i + \sum P_d(a) \cdot C_d(a) + C_m(DI,a) + UC(DI,a) \qquad (1)$$

$$P_d(a) = P_h \cdot P(DI,a) \qquad (2)$$

where C_i: initial construction cost, $P_d(a)$: probability of seismic damage occurrence, $C_d(a)$: seismic loss, $C_m(DI, a)$: maintenance cost, $UC(DI, a)$: user cost, $P_h(a)$:earthquake occurrence probability, $P(DI, a)$: seismic damage probability, a: maximum acceleration, DI: damage index.

Figure 1. Seismic hazard curve

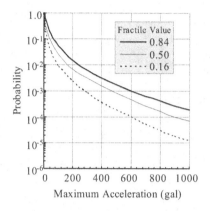

Equation (2) provides the probability of damage occurrence due to the earthquake, which is the multiplication of earthquake occurrence probability with damage probability. In this chapter, the earthquake occurrence probability is calculated by using seismic hazard curve (Shinozuka, Feng, & Naganuma, 2000; Shoji, Fujino, & Abe, 1997). In the seismic hazard curve, the annual occurrence probability of earthquake is calculated by considering the distribution of distance from epicenter, historical earthquake records, horizontal maximum acceleration and active fault. The seismic hazard curve used here is shown in Figure 1, in which the vertical axis is the annual occurrence probability of earthquake and the horizontal axis is the maximum acceleration of earthquake. The damage probability is calculated by using the damage curve.

Definition of Damage Degree and Calculation of Damage Curve

Damage probability is defined in terms of the probability that a bridge pier shows each damage degree among the prescribed damage ranges. The damage curve is calculated here by using the dynamic analysis of the bridge pier.

Bridge Pier Model

Reinforced concrete (RC) bridge pier is used to obtain the damage curve (Furuta & Koyama, 2003). The RC pier is designed according to Design Specification of Highway Bridges, V: Seismic Design (Ministry of Land, Infrastructure and Transportation; MLIT, 2002). The dimensions of the piers designed are presented in Table 1. Figure 2 shows the outline of a

Table 1. The dimensions of RC bridge piers

Model No.	Height of Bridge Pier (mm)	Cross Section of Pier (mm)		Longitudinal Reinforcement	
		Height	Width	Reinforcement	Number of Reinforcement Stage
1	12000	2500	5000	D32	2
2	12000	2500	4500	D32	2
3	12000	2500	4000	D32	2
4	12000	2000	5000	D32	2
5	9000	2000	4500	D29	2
6	9000	2000	4000	D29	2
7	9000	2000	3500	D29	2
8	6000	2200	3000	D29	2
9	6000	2200	2500	D29	1
10	6000	1800	3000	D29	1

Table 2. Damage degrees

Damage Index	Damage Condition of Pier
A_s	Collapse, excessive deformation
A	Severe damage as crack, buckling, break of reinforcement, or large deformation
B	Partial buckling and deformation Partial break of reinforcement and lamination and crack of concrete
C	Partial or slight buckling and deformation
D	No lamination and crack of concrete No damage or minor damage affecting no load capacity

representative bridge pier. Then, the design strength, bending tensile strength and modulus of elasticity of reinforcing bar (SD295) are 21 (N/mm^2), 1.8 (N/mm^2), and 25 (KN/mm^2), respectively. The modulus of elasticity of concrete is 200 (KN/mm^2).

In the seismic design of bridge structure, it is necessary to assure the seismic performance by taking into account the design earthquake intensity, importance of the bridge, and target safety. Two kinds of earthquakes, type 1 earthquake and type 2 earthquake, are considered. Type 1 earthquake means a strong earthquake with the maximum acceleration $a = 800$ gal; whereas type 2 earthquake means a medium earthquake, with the maximum acceleration $a = 400$ gal (MLIT, 2002). The importance of bridge is classified into two categories, A and B, which mean "important" and "not so important," respectively.

Figure 2. A representative RC bridge pier: (a) Front view, (b) side view, (c) cross section of base part of tier

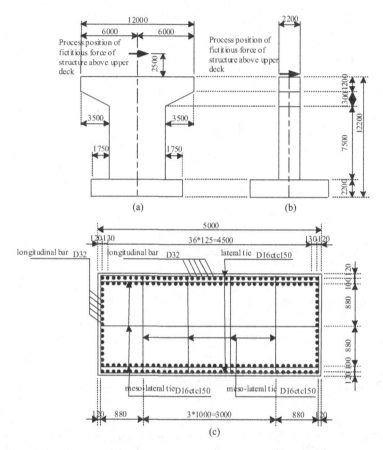

Then, it is assumed that the ground condition is type II that is a normal soil condition in Japan and the importance of the bridge is *B* categorized as "important" in the design specification. Several damage indexes have been proposed so far (Park & Ang, 1985; Furuta, Kameda, & Frangopol, 2004). However, in this chapter, "damage degree" is used and defined in terms of the maximum response displacement, horizontal force, and horizontal displacement of pier. The damage degree is categorized into five ranks such as A_s, *A, B, C,* and *D* as shown in Table 2.

Figure 3 presents the relation between damage degree and horizontal displacement δ. Using the relations, it is possible to estimate the damage degree by the maximum response displacement calculated by the dynamic analysis.

Figure 3. Damage degree and horizontal displacement

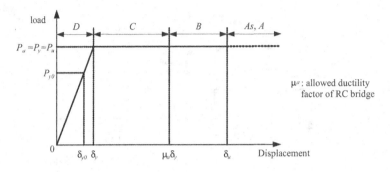

Table 3. Parameters for dynamic analysis

Damping Ratio	0.05
Time period of Newmark β method	0.01 second
Hysteretic model	Takeda model
Earthquake-wave forced direction	Axial direction

Dynamic Analysis of RC Bridge Pier

Dynamic analysis is performed for ten bridges, by using the parameters presented in Table 3, in which a single freedom model and Newmark β method are employed to do the dynamic analysis. It is assumed that the compression strength of concrete is $fc = 21$ (N/mm²) and the reinforcing bar is SD295. As input earthquake wave, type 1 and type 2 earthquakes are used and the ground condition is assumed to be type II. Table 3 presents the parameters for the dynamic analysis, and Table 4 presents six input earthquakes, in which M is the magnitude of earthquake. Using these conditions, the dynamic analysis is performed for 600 times for a RC pier. Figure 4 shows the calculation results, in which the vertical axis is the ratio of maximum response displacement δ_{max} and the yield response displacement δ_y, and the horizontal axis is the maximum acceleration of input earthquake.

Calculation of Damage Curve

In order to calculate the damage probability, it is necessary to determine the distribution function corresponding to the maximum earthquake acceleration. In this chapter, the log-normal distribution is assumed.

Table 4. Input data of six earthquake waves

Earthquake Type	Name	M	Direction
Type I	1968, Hyuga-Nada Earthquake	7.5	LG
			TR
	1994, Hokkaido-Toho Earthquake	8.1	TR
Type II	1995, Hyogo-Nanbu Earthquake	7.2	N-S
			E-W
			N27W

Figure 4. Results of dynamic analysis

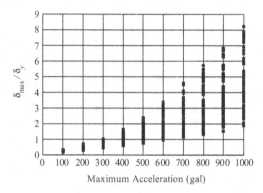

When the distribution of damage degree is given, the damage probability can be calculated as:

$$P(DI,a) = \int_{a}^{b} f_{DI}(x,a)dx \qquad (3)$$

where $[a, b]$ is the interval of each damage degree.

The damage probability is calculated for each damage degree and the results are plotted on a graph with the excess probability as the vertical axis and the maximum acceleration as the horizontal axis. Then, the damage curve can be obtained by combining them. Figure 5 shows the damage curve calculated here.

Figure 5. Seismic damage curve

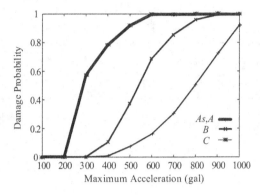

Calculation of User Cost

Here, user cost is defined in terms of the sum of the time cost and energy consumption cost due to the detour or closure of road. The cost by increasing the driving time C_{UT} is calculated as the subtraction of the usual running cost and the running cost for the detour and closure (Nihon Sogo Research Institute, 1998).

$$UC_T = a \cdot \{(Q \cdot T) - (Q_0 \cdot T_0)\} \qquad (4)$$

$$UC_C = \beta \cdot \{(Q \cdot L) - (Q_0 \cdot L_0)\} \qquad (5)$$

Figure 6. Relation between traffic volume and velocity

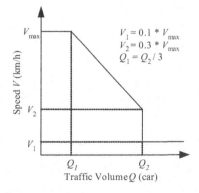

Table 5. Unit driving cost

Speed	Unit Driving Cost
V(km/h)	(Yen/Car/km)
5	35
10	28
20	21
30	18
40	18

where α: unit time cost, β: unit running cost, Q,T,L: detour traffic volume, running time, and link length at the time of road closure, Q_0, T_0, L_0: initial traffic volume, running time, and link length.

Using the data given in (Kinki Branch, MLIT, 1999) and assuming the ratio of small and medium trucks to be 10%, the unit time cost α is estimated as 82 Yen/car/min, and β is assumed as shown in Table 5. The restoring periods are assumed to be two months and two weeks for the damage (A_s, A) and B, respectively.

Maintenance Cost

Taking into account the discount rate, LCC is calculated as:

$$LCC = C_i + \sum P_f(a) \cdot P(DI,a) \cdot \left\{ \frac{C_m(DI,a) + UC(DI,a)}{(1+i)^T} \right\}$$ (6)

where $C_m(DI,a)$: maintenance cost for each damage rank, $UC(DI,a)$: user cost for each damage degree, i: discount rate, T: service life. For each damage degree, repair methods and costs are presented in Table 6. For the simplicity, damage degrees A_s and A are treated equally.

Table 6. Repair methods and costs

Damage Index	Reconstruction Method	Repair Cost	Repair Time
A_s, A	Rebuild	120% of initial construction cost	2 months
B	Repair	73,000 Yen/lm2	1 month
C	Repair	35,000 Yen/lm2	2 weeks
D	No repair		

Reliability Analysis of
RC Pier Considering Seismic Risk

In order to obtain an optimal design based on LCC, it is necessary to define the seismic performance during the service life. In this chapter, the seismic performance during the service life is calculated by reliability index (Melchers, 1999).

Model of Yield Strength

A probabilistic model of required yield strength is adopted. The required yield strength is defined as the minimum yield strength of a structure required to suppress the damage to a prescribed value when the structure is under seismic load.

On the other hand, the limit state of reinforced concrete bridge piers should be determined. Referring to the experimental results of RC bridge piers given by the Japanese Specification of Highway Bridges (MLIT, 2002), the ductility factor for the limit state of reinforced concrete bridge pier is $\mu_T = 4.0$. In this chapter, only type II soil (normal soil in Japan) is taken into consideration. Type 2 earthquake defined in the Japanese Specification of Highway Bridges is employed, which is used to check for strong earthquakes. Here, the required yield strength spectrum and standard required yield strength spectrum adopted are shown in Figure 7 and Figure 8 (Nishimura & Murono, 1999).

In this chapter a lognormal distribution is assumed for the distribution type of the required yield strength spectrum.

The following conditions are assumed; $\mu_T = 4.0$, type 2 earthquake, maximum acceleration = 800 gal, the type of soil = II, mean value and standard deviation of required yield strength are 0.69 and 0.203, respectively.

Figure 7. Spectra of required yield strength

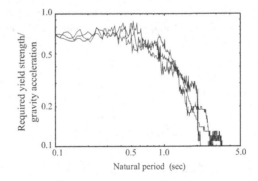

Figure 8. Spectrum of required yield strength (standard value)

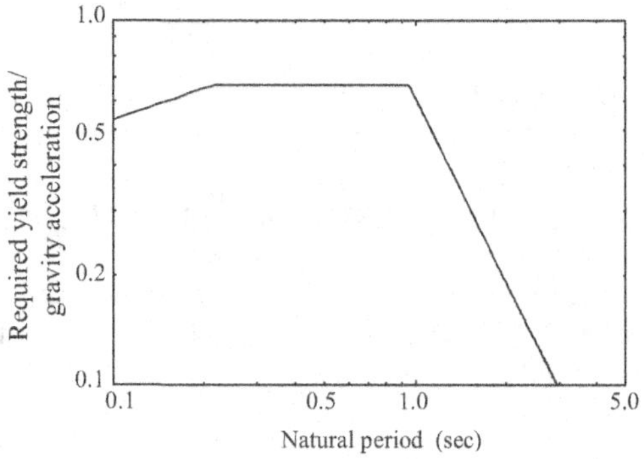

Then, the ultimate limit state is considered as the limit state of RC bridge pier. The limit state function is as follows:

$$Z = P_a - P_y = 0 \tag{7}$$

where P_a = ultimate lateral strength and P_y = required yield strength.

Deterioration Model of RC Bridge Pier

The corrosion of reinforcing bars in chloride environment is considered as the damage factor of RC bridge pier. Usually, the corrosion of reinforcing bars starts by the invasion of chloride ions into concrete. Once the corrosion starts, the cross sectional area of reinforcing bars decreases at a certain rate. Here, the reduction rate of the cross sectional area of reinforcing bars is modeled as follows:

$$A_s(t) = \frac{\pi \{D(t)\}^2}{4A_{s0}} \tag{8}$$

$$D(t) = \begin{cases} D_0 & ; 0 < t \le T_0 \\ D_0 - v_1(t - T_0) & ; T_0 < t \le T_1 \\ D_1 - v_2(t - T_1) & ; T_1 < t \le T_2 \end{cases} \tag{9}$$

where $A_s(t)$: deterioration rate in t years after corrosion starts (cross sectional area of reinforcement after t years / initial cross sectional area of reinforcement), $D(t)$: diameter of reinforcing bar in t years after corrosion, D_0: initial diameter of reinforcing bar, and A_{s0}: initial cross sectional area.

Calculation of Reliability Index

Strength of reinforced concrete bridge piers can not be estimated by only the rate of deterioration. Namely, it is necessary to provide the relation between the deterioration and strength. It is also necessary to clarify the relation between the strength and earthquake intensity. The seismic performance of RC bridge piers during the service time is estimated by using time-dependent reliability index (Melchers, 1999).

- **Step 1:** The diameter of reinforcing bar $D(t)$ at t years after corrosion is computed.
- **Step 2:** Nonlinear analysis by a fiber model is performed and the ultimate lateral strength is computed. Then, the diameter $D(t)$ is calculated according to the Japanese Highway Bridge Design Specification (Japan Society of Civil Engineers, 2001).
- **Step 3:** Using the ultimate lateral strength computed at Step 2, the reliability analysis is performed by using the following limit state function:

$$Z(t) = P_a(t) - P_y = 0 \tag{10}$$

  where $P_a(t)$ = ultimate lateral strength at t years after corrosion starts.
- **Step 4:** The reliability index during the service life is computed by repeating Steps 1 to 3 until t reaches the end of service life.

Optimal Maintenance Planning: Application of Genetic Algorithm

In this chapter, genetic algorithm (GA) is used for the optimization. GA is an algorithm of imitating the natural evolution process following Darwinian survival-of-the-fitness principle (Goldberg, 1989). GA does not need any prior knowledge and is able to understand the algorithm with ease. GA can solve a combinatorial optimization problem with a certain scale through the evolution process in which a gene is encoded according to the problem, and the processes of selection, crossover, and mutation are repeated until the satisfactory convergence is obtained. On the other hand, GA can not necessarily provide the global optimal solution but the quasi-optimal solution (Furuta, Kameda, Nakahara, Takahashi, & Frangopol, 2006).

When the maintenance planning of a structure is decided, the reduction of structural performance should be predicted and appropriate timing and repair/reinforcement methods should be chosen from a viewpoint of safety and economical efficiency (i.e., LCC). Even assuming only two kinds of repair/reinforcement methods, there are more than 1,030 patterns for maintenance planning in the service life. Then, it is difficult to choose an optimal plan from these enormous combinations, considering the safety and economical efficiency. Therefore, GA is applied in this research to decide the optimal maintenance plan.

LCC Evaluation of Road Network

For road networks, three network models (Model 1, Model 2, and Model 3) are employed, which are presented in Figure 9. In these models, it is assumed that each network model includes a road passing over a river, and that traffics reach the destination through detours when some bridges can not be passed. Moreover, it is assumed that the traffic volume and the velocity have a relation shown in Figure 6. Here, user cost is defined in terms of the sum of the time cost and energy consumption cost due to the detour or closure of road. The cost by increasing the driving time C_{UT} is calculated by equations (4) and (5).

For the three road networks, LCC is calculated by assuming that the fractile value in the hazard curve is 0.5, discount rate is 0, and service life is 100 years. Figure 10 shows the calculated results, which indicate that there are big differences among the three networks of Model 1, Model 2, and Model 3, because of the differences in distances of detour and the initial traffic volumes. In the network with high traffics, seismic risk becomes 104,559 (thousand yen) that are 11 times of the initial construction cost and 30 times of the maintenance cost.

Comparing the case involving the user cost in the seismic risk and the case without involving it, the seismic risk is only one third of the initial cost when the user cost is not considered.

Figure 9. Three road network models

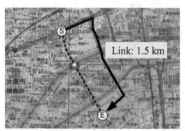

Model 1. Tamae-bashi
Pukushima-ku, Osaka city

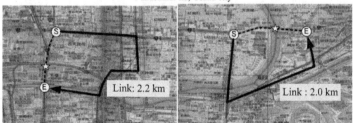

Model 2. Ohe-bashi   Model 3. Sakuranomiya-bashi
Kita-ku, Osaka city   Tyuo-ku, Osaka city

Figure 10. LCC of each model

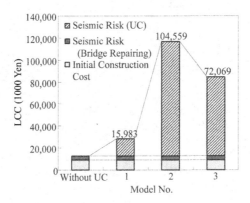

Figure 11. Seismic risk of each maximum acceleration

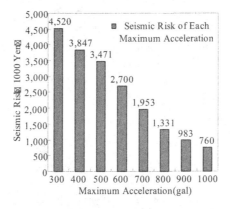

Figure 11 shows the relation between the seismic risk and the maximum acceleration. From this figure, it is obtained that the seismic risk decreases as the maximum acceleration increases. This is due to the fact that since the seismic risk is defined in terms of the multiplication of the seismic loss with the earthquake occurrence probability, the seismic risk decreases rapidly with the increase of the maximum acceleration. Moreover, the bridge pier was designed to satisfy the requirement that the damage should be minor and the bridge function can be recovered in a short period. Therefore, the probabilities of damage *B*, *C*, and *D* become high.

Since it is difficult to determine an appropriate discount rate, only 2% and 5% discount rates are considered. Figure 12 shows the comparison among 0%, 2%, and 5% discount

Figure 12. LCC for each discount rate

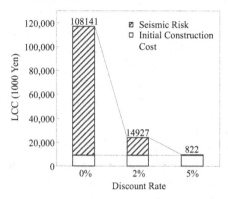

Figure 13. Ratio of seismic risk

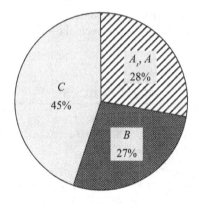

rates. When the discount rate is considered to be 5%, the seismic risk becomes small such as 1/10 of the initial cost. This may be due to the fact that the service life of the bridge structures is very long as 100 years. This implies that the discount rate greatly affects the estimation of LCC.

The effects of damage degree on the seismic risk are examined. Figure 13 presents the ratio of seismic risk corresponding to each damage degree, which implies that the damage degree C is 54 % being the largest and the severe damages As and A have small portions of 28 %. This is due to the fact that while the occurrence probabilities of As and A become larger as the maximum acceleration becomes larger, or the earthquake occurrence probability becomes smaller.

Conclusion

In this study, an attempt was made to propose a calculation method of LCC considering seismic risk based on the reliability theory. Through several numerical calculations, the following conclusions were derived:

Introducing the expected failure cost of structure due to the earthquake excitations, LCC could be evaluated for the case with seismic risk. Applying genetic algorithm, it is possible to obtain the relation between LCC and the safety of bridge considering seismic risk. The damage degree is defined by using the maximum response displacement obtained by the dynamic analysis and the horizontal force and displacement of RC bridge pier.

The seismic performance of RC bridge piers has to be evaluated as a time-dependent process. Through the LCC calculation of several representative road networks, it is obtained that the difference of road network greatly influences on the seismic risk.

Paying attention to the change of LCC according to the change of the maximum acceleration of earthquake, the seismic risk decreases as the maximum acceleration increases. The effect of discount rate is examined by changing its value. As a result, it is obtained that the discount rate has a large influence on the estimation of LCC. This implies that it is very important to determine an appropriate discount rate in the calculation of LCC.

Examining the effect of damage degree, the medium damage level shows a big ratio of 54%, whereas the severe and rather severe damage levels show 28% and 27%, respectively.

References

Frangopol, D. M., & Furuta, H. (Eds.). (2001). *Life-cycle cost analysis and design of civil infrastructure systems.* Reston, VA: ASCE.

Furuta, H., Kameda, T., & Frangopol, D. M. (2004). Balance of structural performance. In *Proceedings of Structures Congress, Nashville, TN* [CD-ROM]. ASCE.

Furuta, H., Kameda, T., Fukuda, Y., & Frangopol, D. M. (2004, March 24-26). Life-cycle cost analysis for infrastructure systems: Life-cycle cost vs. safety level vs. service life. In *Proceedings of Joint International Workshops LCC03/IABMAS and fip/JCSS, EPFL,* Lausanne, Switzerland.

Furuta, H., Kameda, T., Nakahara, K., Takahashi, Y., & Frangopol, D. M. (2006). Optimal bridge maintenance planning using improved multi-objective genetic algorithm. *Structure and Infrastructure Engineering, 2*(1), 33-41.

Furuta, H., & Koyama, K. (2003). Optimal maintenance planning of bridge structures considering earthquake effects. In *Proceedings of the IFIP TC7 Conference*, Antipolis, France.

Furuta, H., Koyama, K., Ohi, M., & Sugimoto, H. (2005). Life-cycle cost evaluation of multiple bridges in road network considering seismic risk. In *Proceedings of Bridge Conference, TRB*, Boston.

Furuta, H., Nose, Y., Dogaki, M., & Frangopol, D. M. (2002). Bridge maintenance system of road network using life-cycle cost and benefit. In *Proceedings of the First IABMAS Conference*, Barcelona, Spain.

Goldberg, D. E. (1989). *Genetic algorithms in search, optimization and machine learning.* Addison Wesley.

Japan Society of Civil Engineers. (1996). *Report on damage by Hanshin Awaji Earthquake* (in Japanese)

Japan Society of Civil Engineers. (2001). *Standard specification for design and construction of concrete in Japan* (*Maintenance*, in Japanese).

Kinki Branch, Ministry of Land, Infrastructure and Transportation. (1999). *Road traffic census* (in Japanese)

Liu, M., & Frangopol, D. M. (2004a). Optimal bridge maintenance planning based on probabilistic performance prediction. *Engineering Structures, 26*(7), 991-1002.

Liu, M., & Frangopol, D.M. (2004b, July 26-28). Probabilistic maintenance prioritization for deteriorating bridges using a multi-objective genetic algorithm. In *Proceedings of the Ninth ASCE Joint Specialty Conference on Probabilistic Mechanics and Structural Reliability.*

Melchers, R. E., (1999). *Structural reliability analysis and prediction* (2nd ed.). Wiley.

Ministry of Land, Infrastructure and Transportation, MLIT. (2001). *Cost estimation standards for civil constructions* (in Japanese).

Ministry of Land, Infrastructure and Transportation, MLIT. (2002). *Standard specification of highway bridges, V: Seismic design, Maruzen* (in Japanese).

Nihon Sogo Research Institute. (1998). *Draft of guideline for evaluation of investment on road* (in Japanese)

Nishimura, A., & Murono, T. (1999). *Calculation of required yield strength spectrum: Report of Railway Research Institute* (in Japanese).

Park, Y. J., & Ang, A. H. S., (1985). Mechanistic seismic damage model for reinforced concrete. *Journal of Structural Engineering, ASCE, 111*(4), 722-739.

Shinozuka, M., Feng, M. Q., & Naganuma, T., (2000). Statistical analysis of fragility curve. *Journal of Engineering Mechanics, ASCE, 126*(12), 1224-1231.

Shoji, M., Fujino, Y., & Abe, M., (1997). Optimization of seismic damage allocation of viaduct systems. In *Proceedings of the JSCE* (Vol. 563, pp. 79-94) (in Japanese).

Chapter XI

Soft Computing Techniques in Probabilistic Seismic Analysis of Structures

Nikos D. Lagaros, University of Thessaly, Greece

Yiannis Tsompanakis, Technical University of Crete, Greece

Michalis Fragiadakis, National Technical University of Athens, Greece

Manolis Papadrakakis, National Technical University of Athens, Greece

Abstract

Earthquake-resistant design of structures using probabilistic analysis is an emerging field in structural engineering. The objective of this chapter is to investigate the efficiency of soft computing methods when incorporated into the solution of computationally intensive earthquake engineering problems. Two methodologies are proposed in this work where limit-state probabilities of exceedance for real world structures are determined. Neural networks based metamodels are used in order to replace a large number of time-consuming structural analyses required for the calculation of a limit-state probability. The Rprop algorithm is employed for the training of the neural networks; using data obtained from appropriately selected structural analyses.

Introduction

The advances in computational hardware and software resources since the early 1990s resulted in the development of new, nonconventional data processing and simulation methods. Among these methods, soft computing has to be mentioned as one of the most eminent approaches to the so-called intelligent methods of information processing. Neural networks (NNs), fuzzy logic, and evolutionary algorithms are the most popular soft-computing techniques. NNs have been widely used in many fields of science and technology and are now becoming popular solution methods for an increasing number of structural engineering problems. NNs constitute a powerful tool that can be used to replace time consuming simulations required in many engineering applications. In the previous years, NNs have been applied mostly for predicting the behavior of a structural system in the context of structural optimal design (Lagaros, Charmpis, & Papadrakakis, 2005; Patnaik, Guptill, & Hopkins, 2005; Salajegheh, Heidari, & Saryazdi, 2005; Zhang & Foschi, 2004), for structural damage assessment (Huang, Hung, Wen, & Tu, 2003; Reda Taha & Lucero, 2005; Sanchez-Silva & García, 2001), on structural reliability analysis (Hurtado & Alvarez, 2002; Nie & Ellingwood, 2004; Papadrakakis & Papadopoulos, 1996), on structural identification (Chakraverty, 2005; Masri, Smyth, Chassiakos, Caughey, & Hunter, 2000; Mróz & Stavroulakis, 2005), for the evaluation of buckling loads of cylindrical shells with geometrical imperfections (Waszczyszyn & Bartczak, 2002) or for the simulation of non-Gaussian stochastic fields (Lagaros, Stefanou, & Papadrakakis, 2005).

Many sources of uncertainty (e.g., material, geometry, loads) are inherent in structural design. Probabilistic analysis of structures leads to safety measures that a design engineer has to take into account due to the aforementioned uncertainties. Probabilistic analysis problems, especially when seismic loading is considered, are highly computationally intensive tasks since, in order to obtain the structural behaviour, a large number of dynamic analyses (e.g., modal response spectrum analysis or nonlinear timehistory analysis) are required (Papadrakakis, Tsompanakis, Lagaros, & Fragiadakis , 2004). In this work, two metamodel based applications are considered in order to reduce the aforementioned computational cost. The efficiency of a trained NN is demonstrated, where a network is used to predict maximum interstory drift values due to different sets of random variables. As soon as the maximum interstory drift is known, the limit-state probabilities are calculated by means of Monte Carlo simulation (MCS). In the first application the probability of exceedance of a limit-state is obtained using proposed by Eurocode 8 (1994). Multimodal Response Spectrum Analysis is. In the second application fragility analysis of a 10-story moment resisting steel frame is evaluated, where limit-state fragilities are determined by means of nonlinear timehistory analysis.

In both applications considered in this chapter, the use of NNs is motivated by the large number of time-consuming simulations required by MCS. The Rprop algorithm is implemented for training the NN, utilizing available information extracted from each record. The trained NN is used to predict maximum interstory drift values, leading to a close prediction of the limit-state probabilities.

Multilayer Perceptron Neural Network

NN metamodels have the ability of learning and accumulating expertise and have found their way into applications in many scientific areas. There is an increasing number of publications that cover a wide range of computational structures technology applications, where most of them are heavily dependent on extensive computer resources that have been investigated or are under development. This trend demonstrates the great potential of NN.

A multilayer perceptron is a feed-forward NN that consists of a number of units, called *neurons*. A neural network is divided to the input layer, one or more hidden layers, and the output layer. In a fully connected network, like those employed in this chapter, each node of a layer is connected to all the nodes of the previous and the next layer. The number of nodes to be used in the hidden layers is not known in advance, for this reason the learning process initiates with a relatively small number of hidden nodes (5 in this study) and gradually increase the number of hidden nodes until achieving the desired convergence (Lagaros & Papadrakakis, 2004). Each layer has its corresponding neurons or nodes and weight connections linked together in order to obtain the desired relation in an input/output set of learning patterns. A single training pattern is an I/O vector of pairs of input-output values in the entire matrix of an I/O training set.

The computed output(s) of the output layer, otherwise known as the observed output(s), are subtracted from the desired or target output(s) to give the error signal:

$$\mathcal{E}(w) = \frac{1}{2m} \left\| E(w) \right\|^2 \tag{1}$$

$$E_i(w) = \sum_{j=1}^{\ell} [out_{k,j} - tar_{k,j}] \tag{2}$$

where m is the number of training pairs, $tar_{k,i}$ and $out_{k,i}$ are the target and the observed output(s) for the node i of the output layer k, respectively. This type of NN training is called supervised learning.

A learning algorithm tries to determine the weights $w_{p,ij}$, in order to achieve the right response for each input vector of the network. Numerical minimization algorithms are used during the training process in order to generate a sequence of weight matrices by means of an iterative procedure. To apply an algorithmic operator $\mathcal{A}$, a starting value of the weight matrix $w^{(0)}$ is needed, while the iteration formula can be written:

$$w^{(t+1)} = \mathcal{A}(w^{(t)}) = w^{(t)} + \Delta w^{(t)} \tag{3}$$

All numerical methods applied in NNs are based on the above formula. The changing part of the algorithm $\Delta w^{(t)}$ is further decomposed into two parts:

$$\Delta w^{(t)} = a_t d^{(t)} \tag{4}$$

where $d^{(t)}$ is a desired search direction and a_t the step size in that direction.

The training methods can be divided into two categories. Algorithms that use global knowledge of the state of the entire network, such as the direction of the overall weight update vector, which are referred to as global techniques. In contrast local adaptation strategies are based on weight specific information only such as the temporal behaviour of the partial derivative of this weight. The local approach is more closely related to the neural network concept of distributed processing in which computations can be made independent to each other. Furthermore, it appears that for many applications local strategies, like Rprop, achieve faster and reliable prediction than global techniques despite the fact that they use less information (Lagaros, Stefanou, & Papadrakakis, 2005). The NN software used in this study has been developed by the authors (Lagaros & Papadrakakis, 2004), while more details about neural networks can be found in Chapter 16 (Kuźniar & Waszczyszyn, 2006).

Seismic Probabilistic Analysis

In the design of structural systems, limiting uncertainties and increasing safety is an important issue. Probabilistic analysis of structures is used as a measure to evaluate the reliability of a structural system, which is defined as the probability that the system will meet some specified demands for a given time period under certain environmental conditions. The performance function of a structural system must be determined to describe the system's behavior and to identify the relationship between the basic parameters of the system. It should be noted that when seismic loading is concerned the uncertainties related to seismic demand and structural capacity are coupled.

Probability of Exceedance

The probability of exceedance p_{exceed} can be determined using a time invariant probabilistic analysis procedure with the following expression:

$$P_{exceed} = Prob[R < S] = \int_{-\infty}^{\infty} F_R(t) f_S(t) dt = 1 - \int_{-\infty}^{\infty} F_S(t) f_R(t) dt \tag{5}$$

where R denotes the structural capacity and S the external loading. The randomness of R and S can be described by known probability density functions $f_R(t)$ and $f_S(t)$, with $F_R(t) = Prob[R < t]$, $F_S(t) = Prob[S < t]$ being the cumulative probability density functions of R and S, respectively.

A limit-state function is defined as $G(R,S) = S-R$ and the probability of exceedance is given by:

$$P_{exceed} = Prob[G(R < S) \geq 0] = \int_{G \geq 0}^{\infty} f_R(R)f_S(S)dRdS \tag{6}$$

It is practically impossible to evaluate p_{exceed} analytically for complex and/or large-scale structures. In such cases the integral of equation (6) can be calculated only approximately using either simulation methods, such as the Monte Carlo simulation, or approximation methods like the first order reliability method (FORM) and the second order reliability method (SORM), or response surface methods (RSM). Despite its high computational cost, MCS is considered as the most efficient method, and is used either for comparison with other approximate methods or as a standalone probabilistic analysis tool.

Fragility Analysis

The seismic fragility of a structure $F_R(x)$ is defined as its limit-state probability, conditioned on a specific peak ground acceleration, spectral velocity, or other control variable that is consistent with the specification of seismic hazard:

$$F_R(x) = Prob[LS_i / PGA \geq x] \tag{7}$$

where LS_i represents the corresponding i^{th} limit-state and the peak ground PGA is the control variable. If the annual probabilities of exceedance $Prob[PGA \geq x]$ of specific levels of earthquake motion are known, then the mean annual frequency of exceedance of the ith limit-state is calculated as follows:

$$P_{LS} = Prob[LS_i] = \int F_R(x)P[PGA \geq x]dx \tag{8}$$

Equation (7) can be used to make decisions about, for example, the adequacy of a design or the need to retrofit. In the present study we seek the fragility $F_R(x)$. Once the fragility is calculated the extension to equation (8) is straightforward. Often $F_R(x)$ is modeled with a lognormal probability distribution, which leads to an analytic calculation. In the present study, Monte Carlo simulation is adopted for the numerical calculation of $F_R(x)$. Numerical calculation of equation (8) provides a more reliable estimate of the limit-state probability, since it is not necessary to make any assumption regarding the seismic data (e.g., that they are lognormally distributed). However, in order to calculate the limit-state probability, a large number of simulations are required for each intensity level, especially when small probabilities are sought.

Monte Carlo Simulation

In probabilistic analysis the MCS method is often employed when the analytical solution is not attainable. This is mainly the case in problems of complex nature with a large number of random variables where all other probabilistic analysis methods are not applicable. Expressing the limit-state function as $G(x) < 0$, where $x = [x_1, x_2, .., x_M]^T$ is the vector of the random variables, the probability of exceedance can be obtained as:

$$P_{LS} = \int\limits_{G(x) \geq 0} f_x(x) dx \tag{9}$$

where $f_x(x)$ denotes the joint probability of failure. Since MCS is based on the theory of large numbers (N_∞) an unbiased estimator of the probability of a limit-state being exceeded:

$$P_{LS} = \frac{1}{N_\infty} \sum_{j=1}^{N_\infty} I(x_j) \tag{10}$$

where $I(x_j)$ is a Boolean vector indicating "successful" and "unsuccessful" simulations. In order to estimate P_{LS} an adequate number of N_{sim} independent random samples is produced using a specific probability density function for the vector x. The response is determined for each random sample x_j and the Monte Carlo estimation of P_{LS} is given in terms of the sample mean by:

$$P_{LS} \cong \frac{N_H}{N_{sim}} \tag{11}$$

N_H is the number of simulations where the maximum interstory drift value exceeds a threshold drift for the limit-state examined. In order to calculate equation (9) N_{sim} analyses have to be performed.

Metamodel Assisted Methodology
for Validating the EC8 Approach

Extreme earthquake events may produce extensive damage to structural systems. It is therefore essential to establish a reliable procedure for assessing the seismic risk of real-world structural systems. The reliability of a steel frame designed to EC8 using the modal response analysis as suggested by Eurocode 8 (1994) is performed. The probability that the life-safety is exceeded is determined, since this is the limit-state that usually controls the design process.

Structural Design Under Seismic Loading Based on EC8

The equations of equilibrium for a linearly elastic system in motion can be written in the usual form:

$$\mathbf{M}\ddot{\mathbf{u}}(t) + \mathbf{C}\dot{\mathbf{u}}(t) + \mathbf{K}\mathbf{u}(t) = \mathbf{R}(t) \tag{12}$$

where $\mathbf{M}$, $\mathbf{C}$, and $\mathbf{K}$ are the mass, damping and stiffness matrices; $\mathbf{R}(t)$ is the external load vector, while $\mathbf{u}(t)$, $\dot{\mathbf{u}}(t)$, and $\ddot{\mathbf{u}}(t)$ are the displacement, velocity, and acceleration vectors of the finite element assemblage, respectively. The multi-modal response spectrum (MmRS) method is a simplified method for the assessment of seismic demand based on the mode superposition approach. Equation (14) is modified according to the modal superposition approach to a system using the following transformation:

$$\bar{\mathbf{M}}_i \ddot{\mathbf{y}}_i(t) + \bar{\mathbf{C}}_i \dot{\mathbf{y}}_i(t) + \bar{\mathbf{K}}_i \mathbf{y}_i(t) = \bar{\mathbf{R}}_i(t) \tag{13}$$

where:

$$\bar{\mathbf{M}}_i = \mathbf{\Phi}_i^T \mathbf{M} \mathbf{\Phi}_i, \ \bar{\mathbf{C}}_i = \mathbf{\Phi}_i^T \mathbf{C} \mathbf{\Phi}_i, \ \bar{\mathbf{K}}_i = \mathbf{\Phi}_i^T \mathbf{K} \mathbf{\Phi}_i \text{ and} (\bar{\mathbf{R}}\ t) = \mathbf{\Phi}_i^T \mathbf{R}(t) \tag{14}$$

are the generalized values of the corresponding matrices and the loading vector, while $\mathbf{\Phi}_i$ is the i-th eigenmode shape matrix. According to the modal superposition approach the system of N differential equations, which are coupled with the off-diagonal terms in the mass, damping and stiffness matrices, is transformed to a set of N independent normal-coordinate equations. The dynamic response can therefore be obtained by solving separately for the response of each normal (modal) coordinate and by superposing the response in the original coordinates.

In the MmRS analysis, a number of different formulas have been proposed to obtain reasonable estimates of the maximum response based on the spectral values. The simplest and most popular formula for combining the modal responses is the square root of sum of squares (SRSS). Thus the maximum total displacement is approximated by:

$$u_{max} = \left(\sum_{i=1}^{N} u_{i,max}^2 \right)^{1/2}$$
$$u_{i,max} = \mathbf{\Phi}_i y_{i,max} \tag{15}$$

where $u_{i,\,max}$ corresponds to the maximum displacement vector corresponding to the i-th eigenmode.

Metamodel Assisted Methodology

In the present implementation the main objective is to investigate the ability of the NN to predict the structural performance in terms of maximum interstory drift. The selection of appropriate I/O training data is an important part of the NN training process. Although the number of training patterns may not be the only concern, the distribution of samples is of greater importance. In the present study the sample space for each random variable is divided into equally spaced distances in order to select suitable training pairs. Having chosen the NN architecture and tested the performance of the trained network, predictions of the probability of exceedance of the design limit-state can be made quickly. The results are then processed by means of MCS to calculate the probability of exceedance p_{exceed} of the design limit-state using equation (6).

The modulus of elasticity, the dimensions (width and height) b and h of the I-shape cross-section and the seismic loading have been considered as random variables. Three alternative test cases are considered depending on the random variables considered: (a) the modulus of elasticity and the earthquake loading are considered to be random variables, (b) the dimensions b and h of the I-shape cross section and earthquake loading are taken as random variables, and (c) all three groups of random variables are considered together. For the implementation of the NN-based metamodel in all three test cases a two-level approximation is employed using two different NN. The first NN predicts the eigenperiod values of the significant modes. The inputs of the NN are the random variables while the outputs are the eigenperiod values. The second NN is used to predict the maximum interstory drift, which is used to determine whether a limit-state has been violated. Therefore, the spectral acceleration values of both X and Y directions are the input values of the NN while the maximum interstory drift is the output. The input and output variables for the two levels of approximation are shown in Table 1.

Table 1. Input and output variables for the two levels of approximation

Test case	1st level of approximation (NN1)		2nd level of approximation (NN2)	
	Inputs	Outputs	Inputs	Outputs
(a)	E	T_i, $i=1,...,8$	$R_{dx}(T_i), R_{dy}(T_i),$ $i=1,...,8$	Max drift θ_{max}
(b)	$b_i, h_i, i=1,...,5$	T_i, $i=1,...,8$	$R_{dx}(T_i), R_{dy}(T_i),$ $i=1,...,8$	Max drift θ_{max}
(c)	$E, b_i, h_i, i=1,...,5$	T_i, $i=1,...,8$	$R_{dx}(T_i), R_{dy}(T_i),$ $i=1,...,8$	Max drift θ_{max}

Figure 1. The six-story space frame

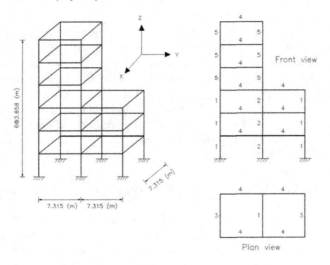

Table 2. List of the natural records

Earthquake	Station	Distance	Site
Tabas	Dayhook	14.0	rock
16 Sept. 1978	Tabas	1.1	rock
Cape Mendocino	Cape Mendocino	6.9	rock
25 April 1992	Petrolia	8.1	soil
	TCU052	1.4	soil
	TCU065	5.0	soil
	TCU067	2.4	soil
	TCU068	0.2	soil
	TCU071	2.9	soil
	TCU072	5.9	soil
	TCU074	12.2	soil
Chi-Chi	TCU075	5.6	soil
20 Sept. 1999	TCU076	5.1	soil
	TCU078	6.9	soil
	TCU079	9.3	soil
	TCU089	7.0	rock
	TCU101	4.9	soil
	TCU102	3.8	soil
	TCU129	3.9	soil

Seismic Probabilistic Analysis Example

The six-story space frame, shown in Figure 1, has been considered to assess the proposed metmodel-assisted structural probabilistic analysis methodology. The space frame consists of 63 members which are divided into five groups having the following cross sections: (a) HEB 650, (b) HEB 650, (c) IPE 450, (d) IPE 400, and (e) HEB 450. The structure is loaded with a permanent action of $G = 3$ kN/m^2 and a live load of $Q = 5$ kN/m^2. In order to take into account that structures may deform inelastically under earthquake loading, the seismic actions are reduced using a behavior factor $q = 4.0$ (Eurocode 8, 1994).

The most common way of defining the seismic loading is by means of a regional design code response spectrum. However, if higher precision is required the use of spectra derived from natural earthquake records is more appropriate. Therefore, a set of 19 natural accelerograms, shown in Table 2, is used. The records are scaled, to the same peak ground acceleration of 0.32g in order to ensure compatibility. Each record corresponds to different earthquake magnitudes and soil properties corresponding to different earthquake events. Two are from the 1992 Cape Mendocino earthquake, two are from the 1978 Tabas, Iran, earthquake and 15 are from the 1999 Chi-chi, Taiwan, earthquake. The response spectra for each scaled record, in X and Y directions, are shown in Figures 2 and 3, respectively. In Table 3, the probability density functions, mean values and standard deviations for all random variables are listed.

Assuming that seismic loading data are distributed lognormally the median spectrum $\hat{x}$ and the standard deviation δ are calculated as follows:

$$\hat{x} = \exp\left[\frac{\sum_{i=1}^{n} \ln(R_{d,i}(T))}{n}\right] \tag{16}$$

$$\delta = \left[\frac{\sum_{i=1}^{n} (\ln(R_{d,i}(T)) - \ln(\hat{x}))^2}{n-1}\right]^{1/2} \tag{17}$$

where $R_{d,i}(T)$ is the response spectrum of the *i-th* record for period value equal to T. The median spectra for both directions are shown in Figures 2 and 3.

Table 3. Characteristics of the random variables

Random variable	Probability density function	Mean value	Standard deviation
E	N	210	10 (%)
b	N	b*	2 (%)
h	N	h*	2 (%)
Seismic load	Log-N	$\hat{x}$ equation (16)	δ equation (17)

Note: Dimensions from the IPE and HEB databases

Figure 2. Natural record response spectra and their median spectrum (longitudinal components)

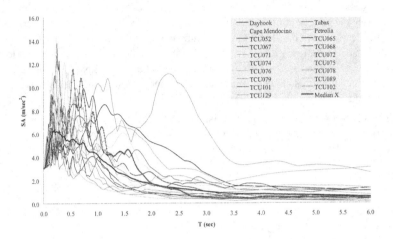

Figure 3. Natural record response spectra and their median spectrum (transverse components)

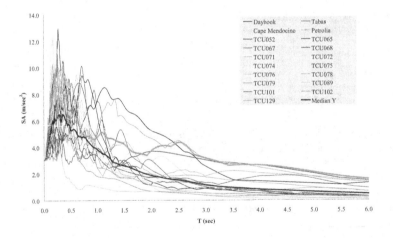

In the first test case one hundred (100) training/testing patterns of the modulus of elasticity are selected in order to train the first NN and one hundred (100) training/testing patterns of the spectral values are selected to training the second NN. Ten out of them are selected to test the efficiency of the trained network. In the second test case the training-testing set was composed by one hundred fifty (150) pairs while in the case of the third test case the

Table 4. "Exact" and predicted values of p_{exceed} and the required CPU time (Adapted from Tsompanakis et al., 2006)

Number of simulations	Test case 1		Test case 2		Test case 3	
	"exact" $P_{exceed}(\%)$	NN $P_{exceed}(\%)$	"exact" $P_{exceed}(\%)$	NN $P_{exceed}(\%)$	"exact" $P_{exceed}(\%)$	NN $P_{exceed}(\%)$
50	0.00	0.00	0.00	0.00	2.00	0.00
100	1.00	2.00	1.00	2.00	2.00	1.00
200	2.00	1.60	1.00	1.00	1.00	3.00
500	1.60	1.26	1.00	1.70	0.40	0.70
1,000	0.90	0.81	0.90	0.67	1.00	0.86
2,000	1.15	1.06	1.45	1.41	1.10	0.97
5,000	1.12	1.04	1.30	1.24	1.32	1.18
10,000	1.09	1.03	1.21	1.14	1.42	1.29
20,000	1.21	1.14	1.25	1.31	1.31	1.19
50,000	1.26	1.16	1.22	1.31	1.36	1.21
100,000	1.28	1.16	1.24	1.31	1.37	1.21
CPU time in seconds						
Pattern selection	-	2	-	3	-	5
Training	-	7	-	9	-	12
Propagation	-	25	-	25	-	25
Total	1154	34	1154	37	1154	42

set was composed by two hundred (200) pairs while in both cases the second NN is trained using one hundred (100) training/testing patterns. For the six story frame of Figure 1, eight modes are required to capture the 90% of the total mass. For each test case a different neural network configuration is used: (1) NN1: 1-10-8, NN2: 16-20-1, (2) NN1: 10-20-8, NN2: 16-20-1 and (3) NN1: 11-20-8, NN2: 16-20-1.

The influence of the three groups of random variables with respect to the number of simulations is show in Figure 4. It can be seen that 2,000 to 5,000 simulations are required in order to calculate accurately the probability of exceedance of the design limit-state. The life safety limit-state is considered violated if the maximum interstory drift exceeds 4.0%.

Once an acceptable trained NN in predicting the maximum drift is obtained, the probability of exceedance for each test case is estimated by means of NN based Monte Carlo simulation. The results for various numbers of simulations are shown in Table 4 for the three test cases examined. It can be seen that the error of the predicted probability of failure with respect to the "exact" one is rather marginal. On the other hand, the computational cost is drastically decreased, approximately 30 times, for all test cases.

Figure 4. Influence of the number of MC simulations on the value of p_{exceed} for the three test cases (Adapted from Tsompanakis, Lagaros, & Stavroulakis, 2006)

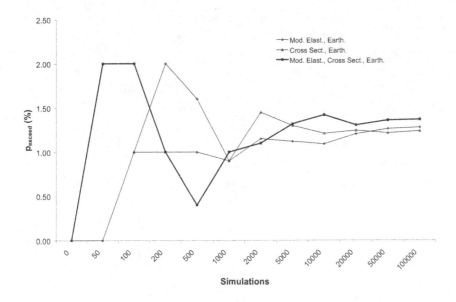

Metamodel Assisted Methodology
for Fragility Analysis

In the second application of this chapter, randomness of ground motion excitation (that influences seismic demand) and of material properties (that affect structural capacity) are also taken into consideration using Monte Carlo simulation. The capacity assessment of steel frames is determined using nonlinear time-history analysis. The probabilistic safety analysis using Monte Carlo simulation and nonlinear time-history analysis results in a computationally intensive problem. In order to reduce the excessive computational cost, techniques based on NN are implemented. For the training of the NN, a number of intensity measures (IMs) are derived from each earthquake record, for the prediction of the level of damage, which is measured by means of maximum interstory drift values θ_{max}.

Random Parameters

The proposed methodology requires that MCS has to be performed at each intensity level. Earthquake records are scaled to a common intensity level that corresponds to the hazard level examined. Scaling is performed using the first mode spectral acceleration of the 5% damped spectrum $Sa(T_1,5\%)$. Therefore, all records are scaled in order to represent the same ground motion intensity in terms of $Sa(T_1, 5\%)$. Earthquake loading is considered as two separate sources of uncertainty, ground motion intensity and the details of ground motion. The first uncertainty refers to the general severity of shaking at a site, which may be measured in terms of any IM such as PGA, $Sa(T_1, 5\%)$, Arias intensity, and so forth. The second source refers to the fact that, although different acceleration time histories can have their amplitudes scaled to a common intensity, there is still uncertainty in the performance, since IMs are imperfect indicators of the structural response. The first source is considered by scaling all records to the same intensity level at each limit-state. The second source is treated by selecting natural records as random variables from a relatively large suite of scenario based records. The concept of considering separately seismic intensity and the details of ground motion is the backbone of the incremental dynamic analysis method (Vamvatsikos & Cornell, 2002), while Porter, Beck, and Shaikhutdinov (2002) have also introduced intensity and different records as two separate uncertain parameters in order to evaluate the sensitivity of structural response to different uncertainties.

The random parameters considered are the material properties and more specifically the modulus of elasticity E and the yield stress f_y, as well as the details of ground motion where a suite of scenario based earthquake records is used. The material properties are assumed to follow the normal distribution while the records are selected randomly from a relatively large bin of natural records.

Predictions of the Seismic Response Using Neural Networks

In the second application, NN are implemented in order to predict the maximum seismic response replacing the time consuming nonlinear timehistory analysis. The NN are trained in order to predict the maximum interstory drift θ_{max} for different earthquake records which the NN identifies using a set of IMs.

The term *intensity measure* is used to denote a number of common ground motion parameters, which represent the amplitude, the frequency content, the duration, or any other ground motion parameter. A number of different IMs have been presented in the literature (Kramer, 1996), while various attempts to relate an IM with a damage measure such as maximum interstory drift values exist (Shome & Cornell, 1999). The IMs adopted can be classified as structure independent (e.g., PGA, Arias intensity) or as both structure and record dependent; for example, $Sa(T_1)$. The complete list of the IMs used in this study is given in Table 5.

It can be seen that the IMs selected, vary from widely used ground motion parameters such as peak ground acceleration (PGA) to more sophisticated measures such as SaC. The definitions and further discussion on the first thirteen measures of Table 5 is given by Kramer (1996). The last two IMs refer to the measure proposed by Cordova, Deirlein, Mehanny, and Cornell (2000), which is defined as:

$$SaC = Sa(T_1)\sqrt{\frac{Sa(c \cdot T_1)}{Sa(T_1)}} \tag{18}$$

The parameter c takes the value 2 and 3 for the 14th and the 15th parameter of Table 5, respectively. These IMs were introduced in order to assist the NN to capture the effects of

Table 5. Intensity measures

No	Intensity Measure
1	*PGA* (g)
2	*PGV* (m)
3	*PGD* (m)
4	*V/A* (sec)
5	Arias intensity (*AI*) (m/sec)
6	Significant duration (*SD*) (5 to 95% of Arias) (sec)
7	RMS acceleration (g)
8	Characteristic intensity (*CI*)
9	CAV
10	Spectral intensity (*SI*)
11	Total duration (*TD*) (sec)
12	$Sa(T_1)$ (g)
13	$Sv(T_1)$ (cm)
14	*SaC*, *c*=2 (g)
15	*SaC*, *c*=3 (g)

Table 6. Intensity measures combinations

ID	IM combinations	NN
A	*PGA*	3-5-1
B	*PGA, PGV*	4-8-1
C	*PGA, PGV, PGD*	5-8-1
D	*PGA, PGV, PGD, AI*	6-12-1
E	*PGA, PGV, PGD, AI, CAV*	7-14-1
F	*PGA, PGV, PGD, AI, CAV, SI*	8-16-1
G	*PGA, PGV, PGD, AI, CAV, SI, $Sv(T_1)$*	9-20-1
H	*PGA, PGV, PGD, AI, CAV, SI, $Sv(T_1)$, $SaC_{c=2}$*	10-21-1
I	*PGA, PGV, PGD, AI, CAV, SI, $Sv(T_1)$, $SaC_{c=2}$, $SaC_{c=3}$*	11-25-1
J	*PGA, PGV, PGD, V/A, AI, SD, RMS, CI, CAV, SI, TD, $Sv(T_1)$, $SaC_{c=2}$, $SaC_{c=3}$*	16-30-1

inelasticity by considering the elastic spectrum at an "effective" period longer than T_1, thus reflecting the reduction in stiffness.

For each intensity level separate training of the NN is performed by means of the IMs of Table 5. The training process is based on the fact that the trained NN will assign small weights to

Figure 5. Ten-story steel moment frame

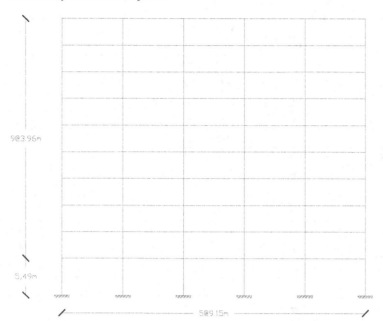

the IMs which have poor correlation with the damage measure selected. Instead of using the whole set, the suitability of using only some of the IMs was examined as indicated in Table 5. The parametric study was performed for various intensity levels since the performance of an IM depends also on the level of nonlinearity that the structure has undergone. The 10 combinations of IMs, shown in Table 6, are compared with respect to their ability to predict the maximum interstory drift in different hazards levels. The number of input nodes in the NNs trained for the ten different IM combinations varies from 3 to 16, which is the modulus of elasticity and the yield stress plus the number of IMs employed, while the output parameters for all combinations is the maximum interstory drift θ_{max} value. The type of the NN used for all combinations can also be found in Table 6, while the number of the hidden nodes has been defined according to the procedure described previously.

Neural Networks Based Fragility Assessment

A suite of 95 scenario-based natural records were used in this application. All records correspond to relatively large magnitudes of 6.0–6.9 and moderate distances, all recorded on firm soil and bearing no marks of directivity. In order to obtain the fragility curves, 16 intensity levels expressed in *PGA* terms ranging from 0.05g to 1.25g were used. For each intensity level, the probability of five limit-states being exceeded is calculated. Each limit-state is defined by means of a corresponding maximum interstory drift θ_{max} value. The limit-states

Table 7. Prediction errors (%) on the maximum interstory drift θ_{max}

					IM Combination					
	A	B	C	D	E	F	G	H	I	J
					PGA = 0.05g					
MAX	49.7	39.6	19.6	32.6	15.9	14.8	4.9	27.0	23.6	9.0
MIN	4.4	0.2	0.9	0.3	0.3	0.6	1.0	1.8	0.6	0.3
MEDIAN	29.1	9.9	4.8	5.2	5.9	5.5	2.9	5.3	4.7	4.0
					PGA = 0.27g					
MAX	32.6	28.2	21.4	26.6	46.7	23.6	9.5	24.5	35.9	9.6
MIN	0.9	0.1	1.1	0.5	1.4	0.1	0.6	0.7	0.3	1.5
MEDIAN	16.9	16.1	9.8	17.2	9.0	7.2	4.9	7.2	4.4	4.4
					PGA = 0.56g					
MAX	62.0	67.2	28.8	42.2	35.4	28.0	33.2	29.3	16.4	9.2
MIN	6.7	3.2	0.2	0.6	0.1	0.2	0.7	1.3	2.2	0.8
MEDIAN	18.5	20.4	9.7	12.6	19.2	15.5	9.0	9.2	7.1	4.3
					PGA = 0.90g					
MAX	72.1	45.2	51.0	23.3	13.3	16.9	12.0	8.7	12.5	9.2
MIN	3.0	6.1	1.4	0.5	1.2	0.9	0.5	0.6	0.2	0.9
MEDIAN	36.5	15.2	15.7	3.0	2.8	7.8	1.8	3.8	2.9	3.5

considered cover the range from serviceability, to life safety and finally to the onset of collapse. The corresponding θ_{max} threshold values range from 0.2% to 6%.

The test example considered to demonstrate the efficiency of the proposed procedure is the five-bay, 10-story moment resisting plane frame of Figure 5. The mean values of the modulus of elasticity is equal to 210GPa and the yield stress is $f_y = 235$MPa. The coefficients of variation for E and f_y are considered as 5% and 10%, respectively, while both variables are assumed to follow the normal distribution. The constitutive law is bilinear with a strain hardening ratio of 0.01, while the frame is assumed to have rigid connections and fixed supports. The permanent load is equal to 5kN/m² and the live load is taken as $Q = 2$kN/m². The gravity loads are contributed from an effective area of 5m. All analyses were performed using a force-based fiber beam-column element that allows the use of a single element per member. The same material properties are used for all members of the frame. Geometric nonlinearities were taken into consideration.

For training the NN both training and testing sets have to be selected for each hazard level. The selection of the sets is based on the requirement that the full range of possible results has to be taken into account in the training step. Therefore, training/testing triads of the material properties and the records are randomly generated. In the case of earthquake records the

Figure 6. Fragility curves

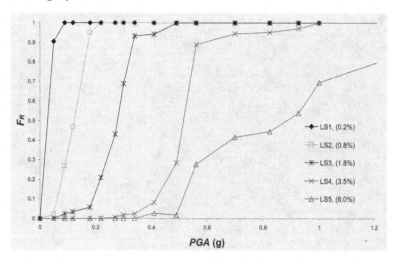

Figure 7. Number of NN simulations required (near collapse limit-state, $\theta_{max} \geq 6.0\%$)

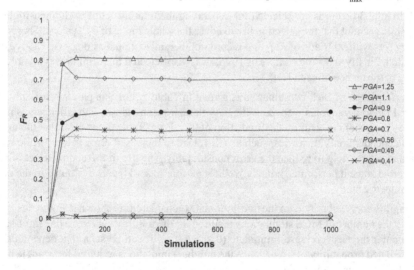

selection has to take into account that the scaling factor should be between 0.2 and 5. This restriction is applied because large scaling factors are likely to produce unrealistic earthquake ground motions. Furthermore, the records selected for generating the training set have to cover the whole range of structural damage for the intensity level in consideration. Thus, nonlinear timehistory analyses were performed, for mean E and f_y values, where the θ_{max} values of each record that satisfy the previous requirement were determined for each hazard

Figure 8. Prediction of θ_{max} *for the testing sample*

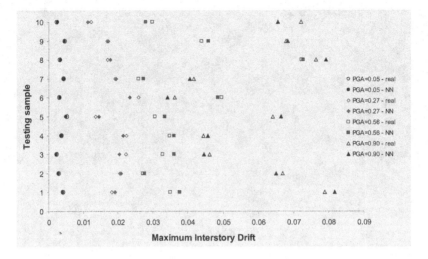

level. In total 30 records are selected for generating the training set of each intensity level taking into account that the selection has to cover the whole range of θ_{max} values. Therefore, training sets with 90 triads of E, f_y and record number, all sampled as discussed above, are generated. Finally, a testing sample of 10 triads is also selected in a similar way in order to test the performance of each NN.

The performance of each combination is shown in Table 7. For this parametric study, the material random variables were considered with their mean values. The efficiency is measured by means of the error on the prediction of θ_{max}. From Table 6 it is clear that the use of record dependent only measures, such as *PGA*, lead to increased error values, while more refined measures help to reduce the error considerably. The use of the complete set of IMs is preferred since it performed equally well for all four hazard levels examined in the present parametric study.

The fragility curves obtained for the five limit-states considered are shown in Figure 6. Figure 7 shows the number of MCS simulations required for the fragility curve of a one limit-state, in particular the near collapse limit-state ($\theta_{max} \geq 6.0\%$). It can be seen that depending on the calculated probability of exceedance the number simulations required for a single point of the fragility curve, ranges from 50 to 1000. The validity of the prediction obtained with the NN is shown in Figure 8. The maximum interstory drift values predicted for the 10 components of the testing set compared to the values obtained with nonlinear timehistory analysis are shown in Figure 8 for four intensity levels. Figure 8, gives the impression that better prediction is obtained for the lower intensity levels, however this occurs because the horizontal axis of Figure 8 corresponds to θ_{max} values and not on the relative error on θ_{max}, which practically remains constant for all four intensity levels.

Conclusion

This chapter presents applications of NN, in computationally demanding tasks in structural mechanics. The computational effort involved in the conventional MCS becomes excessive in large-scale problems, especially when earthquake loading is considered because of the large sample size and the computing time required for each Monte Carlo run. The use of NN can practically eliminate any limitation on the scale of the problem and the sample size used for MCS.

On the other hand, a very efficient procedure for the fragility analysis of structures based on properly trained neural networks is presented. The NN are trained by means of a set of intensity measures that can be easily extracted from the ground motion records. The proposed methodology allows the use of Monte Carlo simulation for the calculation of the limit-state fragility curves, where the only simplifying assumptions made are the distributions of the random variables. The proposed formulation offers a different approach to an emerging problem in earthquake engineering leading to reduction of the computational cost. The results obtained once combined with regional hazard curves can be directly applied to the performance-based design of steel frames.

References

Chakraverty, S. (2005). Identification of structural parameters of multistory shear buildings from modal data. *Earthquake Engineering and Structural Dynamics, 34*, 543-554.

Cordova, P. P., Deirlein, G. G., Mehanny, S. S., & Cornell, C. A., (2000, September 11-13). *Development of a two-parameter seismic intensity measure and probabilistic assessment procedure.* Paper presented at the Second U.S.-Japan Workshop on Performance-Based Earthquake Engineering of Reinforced Concrete Building Structures, Saporo, Japan.

Eurocode 8. (1994). Design provisions for earthquake resistant structures. *CEN, ENV, European Committee for Standardization,* Brussels.

Huang, C. S., Hung, S. L., Wen, C. M., & Tu, T. T. (2003). A neural network approach for structural identification and diagnosis of a building from seismic response data. *Earthquake Engineering and Structural Dynamics, 32*(2), 187-206.

Hurtado, J. E., & Alvarez. D. A., (2002). Neural-network-based reliability analysis: A comparative study. *Computers Methods in Applied Machines and Engineering, 191*, 113-132.

Kramer, S. L. (1996). *Geotechnical earthquake engineering.* Englewood Cliffs, NJ: Prentice Hall.

Kuźniar, K., & Waszczyszyn, Z. (2006). Neural networks for the simulation and identification analysis of buildings subjected to paraseismic excitations. In N.D. Lagaros & Y. Tsompanakis (Eds.), *Intelligent computational paradigms in earthquake engineering.* Hershey, PA: Idea Group Publishing.

Lagaros, N. D., Charmpis, D. C., Papadrakakis, M. (2005). An adaptive neural network strategy for improving the computational performance of evolutionary structural optimization. *Computer Methods in Applied Mechanics and Engineering, 194*(30/33), 3374-3393.

Lagaros, N. D., & Papadrakakis, M. (2004). Learning improvement of neural networks used in structural optimization. *Advances in Engineering Software, 35,* 9-25.

Lagaros, N. D., Stefanou G., & Papadrakakis, M. (2005). Soft computing hybrid simulation of highly skewed non-gaussian stochastic fields. *Computers Methods in Applied Machines and Engineering, 194*(45/47), 4824-4844.

Masri, S. F., Smyth, A. W., Chassiakos, A. G., Caughey, T. K., & Hunter, N. F. (2000). Application of neural networks for detection of changes in nonlinear systems. *Journal of Engineering Mechanics, 126*(7), 666-676.

Mróz, Z., & Stavroulakis, G. (Eds.). (2005). *Parameter identification of materials and structures* (CISM Courses and Notes, No. 469). Springer.

Nie, J., & Ellingwood, B. R. (2004). A new directional simulation method for system reliability. Part II: application of neural networks. *Probabilistic Engineering Mechanics, 19*(4), 437-447.

Papadrakakis, M, Papadopoulos, V., & Lagaros, N. D. (1996). Structural reliability analysis of elastic-plastic structures using neural networks and Monte Carlo simulation. *Computers Methods in Applied Machines and Engineering, 136,* 145-163.

Papadrakakis, M., Tsompanakis, Y., Lagaros, N. D., & Fragiadakis, M. (2004). Reliability based optimization of steel frames under seismic loading conditions using evolutionary computation. *Special Issue of the Journal of Theoretical and Applied Mechanics on Computational Intelligence in Mechanics, 42*(3), 585-608.

Patnaik, S. N., Guptill, J. D., & Hopkins, D. A. (2005). Subproblem optimization with regression and neural network approximators. *Computer Methods in Applied Mechanics and Engineering, 194*(30/33), 3359-3373.

Porter, K. A., Beck, J. L., & Shaikhutdinov, R. V. (2002). Sensitivity of building loss estimates to major uncertain variables. *Earthquake Spectra, 18,* 719-743.

Reda Taha, M. M., & Lucero, J. (2005). Damage identification for structural health monitoring using fuzzy pattern recognition. *Engineering Structures, 27*(12), 1774-1783.

Salajegheh, E., Heidari, A., & Saryazdi S. (2005). Optimum design of structures against earthquake by a modified genetic algorithm using discrete wavelet transforms. *International Journal for Numerical Methods in Engineering, 62,* 2178-2192.

Sanchez-Silva, M., & García, L. (2001). Earthquake damage assessment based on fuzzy logic and neural networks. *Earthquake Spectra, 17*(1), 89-112.

Shome, N., & Cornell, C. A. (1999). *Probabilistic seismic demand analysis of non-linear structures* (Tech. Rep. No. RMS-35). Stanford, CA: Stanford University.

Tsompanakis, Y., Lagaros, N. D., & Stavroulakis, G. (2006). Soft computing techniques in parameter identification and probabilistic seismic analysis of structures. *Advances in Engineering Software.* (under review).

Vamvatsikos, D., & Cornell, C.A. (2002). Incremental dynamic analysis. *Earthquake Engineering and Structural Dynamics, 31*, 491-514.

Waszczyszyn, Z., & Bartczak, M. (2002). Neural prediction of buckling loads of cylindrical shells with geometrical imperfections. *International Journal of Non-linear Mechanics, 37*(4), 763-776.

Zhang, J., & Foschi, R. O. (2004). Performance-based design and seismic reliability analysis using designed experiments and neural networks. *Probabilistic Engineering Mechanics, 19*, 259-267.

Section III

Structural Identification
Applications

Chapter XII

Inverse Analysis of Weak and Strong Motion Downhole Array Data:
A Hybrid Optimization Algorithm

Dominic Assimaki, Georgia Institute of Technology, USA

Abstract

A seismic waveform inversion algorithm is proposed for the estimation of elastic soil properties using low amplitude, downhole array recordings. Based on a global optimization scheme in the wavelet domain, complemented by a local least-square fit operator in the frequency domain, the hybrid scheme can efficiently identify the optimal solution vicinity in the stochastic search space, whereas the best fit model detection is substantially accelerated through the local deterministic inversion. The applicability of the algorithm is next illustrated using downhole array data obtained by the Kik-net strong motion network during the $M_w 7.0$ Sanriku-Minami earthquake. Inversion of low-amplitude waveforms is first employed for the estimation of low-strain dynamic soil properties at five stations. Successively, inversion of the mainshock empirical site response is employed to extract the equivalent linear dynamic soil properties at the same locations. The inversion algorithm is shown to provide robust estimates of the linear and equivalent linear impedance profiles, while the attenuation structures are strongly affected by scattering effects in the near-surficial heterogeneous layers.

Introduction

Current state-of-practice site response methodologies primarily rely on geotechnical and geophysical investigation for the necessary information on density and low-strain shear wave velocity variation with depth. Even further, *attenuation*, a critical yet least explored mechanism of seismic energy dissipation and redistribution, is either approximated by means of empirical correlations or inferred based on limited laboratory data. At larger strains, which the material is anticipated to experience during strong motion events, soil properties are mainly evaluated through laboratory testing. Nonetheless, even the applicability of laboratory testing is limited, due to sample disturbance and difficulties in reproducing the in-situ stress-state and seismic loading.

The scarcity of near-surface geotechnical information, the error propagation of laboratory and in-situ measurement techniques, and the limited resolution of the continuum, usually result in predictions of surface ground motion that compare poorly with low amplitude observations. This discrepancy is even further aggravated for strong ground motion, associated with hysteretic, nonlinear, and potentially irreversible material deformations.

Seismic observations of site response may be a valuable complement to in-situ and laboratory geotechnical investigation techniques. Among others, ground motion recordings at various depths acquired through downhole instrumentation—increasingly deployed in seismically active areas over the past years—provides critical constraints on interpretation and prediction methodologies for site response assessment, as well as information on the real material behavior and overall site response over a wide-range of loading conditions.

In this chapter, a seismogram *inversion* algorithm is developed for the estimation of dynamic soil properties using downhole array data. Comprising a genetic algorithm in the wavelet domain complemented by a local least-square fit operator in the frequency domain, the hybrid scheme can efficiently identify the optimal solution vicinity in the stochastic search space, while the best fit model detection is substantially accelerated through the local deterministic inversion. Results, illustrated for multiple weak and strong motion recordings from the M_w 7.0 Miyagi-Oki earthquake, obtained at selected stations of the Japanese strong motion network Kik-Net, highlight the role of efficient numerical techniques in understanding complex physical phenomena, such as strong motion seismic wave reverberation and scattering in the near surface, leading to the evaluation and improvement of current site response methodologies.

Background:
Observational Evidence Of Site Effects

The destructiveness of ground shaking during earthquakes can be significantly enhanced by *local soil conditions*, a term that refers to the mechanical properties of the surficial geological formations. In particular, documented evidence during past events reveals that the variability in seismic intensity and structural damage severity correlates strongly to the variability of soil stratigraphy at a given area, and examples include—among others—the nonuniform distribution of damage in Tokyo during the 1923 Kanto Earthquake (Ohsaki, 1969); in Caracas during the 1967 Venezuelan earthquake (Seed et al., 1972); in Bucharest during the 1977 Vranehla earthquake (Tezcan, Yerlici, & Durgunoglou, 1979); in Mexico City during the earthquakes of 1957 and especially of 1985 (Rosenblueth, 1960; Seed & Romo, 1987); in San Francisco and Oakland during the 1989 Loma Prieta earthquake (Housner, 1990); in Kobe during the 1995 earthquake (*Soils and Foundations*, 1996); and in Adapazari during the Kocaeli earthquake (Earthquake Spectra, 2000).

Based on ground motion time-histories recorded in the last 3 decades, the influence of subsoil characteristics involves the amplitude level, frequency composition, and duration of surface ground shaking, rendering the detailed description of local soil conditions at any site critical for the assessment of seismic risk, for microzonation studies and for the seismic design and retrofit of important facilities and long structures. Towards understanding the critical mechanisms that govern the seismic response of near-surface geological formations and evaluating currently employed site response methodologies, downhole array recordings obtained in seismically active areas over the past years (e.g., Japan, United States, Taiwan, Mexico, Greece), have been shown to be a valuable complement to existing in-situ and laboratory techniques. Among others, borehole measurements provided direct *in situ* evidence of nonlinearity (e.g., Aguirre & Irikura, 1997; Iai, Morita, Kaeoka, Matsunaga, & Abiko, 1995; Sato, Kokusho, Matsumoto, & Yamada, 1996; Satoh, Fushimi, & Tatsumi, 2001; Seed & Idriss, 1970; Wen, Beresnev, & Yeh, 1994; Zeghal & Elgamal, 1994); they have invited a reevaluation of the use of surface-rock recordings as input motion to soil columns (e.g., Boore & Joyner, 1997; Satoh, Kawase, & Sato, 1995; Steidl, Tumarkin, & Archuleta, 1996), and they have provided basic information about scaling and alluvium sites are located at the surface (e.g., Bonilla, Steidl, Lindley, Tumarkin, & Archuleta, 1997; Borcherdt, 1970; Field, 1996; Field & Jacob, 1995; Hartzell, 1992; Hartzell et al., 1996b; Kato, Aki, & Takemura, 1995; Margheriti, Wennerberg, & Boatwright, 1994; Su, Anderson, Brune, & Zeng, 1996;).

In the ensuing, an efficient seismic waveform *inversion* algorithm is presented in the "Downhole Array Seismogram Inversion" section. Successively, the algorithm is employed for the estimation of low-strain velocity, attenuation and density soil profiles in the near surface using weak motion recordings from the Sanriku-Minami Earthquake (see section "The Sanriku-Minami Earthquake") in section "Weak Motion Data Inversion" These results are also used in "Sensitivity Analysis of the Inverted Attenuation Structure" section to investigate phenomena related to the strongly heterogeneous nature of sedimentary formations. Finally, the "Strong Motion Data Inversion" section presents the value of downhole array seismogram recordings in evaluating current site response methodologies, by using strong ground motion inversion to estimate the extent of nonlinearity exerted by soft soils during strong shaking.

Downhole Array Seismogram Inversion

Seismogram inversion is a nonlinear multiparametric problem, where one attempts to estimate physical parameters from available data. In ideal cases, there exists an exact theory that prescribes how the data should be transformed to reconstruct the model, usually provided that infinitely many and noise-free data sets are available. Therefore, despite the mathematical elegance of exact nonlinear inversion schemes, they have the following drawbacks: (a) they are applicable to ideal conditions, (b) they are often unstable, and (c) they usually predict continuous functions of space variables based on a finite number of available data.

The fact that in realistic experiments, a finite amount of data is available to reconstruct a model with infinitely many degrees of freedom, necessarily implies that the inverse problem is not unique, namely there exist several (usually infinitely many) models that explain the data equally well. Clearly, in this case, the model *estimated* by means of data inversion is not necessarily equal to the true model, and the inversion process needs also to *appraise* what properties of the true model are recovered by the best-fit model, and what errors are attached to it; in fact, acknowledging the existence of errors and limited resolution is necessary for the physical interpretation of a model (Snieder & Trampert, 1999; Trampert, 1998). Nonlinearity introduces additional complexity to the inverse analysis, affecting both the estimation and the appraisal problem, and—in practical problems—it is usually treated as a nonlinear optimization problem, where a suitably chosen measure of the data misfit is reduced as a function of the model parameters.

In seismic applications, two data misfit measures are commonly used. One is the error defined as $E(m) = d - g(m)$, where m and d are model and data vectors respectively, and g is the forward modeling operator. The optimization process determines the m's that minimize the error energy. A second measure is the normalized correlation function, defined as follows:

$$C(m) = \frac{d \otimes g(m)}{(d \otimes d)^{1/2} (g(m) \otimes g(m))^{1/2}} \tag{1}$$

where $\otimes$ represents the cross-correlation; in this case, the optimization scheme estimates the m's that maximize $C(m)$. Both these functions are referred to as "objective" or "fitness" functions.

In the forward problem, the model parameters are nonlinearly related to the measured process, reflected on $C(m)$ and $E(m)$ as multiple local maxima and minima respectively. As a result, traditional search techniques based on local linearization, which use characteristics of the problem to determine the next sampling point (e.g. gradients, Hessians, linearity and continuity) are computationally efficient, but fail to identify the *best fit* solution, when the starting model is too far from the global optimal solution. On the other hand, stochastic search techniques (e.g., genetic algorithms, simulated annealing) have been shown to efficiently identify promising regions in the search space, but perform very poorly in a localized search (Bersini & Renders, 1994; Davis, 1991; Houck, Joines, & Kay, 1996; Michalewicz, 1994).

The proposed optimization technique is a two-step process, namely a genetic algorithm in the wavelet domain coupled to a nonlinear least-square fit in the frequency domain, thus improving the computational efficiency of the former, while avoiding the pitfalls of using local linearization techniques—such as the latter—for the optimization of multimodal, discontinuous and nondifferentiable functions. The parameters to be estimated are stepwise variations of the shear wave velocity, attenuation and density with depth, for horizontally

Figure 1. (a) Schematic representation of forward numerical model used in the optimization scheme, and (b) horizontally stratified layered structure overlying rigid bedrock (Haskell-Thompson model)

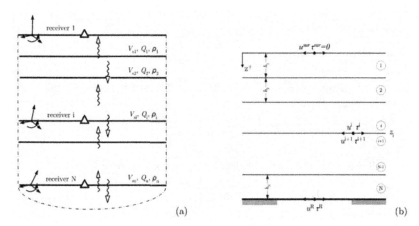

(a)   (b)

layered media with predefined layer thickness. Equality constrains are imposed on the vector of unknowns to bound the search space, based on independent geological and geotechnical site characterization data. Note that for the discretization of the continuum, a preliminary analysis precedes the inversion scheme to ensure that the layer thicknesses are at least equal to the quarter length of propagating waveforms ($h_i \geq V_{si} / 4f_{max}$); the satisfaction of this criterion is also verified at convergence of the optimization algorithm.

Numerical Simulation and Data Processing

One-dimensional downhole seismograms can be simulated by means of standard frequency domain analyses, as proposed by Brune (1970) and revised by Madariaga (1976). Accounting both for upgoing and downgoing waves, we here use the Haskell-Thompson transfer function for horizontally layered media overlying rigid bedrock and subjected to antiplane (SH) incident motion. In this representation, each layer of soil behaves as a Kelvin-Voigt solid.

The forward model schematically illustrated in Figure 1, comprises n homogeneous, horizontal soil layers overlying bedrock, where the total motion is prescribed and equal to u_R. The soil layers have the following characteristics:

$$h_i, \ \rho_i, \ V_{s_i}, \ Q_i, \ i = 1, 2, \dots N \tag{2}$$

where: h = thickness (m), ρ = density (Mg/m³), V_s = shear wave velocity (m/sec), and $1/2\,Q$ = hysteretic damping. At any given layer, i, at a given input frequency, f, we consider the following complex quantities:

$$V_{s_i}^* = V_{s_i} \sqrt{1 + \frac{i}{Q_i}} \qquad G_i^* = \rho_i V_{s_i}^2 \left(1 + \frac{i}{Q_i}\right)$$

$$k_i^* = \frac{\omega}{V_{s_i} \sqrt{1 + \dfrac{i}{Q_i}}} = \frac{2\pi f}{V_{s_i} \sqrt{1 + \dfrac{i}{Q_i}}} \tag{3}$$

where $V_{s_i}^*$ is the complex shear wave velocity, G_i^* is the complex shear modulus, and k_i^* is the complex wavenumber.

We also define:

1. $z = 0.0$ at the free surface (stress-free boundary) of the profile,

2. $z_i = \sum_{j=1}^{i} h_i$ as the total depth of the $i\text{-}th$ layer,

3. A_i as the amplitude of the down-going waves at each layer, and

4. B_i as the amplitude of up-going waves at each layer.

The total motion in each layer (summation of up-going and down-going waves) is:

$$u_i = A_i\, e^{i\left(\omega t + k_i^* z_i\right)} + B_i\, e^{i\left(\omega t - k_i^* z_i\right)} \tag{4}$$

Therefore, the total shear stress in each layer is defined as:

$$\tau_i = i k_i^* G_i^* A_i\, e^{i\left(\omega t + k_i^* z_i\right)} - i k_i^* G_i^* B_i\, e^{i\left(\omega t - k_i^* z_i\right)} \tag{5}$$

We now define the matrix of displacement-stress coefficients as follows:

$$\begin{Bmatrix} u_i \\ \tau_i \end{Bmatrix} = \underbrace{\begin{bmatrix} e^{i k_i^* z_i} & e^{-i k_i^* z_i} \\ i k_i^* G_i^* e^{i k_i^* z_i} & -i k_i^* G_i^* e^{i k_i^* z_i} \end{bmatrix}}_{\Delta_i(z_i)} \begin{Bmatrix} A_i \\ B_i \end{Bmatrix} \tag{6}$$

Applying continuity of displacements and stresses at the soil layer interfaces, these conditions can be expressed as follows:

$$\begin{Bmatrix} A_{i+1} \\ B_{i+1} \end{Bmatrix} = \underbrace{\begin{bmatrix} e^{i k_{i+1}^* z_i} & e^{-i k_{i+1}^* z_i} \\ i k_{i+1}^* G_{i+1}^* e^{i k_{i+1}^* z_i} & -i k_{i+1}^* G_{i+1}^* e^{i k_{i+1}^* z_i} \end{bmatrix}^{-1} \begin{bmatrix} e^{i k_i^* z_i} & e^{-i k_i^* z_i} \\ i k_i^* G_i^* e^{i k_i^* z_i} & -i k_i^* G_i^* e^{i k_i^* z_i} \end{bmatrix}}_{\Delta\Delta_i(z_i) = \Re_i} \begin{Bmatrix} A_i \\ B_i \end{Bmatrix} \tag{7}$$

Applying equation (7) from the *N-th* to the 1ˢᵗ interface, we have:

$$A_N = \Re_{N-1}\, A_{N-1} = \Re_{N-1}\, \Re_{N-2}\, A_{N-2} = \prod_{i=N-1}^{1} \Re_i\, A_1 \tag{8}$$

Therefore, at the soil/bedrock interface we have:

$$\begin{Bmatrix} u_R \\ \tau_R \end{Bmatrix} = \Delta_{N(z=H)}\, A_N = \Delta_{N(z=H)} \prod_{i=N-1}^{1} \Re_i\, A_1 = \Delta_{N(z=H)} \prod_{i=N-1}^{1} \Re_i\, \Delta_{1(z=0)}^{-1} \begin{Bmatrix} u_1(0) \\ \tau_1(0) \end{Bmatrix} \tag{9}$$

where $u_1(0)$ and $\tau_1(0)$ are the total surface displacement and stress respectively.

Finally, applying $\tau_1(0) = 0$ at the stress-free boundary $(z = 0)$, and $u_N(H) = u_R$ at the soil/bedrock interface, equation (9) is transformed as follows:

$$\begin{Bmatrix} u_R \\ \tau_R \end{Bmatrix} = \underbrace{\Delta_{N(z=H)} \prod_{i=N-1}^{1} \Re_i\, \Delta_{1(z=0)}^{-1}}_{\begin{bmatrix} \Im_{11} & \Im_{12} \\ \Im_{21} & \Im_{22} \end{bmatrix}} \begin{Bmatrix} u_{sur} \\ 0 \end{Bmatrix} \tag{10}$$

$$\frac{u_{sur}}{u_R} = \Im_{11}^{-1}$$

In the ensuing, we refer to $\Im_{11}^{-1}$ as the theoretical surface-to-bedrock transfer function. It should be noted that equation (10) describes the frequency response of layered media to upgoing and downgoing SH waves, prescribed at any given depth within the profile, irrespective of the soil conditions at larger depths. For the interpretation of downhole array seismic data in particular, equation (10) describes the frequency response of the soil structure between any receiver within the profile and the receiver located at ground surface.

To simulate the anti-plane wave propagation problem using downhole array recordings, we compute the transverse motion at all receiver depths by rotating the NS and EW seismogram components through the *great circle path*, based on the event and receiver coordinates. As can be readily seen, the mathematical representation (Figure 1) approximates the physical problem, provided that the angle of incidence at the borehole level is small enough, for the seismic waves to be considered "vertically propagating".

Subsequently, the rotated components are de-noised (applying an energy threshold equal to 5% of the maximum spectral amplitude level) and filtered, using a Butterworth filter with pass-band (1-15Hz). In particular, we implement a noncausal infinite-duration impulse response filter (IIR) by applying one causal filter to the signal forward in time and successively, an anti-causal filter backwards on the filtered signal (Gustafsson, 1996).

Finally, accounting for the fact that large events have longer source durations than smaller ones, we define the appropriate amount of digital information (i.e., seismogram time-window) to be used in the optimization scheme as a function of the event's magnitude. Following

Abercrombie (1995), we use 1-second (1.0 s) windows for small events ($M_L < 3$), 2-second (2.0s) windows for events $3 < M_L < 4$, and 4-second (4.0 s) windows for the larger events.

It should be noted that the selection of the appropriate time-window is a trade-off between control points in the inversion scheme, and excess information that cannot be reproduced by the mathematical model. While using long time-windows ensures the stable estimation of the average empirical site response, complex phenomena—such as small-scale scattering—are *bound* on the same time to be simulated though a forward numerical operator that describes purely vertically propagating upgoing and downgoing waves. A sensitivity analysis, illustrating the dependence of the inverted soil structure on the duration of the time window used, is presented in the section "The Sanriku-Minami Earthquake" of this chapter.

Genetic Algorithm Optimization in the Wavelet Domain

Genetic algorithms have been traditionally used to solve difficult problems with objective functions that do not possess properties such as continuity, differentiability, satisfaction of the Lipschitz condition, and so forth (Davis 1991; Goldberg 1989; Holland 1975; Michalewicz, 1994). These algorithms maintain and transform a family or population of solutions, and implement a "survival of the fittest" strategy in their search for better solutions.

In general, the fittest individuals of any population tend to reproduce and survive to the next generation, thus improving successive generations. Nonetheless, inferior individuals can—by chance—also survive and reproduce. Genetic algorithms have been shown to solve linear and nonlinear problems by exploring all regions of the state space and exponentially exploiting promising areas through mutation, crossover, and selection operations applied to individuals in the population (Michalewicz, 1994). For a more complete discussion of genetic algorithms, including extensions and related topics, the reader is also referred to Davis (1991), Goldberg (1989), and Holland (1975).

The use of a genetic algorithm requires the determination of six fundamental issues: (1) the chromosome representation, (2) the selection function, (3) the genetic operators evaluating the reproduction function, (4) the creation of the initial population, (5) the termination criteria, and (6) the objective function.

A chromosome representation is needed to describe each individual in the population of interest. The representation scheme determines both the problem's structure in the genetic algorithm, and the genetic operators used. Each individual or chromosome comprises a sequence of genes from a certain alphabet. In Holland's original design, the alphabet was limited to binary digits. Since then, problem representation has been the subject of much investigation. It has been shown that more natural representations are more efficient and produce better solutions (Michalewicz, 1994). One useful representation of an individual or chromosome for function optimization involves genes or variables from an alphabet of floating point numbers with values within the variables upper and lower bounds. Michalewicz (1994) has done extensive experimentation comparing real valued and binary genetic algorithms, and shows that real valued genetic algorithms are an order of magnitude more efficient in terms of CPU time. He also shows that a real valued representation moves the problem closer to the problem representation which offers higher precision with more consistent results across replications. Based on the aforementioned studies, a floating number representation was adopted for the purpose of this study.

The genetic algorithm must be supplied with an initial population. The most common method is to randomly generate solutions for the entire population. Nonetheless, since genetic algorithms can iteratively improve existing solutions (i.e., solutions from other heuristic and/or current practices), the first parental population can be seeded with potentially good solutions while randomly generating solutions for the remainder of the population. In this study, random generation of the initial population is allowed within the parameter vector boundaries, seeded upon availability of on-site investigation data with estimates of the shear-wave velocity, attenuation and density profiles. Successively, the genetic algorithm moves from generation to generation selecting and reproducing parents until a termination criterion is met. We here implement a scheme of multiple termination criteria, namely minimum thresholds on (a) the summation of population deviations, and (b) the best solution improvement over a specified number of generations.

Finally, there exist multiple evaluation functions that can be used in a genetic algorithm, provided that they allow the population to be mapped into a partially ordered set. Beyond this constraint, the appropriate function is independent of the stochastic search process, and is selected to ensure efficient convergence of the optimization problem studied. The following section presents the objective function used for the purpose of this study.

Wavelet Domain Objective Function

For the global optimization scheme, the objective function is here defined as the normalized correlation between observed data and synthetics, as follows (Hartzell et al., 1996; Stoffa & Sen, 1991):

$$C(m) = \frac{1}{N_p} \sum_{1}^{N_b} \frac{2 \sum_{1}^{N_{TS}} a_0 \, a_s^*(m)}{\left[\sum_{1}^{N_{TS}} a_0 \, a_0^* \right] + \left[\sum_{1}^{N_{TS}} a_s(m) \, a_s^*(m) \right]} \tag{11}$$

where a_0, $a_s^*(m)$ stands for the observed and synthetic seismograms respectively, N_{TS} is the number of time steps, and N_p is the number of wavelet decomposition bands of the signal.

In this case, the mathematical representation of the forward problem propagates the measured total motion at the borehole depth to the surface through an idealized medium. Successively, the coherency between measured and predicted processes at the surface station of the array, maps the similarity between the idealized soil configuration and the real soil structure. Therefore, the objective of the optimization scheme is to maximize the normalized cross-correlation, identifying the so-called best-fit soil configuration. It should be noted herein that, upon availability of multiple downhole instruments, the objective function can be modified so as the optimization process to maximize the average cross-correlation over all available downhole-surface pairs.

Decomposing the signal in the wavelet domain, and normalizing the approximation and details—as opposed to the original signal—in the objective function definition, allows for equal weighting of the information across all frequency bands. This approach is preferable to a time-domain representation, which would inevitably emphasize the larger amplitude signals of the nonstationary ground motion (in time and frequency). We perform the signal representation (expansion) using Meyer orthogonal wavelets (Daubechies, 1992). In the following section, we briefly describe the advantages of using wavelet analysis in geophysical applications and provide details on the selected wavelet function. For further information, the reader is referred to Foufoula-Georgiou and Kumar (1994), a collection of papers describing the advantages of wavelet transforms in the analysis of geophysical processes.

Discrete Meyer Wavelet Decomposition

The wavelet transform originated in geophysics for the analysis of seismic signals (Morlet Arens,Fourgeau, & Giard, 1982a, 1982b) and was later formalized by Grossmann & Morlet (1984) and Goupillaud Grossmann, & Morlet, (1984). Successively, important advances were introduced by Meyer (1992), Mallat (1989a, 1989b), Daubechies (1988, 1992), Chui (1992), Wornell (1995), and Holschneider (1995), among others. Wavelets have been extensively used in studies of geophysical processes or signals, both as integration kernels to extract information about the processes, and as bases for their representation or characterization. In the form of analyzing kernels, wavelets enable the localized study of a signal by means of a scale-dependent detail description. By means of this process, referred to as *time-frequency localization*, broad and fine signal features can be separately analyzed on large and small scales correspondingly. This property of wavelet analysis is especially useful for the analysis of nonstationary signals and signals with short transient components, namely features at different scales or singularities.

Wavelets can also be used for the description of a process, in the form of elementary building blocks in a decomposition series or expansion. Similarly to the well-known Fourier series, a signal wavelet representation is provided by an infinite series expansion of dilated (or contracted) and translated versions of a fundamental wavelet, each multiplied by an appropriate coefficient. For a particular geophysical application, decision on the appropriate expansion (wavelet, Fourier or spline), and selection of the optimum wavelet representation, depends on the purpose of the analysis.

For the decomposition of nonstationary signals in particular, or for signals with time-dependent frequency content, an *orthogonal, local* and *universal* basis is usually selected. By means of orthogonal wavelet transforms, discrete signals can be represented at various resolutions by means of the so-called multiresolution analysis (Daubechies, 1988; Mallat, 1989a), a process that describes both the signal decomposition and to the development of efficient mechanisms that govern the transition from one level of resolution to another. For example, for the DWT of a signal $f(t)$, the first step produces, starting from f, two sets of coefficients: approximation coefficients cA_1, and detail coefficients cD_1. These vectors are obtained by convolving f with a low-pass filter L0-D for the approximations, and a high-pass filter Hi-D for the details, followed by a dyadic decimation (i.e. downsampling). The next step splits the approximation coefficients cA_1 in two parts using the same scheme, replacing f by cA_1, and producing cA_2 and cD_2, and so on. Once the process is completed, the wavelet decomposition of the signal f analyzed at resolution level i, has the following structure: $(cA_i, cD_i,..., cD_1)$.

In the proposed optimization scheme, wavelet multiresolution analysis is used to describe the nonstationary seismic signals at various frequency bands, and successively, the signal amplitudes are normalized at the resolution level thus being assigned equal weight across the frequency range of interest. In particular, the measured and computed signals are decomposed using an orthogonal, discrete wavelet basis at five levels of resolution, and the objective function of the genetic algorithm is defined as the average normalized cross

Figure 2. Selected time-window of trial synthetic and surface observation (top), wavelet decomposition (bottom left) and net error function distribution (bottom right) of sample time history recorded at station IWTH04 of the Japanese Kik-Net seismic network, during the M_L = 4.8 aftershock (06/10/2003 16:24) of the Sanriku-Minami earthquake

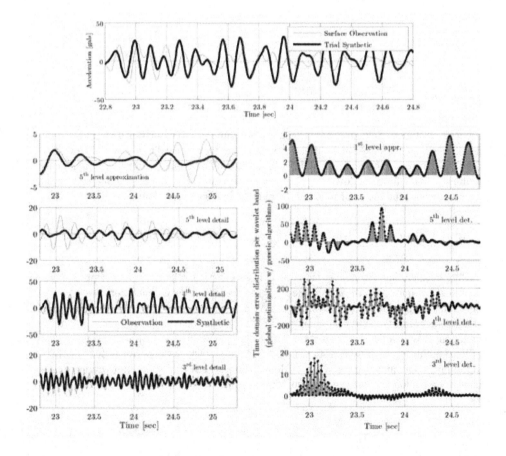

correlation between synthetics and observations across all wavelet bands of interest. The so-called Meyer wavelet, $\hat{\psi}$ (ω), and scaling, $\hat{\lambda}$ (ω), functions are described in the frequency domain by equations (12) and (13):

Wavelet function

$$
\begin{cases}
\hat{\psi}\,(\omega) = (2\,\pi)^{-1/2}\,e^{i\omega/2}\,\sin\left(\dfrac{\pi}{2}\,\mathbf{v}\left(\dfrac{3}{2\pi}\,|\omega| - 1\right)\right) & \text{if} & \dfrac{2\pi}{3} \le |\omega| \le \dfrac{4\pi}{3} \\[3ex]
\hat{\psi}\,(\omega) = (2\,\pi)^{-1/2}\,e^{i\omega/2}\,\sin\left(\dfrac{\pi}{2}\,\mathbf{v}\left(\dfrac{3}{4\pi}\,|\omega| - 1\right)\right) & \text{if} & \dfrac{4\pi}{3} \le |\omega| \le \dfrac{8\pi}{3} \\[3ex]
\hat{\psi}\,(\omega) = 0 & \text{if} & |\omega| \notin \left[\dfrac{2\pi}{3}, \dfrac{8\pi}{3}\right] \\[3ex]
\text{where} \\[1ex]
\mathbf{v}(a) = a^4\left(35 - 84a + 70a^2 - 20a^3\right), \quad a \in [0,1] & & \text{(12)}
\end{cases}
$$

Scaling function

$$
\begin{cases}
\hat{\lambda}\,(\omega) = (2\pi)^{-1/2} & \text{if} & |\omega| \le \dfrac{2\pi}{3} \\[3ex]
\hat{\lambda}\,(\omega) = (2\pi)^{-1/2}\cos\left(\dfrac{\pi}{2}\,\mathbf{v}\left(\dfrac{3}{2\pi}\,|\omega| - 1\right)\right) & \text{if} & \dfrac{2\pi}{3} \le |\omega| \le \dfrac{4\pi}{3} \\[3ex]
\hat{\lambda}\,(\omega) = 0 & \text{if} & |\omega| > \dfrac{4\pi}{3} \quad\quad \text{(13)}
\end{cases}
$$

Figure 2 illustrates an example of the genetic algorithm objective function definition, formulated for a single surface-borehole station pair, where the recorded total borehole motion is used to estimate surface synthetic motions based on trial soil configurations. In particular, the figure shows the time window of the signal under investigation, the corresponding wavelet decomposition, and the error function distribution in the wavelet domain.

Local Hill-Climbing in the Frequency Domain

Further accelerating the convergence of the optimization scheme, a local improvement operator is employed at the end of the selection process of each generation. In particular, once the best fit solutions are identified, a nonlinear Gauss-Newton scheme is used, opting at convergence of the active parental generation towards local minima or maxima prior to mutation, cross-over and reproduction. This technique, referred to as *hill-climbing* method of local optimization, has been shown to significantly enhance the performance of genetic algorithms (Bersini & Renders, 1994; Houck, Joines, & Kay, 1996). An overview of the local optimization scheme is presented in the ensuing.

Gauss-Newton Algorithms for Nonlinear Least-Square Optimization

In unconstrained optimization problems, one seeks a local minimum of a real-valued function, $f(x)$, where x is a vector of n real variables. The problem can be mathematically stated as follows:

$$\text{minimize}_x \ f(\overline{x}) \quad \overline{x} \in \mathbb{R}^n \ \text{i.e.} \ \overline{x} = (x_1, x_2, ..., x_n)^T \tag{14}$$

By contrast, global optimization algorithms attempt to identify a solution x^*, which minimizes f over all possible vectors $(\overline{x})$, a substantially more cumbersome and computationally expensive process. As a result, local optimization methods are selected for multiple applications, and can yield satisfactory estimates of the solution; the efficiency of the algorithm, however, depends strongly on the user-provided starting trial vector.

The general formulation of the nonlinear least-squares problem can be expressed as follows:

$$\min \left\{ r(\overline{x}) : \overline{x} \in \mathbb{R}^n \right\} \tag{15}$$

where r is the function defined by $r(\overline{x}) = \dfrac{1}{2} \left\| f(\overline{x}) \right\|_2^2$ and f is a vector-valued function mapping $\mathbb{R}^n$ to $\mathbb{R}^m$.

For a physical process, modeled by a nonlinear function ϕ that depends on a parameter vector $\overline{x}$ and time t, if b_i is the actual output of the system at time t_i, the residual:

$$\phi(\overline{x}, t_i) - b_i \tag{16}$$

provides a measure of the discrepancy between the predicted and observed outputs of the system at time t_i. A reasonable estimate for the parameter $\overline{x}$ may be obtained by defining the $i\text{-}th$ component of f as:

$$f_i(\overline{x}) = \phi(\overline{x}, t_i) - b_i \tag{17}$$

and solving the least-squares problem based on the aforementioned definition of f.

From an algorithm view point, the feature that distinguishes least-square problems from the general formulation of an unconstrained optimization problem is the structure of the Hessian matrix of r. In particular, the Jacobian matrix of f, namely:

$$f'(\overline{x}) = (\partial_1 f(\overline{x}), \partial_2 f(\overline{x}), ... \partial_n f(\overline{x})), \tag{18}$$

can be used to express the gradient of r since $\nabla r(\overline{x}) = f'(\overline{x})^T f(\overline{x})$. Similarly, $f'(\overline{x})$ is a part of the Hessian matrix, as shown as follows.

$$\nabla^2 r(\overline{x}) = f'(\overline{x})^T f'(\overline{x}) + \sum_{i=1}^{m} f_i(\overline{x}) \nabla^2 f_i(\overline{x}). \tag{19}$$

Calculation of the gradient of r relies therefore on evaluating the Jacobian matrix $f'(\overline{x})$. Once completed, the first term of the Hessian matrix $\nabla^2 r(\overline{x})$ is known without further computations. Nonlinear least-squares algorithms exploit this structure.

For a large number of practical applications, the first term in $\nabla^2 r(\overline{x})$ is substantially larger than the second term, and the relative magnitude of the first term increases for small values of residuals $f_i(\overline{x})$, that is, in the vicinity of the local solution. More specifically, a problem is considered to have small residuals, if for all $\overline{x}$ near the solution, the quantities:

$$\left| f_i(\overline{x}) \right| \left\| \nabla^2 f_i(\overline{x}) \right\| \ll \min \left(\text{eigenvalue} \left[f'(\overline{x})^T f'(\overline{x}) \right] \right), \quad i = 1, 2, ..., n. \tag{20}$$

An algorithm, particularly suited for the small-residual case, is the Gauss-Newton algorithm, where the Hessian is approximated by its first term. In a line-search version of the Gauss-Newton algorithm, the search direction d_k from the current iteration satisfies the linear system:

$$(f'(\overline{x}_k)^T f'(\overline{x}_k)) d_k = -f'(\overline{x}_k)^T f(\overline{x}_k) \tag{21}$$

Note than any solution of this equation is a descent direction, since:

$$d_k^T \nabla r(\overline{x}_k) = -\left\| f'(\overline{x}_k) d_k \right\|_2^2 < 0, \text{ unless } \nabla r(\overline{x}_k) = 0. \tag{22}$$

Newton-Gauss algorithms perform a line-search along the direction to obtain the new trial vector. The suitability of a candidate's step length can be determined, as in the general case of unconstrained minimization, by enforcing the sufficient decrease condition and the curvature condition.

Frequency-Domain Objective Function

For the purpose of this study, the objective function of the nonlinear least-square optimization is defined in the frequency domain, as the energy error between the model and data vectors:

$$E(m) = \frac{\sum_{1}^{N_\omega}(A_o - A_S^*(m))^2}{\sum_{1}^{N_\omega}A_o^2 + \sum_{1}^{N_\omega}(A_S^*(m))^2}$$ (23)

where A_0, $A_S^*(m)$ stand for the empirical and theoretical transfer functions respectively, and N_ω is the number of frequencies. As can readily be seen from equation (17), the functions ϕ and b_i are:

$$\phi = \Re_{11}^{-1}\left(\{V_s, Q, \rho\}, \omega\right) = A_s\left(V_s, Q, \rho\right)$$ (24)

$$b_i = \frac{FFT(a^{sur})}{FFT(a^{bor})} = A_0$$

where V_s, Q, ρ are the shear wave velocity, attenuation and density vectors correspondingly, of n soil layers with predefined thickness, ω is the circular frequency, and a is the recorded acceleration time history.

To ensure that the empirical transfer function is computed using identical volume of information from the borehole and surface motion, the borehole record is shifted in time, so that

Figure 3. Sample acceleration time-histories at station iwth04 of the Japanese Kik-Net seismic network, during the M_L= 4.8 aftershock (06/10/2003 16:24) of the Sanriku-Minami earthquake (top) and corresponding cross-correlation function, where the borehole to surface travel time is interpreted as the time lag where the function is maximized.

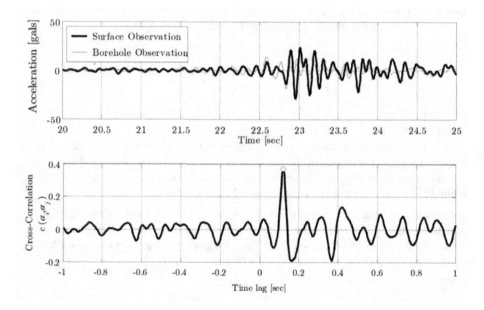

Figure 4. Selected time-window for surface to borehole empirical transfer function estimation (top), theoretical and empirical transfer function (bottom left) and estimation residuals (bottom right) of sample time history recorded at station iwth04 of the Japanese Kik-Net seismic network, during the $M_L = 4.8$ aftershock (06/10/2003 16:24) of the Sanriku-Minami earthquake

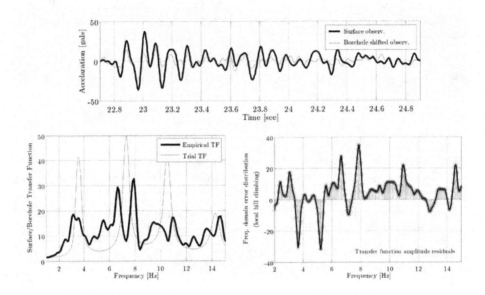

the maximum surface-to-borehole cross-correlation coefficient occurs at zero time lag. In particular, given two acceleration time histories $a_i(t)$ and $a_j(t)$ recorded at stations i and j, the cross-correlation function is expressed as (Bendat & Piersol, 1996):

$$c_{a_i,a_j}(t) = \frac{\frac{1}{N-m}\sum_{n=1}^{N-m} a_i(n\Delta t) a_j((n+m)\Delta t)}{\sqrt{\frac{1}{N}\sum_{n=1}^{N} a_i^2(n\Delta t)}\sqrt{\frac{1}{N}\sum_{n=1}^{N} a_j^2(n\Delta t)}}, \quad m = 1, 2, \ldots N \tag{25}$$

where Δt is the time step of digital data, $t = m\Delta t$ is the time delay and $N\Delta t$ is the acceleration record length. The cross-correlation function reaches a major peak at time lag $t = t_d$, which corresponds to the travel time from station i to j (see Figure 3). Therefore, shifting the borehole data ensures optimal coherency of the two time-histories.

Successively, the empirical transfer function is defined as the amplitude of the complex ratio between the Fourier surface and shifted borehole motion spectra, which corresponds to the same tapered time window used in the time-domain optimization process. A schematic representation of the error distribution in the frequency domain is illustrated in Figure 4.

Figure 5. Schematic representation of the proposed hybrid optimization scheme, illustrated for population j of the genetic algorithm. The process is repeated in series for k seismograms, and the global optimum solution is obtained by averaging k best-fit models.

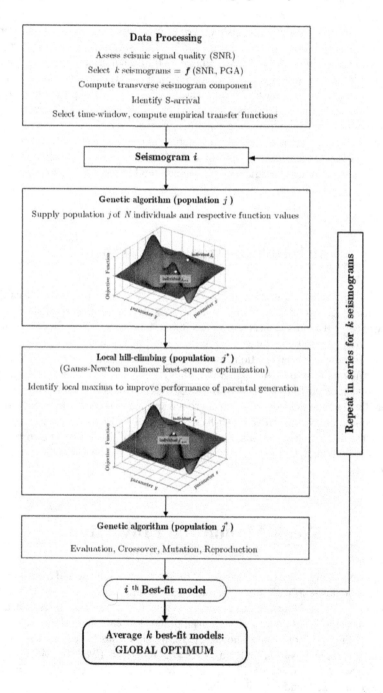

Overview of the Global-Local Optimization Scheme

The proposed optimization algorithm, namely a stochastic search technique combined with a nonlinear least-square scheme operating at the parental level of each generation, is repeated in series for multiple borehole and surface waveform pairs. Among the total number of available motions recorded at a certain station, a subset is selected on the basis of the available signal-to-noise ratio (SNR). By averaging the optimal solution for multiple events, we minimize both the error propagation arising both from the measured process and from the limitations of the forward numerical model, thus obtaining the most probable *best-fit* solution to the inverse problem. The proposed algorithm is schematically illustrated in Figure 5. The global-local inversion technique can efficiently identify the *optimal solution vicinity* in the search space by means of the hybrid genetic algorithm, whereas the use of nonlinear least-square fit accelerates substantially the detection of the *best fit* model. The algorithm has been implemented in the mathematical computer code MATLAB 7, and typical results are presented in the following section of this chapter.

The Sanriku-Minami Earthquake

The Sanriku-Minami earthquake (5/26/2003, 18:24GMT, MW 7.0), an intraslab earthquake of the Pacific Plate, occurred at about 70 km depth, several kilometers offshore, near the boundary of Miyagi and Iwate Prefectures (38.806N, 141.685E; NIED, Hi-Net, Figure 6b). The ground motion was widely felt in the northern half of Japan and recorded at about 800 stations of nationwide strong motion networks K-Net (Kinoshita, 1998) and Kik-Net. At five stations, including two JMA (Japan Meteorological Agency) stations, PGA (peak ground acceleration) exceeded 1g (980 gals) within an epicentral distance of about 100 km. Figures 6a and 6b illustrate the PGA distribution at ground surface as recorded by Kik-Net, and the aftershock distribution as recorded by Hi-Net, respectively. Also shown on Figure 6a, is the location of the five Kik-Net stations used for the purpose of this study. The corresponding station codes, longitude, latitude, ground surface elevation and borehole instrument depths are listed in Table 1.

Weak Motion Data Inversion

Among approximately 240 events recorded at these stations in the period May 3-July 3, we identified 18 borehole-surface 3-component acceleration time history pairs with characteristics $M_W > 3.5$ and $a_{max} < 0.1g$. Note that this observation selection criterion ensures both adequate signal-to-noise ratio (SNR), and acceleration amplitude low enough to elicit elastic material behavior. For the inversion results shown, the following events have been selected on the basis of the surface observation SNR: (a) M4.9 16:24GMT 06/10/03, (b) M4.4 21:12GMT 05/27/03, (c) M4.9 00:44GMT 05/27/03, and (d) M4.4 07:41GMT 05/27/03, denoted heretofore as events 1, 2, 3, and 4 respectively.

Figure 6. (a) Distribution of PGA at the surface, as recorded by the Japanese Strong Motion Seismogram Network, Kik-Net. Depths of the boreholes are larger than 100m at most stations, whereas shear wave velocities are larger than 500 m/s at the depth of the borehole instruments. (b) Aftershock distribution by manual picking of seismogram data, recorded by the Japanese High Sensitivity Seismogram Network Kik-Net.

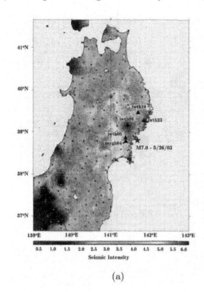

(a)

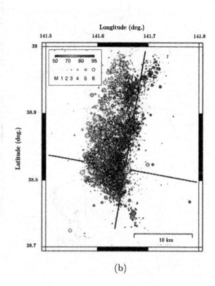

(b)

Table 1. Station codes, names, longitude, latitude, ground surface elevation and borehole instrument depths of the Kik-Net stations used for the purpose of this study

Station Code	Station Name	Latitude [deg]	Longitude [deg]	Surface Alt. [m]	Borehole Depth [m]
IWTH04	SUMITA	39.178	141.394	620	106
IWTH05	FUJISAWA	38.863	141.355	120	100
IWTH18	KAWAI-S	39.460	141.681	552	100
IWTH23	KAMAISHI	39.272	141.827	44	103
MYGH04	TOUWA	38.783	141.329	35	100

Figure 7. Shear wave velocity structure (left) from geophysical site investigation at site iwth04; Attenuation (middle) and density (right) structures, estimated using empirical correlations.

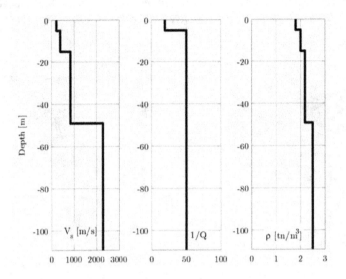

To illustrate the inaccuracy arising in weak motion site response assessment due to the limited amount of geotechnical information, we here illustrate an example of elastic wave propagation simulation at station *iwth*04, where the borehole recorded seismogram is used as total input motion and the ground surface synthetic is evaluated based on the available low-strain dynamic soil properties. In particular, Figure 7 shows the shear wave velocity (V_s) structure at station *iwth*04, based on the geophysical site investigation, and reported at the Kik-Net Strong Motion Network Web site (www.kik.bosai.go.jp/kik/). Also shown, are estimates of the local attenuation and density structures, as reported by the National Institute for Earthquake Disaster Prevention (NIED) for the broader geographical area under investigation.

Using the available geotechnical information, a forward prediction of the site response for event 1 is shown in Figure 8. In particular, the figure illustrates the surface synthetic and observation acceleration time histories, and the theoretical transfer function. The latter is compared to (a) the empirical transfer function, extracted from event 1 using a 2.0s, 5% tapered time window; and (b) the average site response, computed by averaging the empirical transfer function over 18 events recorded at site *iwth*04. Clearly, the comparison between synthetics and observations at ground surface is very poor, highlighting both the coarse discretization of the available geotechnical information near the surface, and the divergence of the estimated attenuation and density structures from the true soil conditions.

Figure 8. (a) Comparison of observation and synthetic surface response to event 1, computed using the available local geotechnical information (Figure 7), and (b) comparison of empirical transfer function corresponding to event 1, theoretical (synthetic) transfer function computed using the available local geotechnical information (Figure 7), and average site response estimated using 28 empirical transfer functions at station iwth04

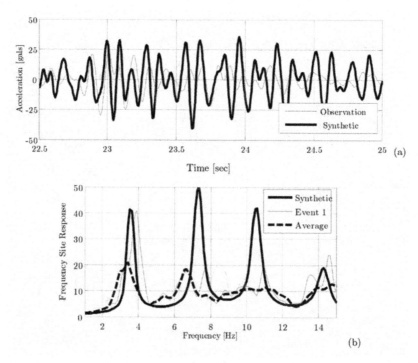

Figure 9. Efficiency of the hybrid optimization scheme with local hill-climbing (LHC) based on the discrete Meyer wavelet decomposition (solid line): comparison with an alternative orthogonal wavelet decomposition (thick dashed line), and a traditional genetic algorithm (thin dashed-dotted line) performance

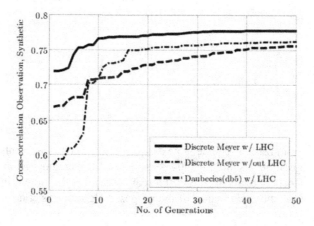

Figure 10. Estimation of travel time at station iwth04, as the average time-lag correspond-ing to the peak borehole-to-surface cross-correlation (events 1-4)

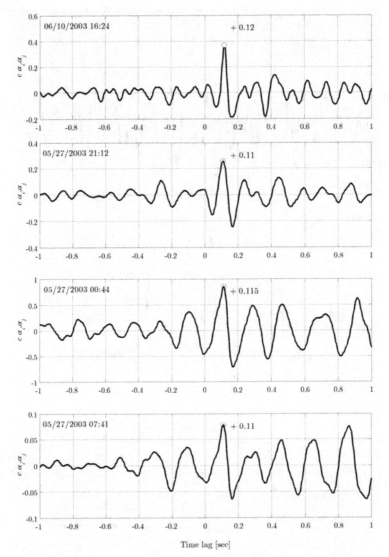

Time lag [sec]

The global-local optimization scheme is next employed in series for events 1-4, and the optimal soil configuration at station *iwth*04 is estimated by averaging the individual best-fit profiles. The effectiveness of the algorithm convergence is illustrated in Figure 9 for event 1. In particular, the performance of the proposed algorithm is compared with (a) a

Figure 11. (a) Mean inverted shear wave velocity structure at stations 1-5 using 4 low intensity events (enumerated 1-4 in decreasing SNR order); also shown the shear wave velocity from geophysical site investigation at the corresponding stations (left). Mean attenuation inverted structure (middle), and mean density inverted structure (right); also shown, are empirical estimates of local attenuation and density, used in conjunction with the local site investigation information to seed the first parental generation with potentially good solutions. (b) Comparison of surface observations and synthetics computed on the basis of the average global optimum profiles at the corresponding stations, for events 1-4.

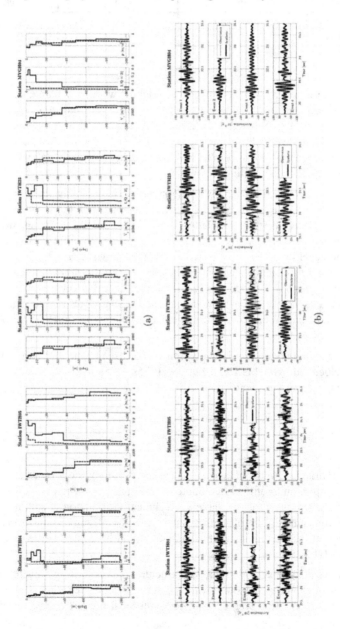

Table 2. Borehole-to-surface travel time at stations 1-5, estimated on the basis of the lag that corresponds to the maximum seismogram cross-correlation: Comparison of empirical and synthetic (computed using the average global optimum profile) ground motion

Station Code	t αsur,αbor (recording)	t αsur,αbor (synthetic)	Error [%]
IWTH04	0.1200	0.1150	4.17
IWTH05	0.1063	0.1038	2.35
IWTH18	0.1250	0.1188	4.96
IWTH23	0.0675	0.0693	2.67
MYGH04	0.1375	0.1365	0.73

global-local scheme based on an alternative orthogonal wavelet decomposition (5th level Daubecies wavelet); and (b) the stochastic (global) part of the algorithm, i.e. a traditional genetic algorithm. As can be readily seen, the selected wavelet decomposition and local improvement operator lead to faster and more effective convergence. Note also that Bersini and Renders (1994) report similar results, when comparing the hybrid optimization scheme with traditional simulated annealing techniques.

Successively, we illustrate the accuracy of the inversion in depicting the exact S-wave arrival, by comparing the apparent velocity at station *iwth*04 with the weighted average of the optimum shear wave velocity profile. As shown in Section 2.3.2, the apparent velocity of propagation, v_a, between stations i and j may be estimated from the surface-to-borehole cross correlation coefficient as: $v_a = d/t_d$, where d is the known separation distance between stations i and j, and t_d is the time lag where the cross-correlation function reaches a major peak. For small angles of incidence with respect to the vertical direction, the apparent shear wave velocity is a very close approximation to the actual quantity, v. Figure 10 illustrates the surface-to-borehole cross-correlation coefficient as a function of time lag, for the selected events at site *iwth*04.

Using the average borehole-to-surface travel time (borehole depth at -109m), the apparent shear wave velocity at *iwth*04 is v_a = 109m /0.11375s = 958 m/s $\approx v$. Next, we define the weighted average shear wave velocity as:

$$\bar{v} = \sum_{i=1}^{nlay} v_i h_i \left/ \sum_{i=1}^{nlay} h_i \right. \qquad (26)$$

where $i = 1 \ldots nlay$ is the number of layers, v_i is the layer velocity and h_i is the layer thickness. For the average optimum profile, we compute $\bar{v}$ = 937m/s $\simeq v_a$, whereas the same quantity is $\bar{v}$ = 1572m/s when using the available geotechnical information at station *iwth*04 (see Figure 7). The robustness of the algorithm was verified for all stations under investigation, and comparison of the average theoretical and empirical borehole-to-surface travel times in presented in Table 2.

Figure 11a shows the average optimum soil structure, obtained using 4 events in series at each station. Also shown is the shear wave velocity structure reported at each site (e.g., Figure 7 for station *iwth*04), as well as estimates of the local attenuation and density profiles.

Figure 12. Weighted average shear wave velocity at stations 1-5, computed on the basis of the average global optimum profiles, and categorization of the site conditions based on the NEHRP site classification system

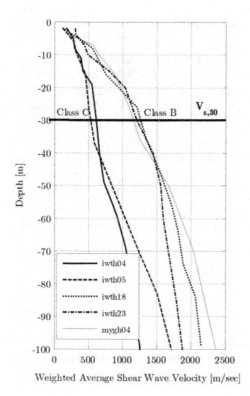

Note that since genetic algorithms can iteratively improve existing models (i.e., estimates obtained through heuristic methods and/or current practices), the available geotechnical information—if any—is also used to seed the first parental population, which is randomly generated within the parameter vector boundaries, with potentially good solutions, and accelerate convergence of the stochastic search.

As can be readily seen, the inverted V_s structure does not deviate substantially from the coarsely discretized reported data, despite the refined discretization of the continuum used in our inverse analysis, typically one fourth of the reported layer thickness. Furthermore, as expected, the density structure is very similar to one estimated on the basis of the available geotechnical information. Nonetheless, the inverted attenuation structure near the surface is shown to be very sensitive to the frequency content of the incident motion, with standard deviation of the results being on the order of 100% at the top 20 m of the profiles. This uncertainty is introduced in the attenuation prediction by the stochastic nature of the physical mechanism. The physical problem simulated contains inherently two mechanisms of attenuation, the material absorption (intrinsic) and the propagating energy redistribution (scattering), whereas the inverted attenuation profile accounts for both phenomena through a single, frequency-independent parameter. Even further, the attenuation values close to the surface are quite substantial when compared to published laboratory measurements of low amplitude dynamic soil properties (typically on the order of $1/Q \approx 0.02 - 0.08$ for stiff soil formations).

It should be noted, however, that the *best-fit* average and corresponding standard deviation of the attenuation structure should have—ideally—been computed on the basis of infinitely many observations. Nonetheless, the relatively small number of events used in this case is based both on the target efficiency of the scheme and on the limited availability of quality digital information.

The average optimum shear wave velocity, attenuation and density profiles shown in Figure 11a, are next subjected to the total motion recorded at the borehole level of the corresponding station, during the 4 events under investigation. Figure 11b compares the surface measured and synthetic waveforms, for the 2sec time-window used in the optimization process.

Successively, Figure 12 illustrates the weighted average shear wave velocity profiles that correspond to the *best-fit* average structures, defined for each configuration at the interface of layers i and $i+1$ as:

Figure 13. Energy partition pattern at ground surface, computed for events 1-4 at station iwth04: Comparison between observations and synthetics, computed using the global optimum solution

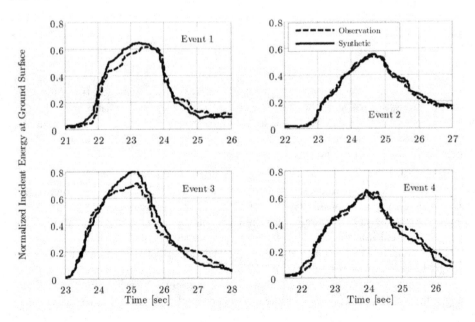

$$\overline{v}\left(d_{i}\right)=\sum_{1}^{i} v_i h_i \bigg/ \sum_{1}^{i} h_i \qquad (27)$$

where d_i is the total depth to the layers interface measured from ground surface, v_i is the layer velocity and h_i is the layer thickness. According to the NEHRP site classification, where categorization of local soil conditions is based on the average shear wave velocity of the surficial 30m ($V_{s,30}$), sites iwth04 and iwth05 are categorized as being Class-C sites (Dense Soil or Soft Rock, 360m/s < V_{S30} < 760m/s), while sites iwth18, iwth23 and myg04 belong to Class-B (Firm-to-Hard Rock, 760m/s < V_{S30} < 1500m/s).

As can be readily seen, the proposed scheme can depict the impedance and attenuation structures, resulting in accurate representation of both the frequency content and amplitude of the recorded surface ground motion. It should be noted, however, that the complexity of the measured process—clearly manifesting in the frequency response of the system—highlights further the stochastic nature of the physical problem, which is not captured by simplified mathematical models. The limitations of our forward numerical simulation, which attempts to approximate 3D strongly heterogeneous systems by means of 1D horizontally stratified homogeneous configurations, can be also illustrated by comparing the synthetic and measured rate of incident seismic energy at ground surface.

In the time-domain, the energy of the direct wave can be expressed as the summation of the squared amplitudes in the window containing the direct arrival (Frankel & Wennerberg, 1987; Korn, 1997). To estimate the normalized seismic energy, we first compute the wave energy in an appropriate time window as follows:

$$E_i\left(t_m\right)=\int_{t_1}^{t_2} A_i^2(t)\, dt \qquad (29)$$

where $A_i^2(t)$ is the square wave amplitude at the i-th station at time t, and t_m is the central time (mid-point) of the window between t_1 and t_2. Successively, the energy estimated in each window is normalized by the total energy of the waveform, starting from the S-wave arrival to the end (90% of the total incident energy) of the waveform under consideration. Therefore, the normalized energy arriving at station i at time t_m is defined as:

$$\overline{E}_i\left(t_m\right)=\frac{\displaystyle\int_{t_1}^{t_2} A_i^2(t)\, dt}{\displaystyle\int_{t_S}^{T_{90}} A_i^2(t)\, dt} \qquad (30)$$

where t_S is the S-wave arrival time, and T_{90} is the lapse time that corresponds to 90% of the total incident wave energy. The normalized energy time history, computed by sliding the window (t_1, t_2) across the waveform, is referred to as the energy partition pattern (Sivaji Nidhizawa, Kitagawa, & Fukushima, 2002).

Figure 13 plots the energy partition pattern for events 1-4 at station iwth04. Comparison of the measured and optimal synthetic surface motions shows good agreement both of the incident energy rate and the maximum energy level at ground surface. Nonetheless, the synthetic

Figure 14. (a) Borehole and (b) surface acceleration spectrograms of event 1, recorded at station iwth04; (c) transfer function amplitude spectrogram at the same location. Note that the S-wave arrival is estimated at t=22.80s, corresponding to the end of the coherent site response region of the spectrogram.

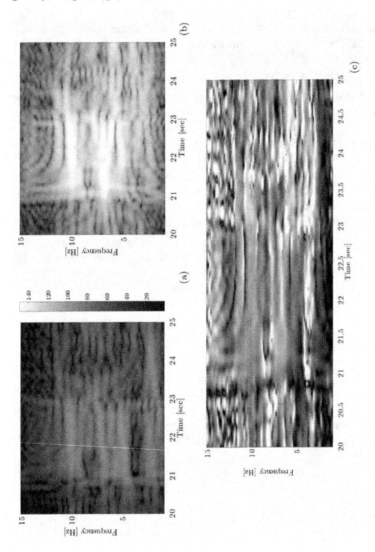

*Figure 15. Surface to borehole cross-correlation coefficient for the event described above;
the coefficient corresponding to the S-arrival window used in the inversion scheme is also
indicated*

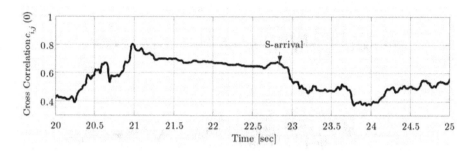

*Figure 16. Inverted shear wave velocity structure V_s [m/sec] (left), attenuation Q (middle)
and density ρ [Mg/m³] (right) at site iwth04 using events 1-4 and a 10s optimization window
(mean (μ) ± standard deviation (σ)); also shown with dashed line is the average inverted
structure at the same station, using a 2 sec optimization window and events 1-4*

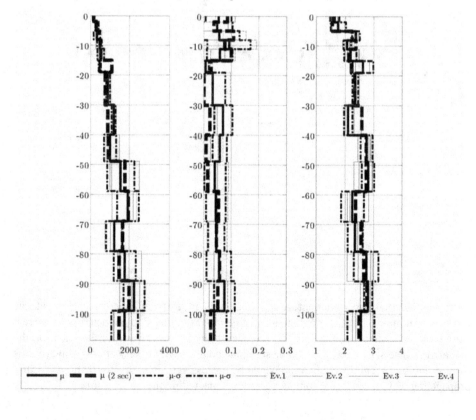

Figure 17. Surface observations and global optimum synthetics at station iwth04, for events 1-4: Comparison of the results obtained using 10s and 2s time-windows in the objective function definition

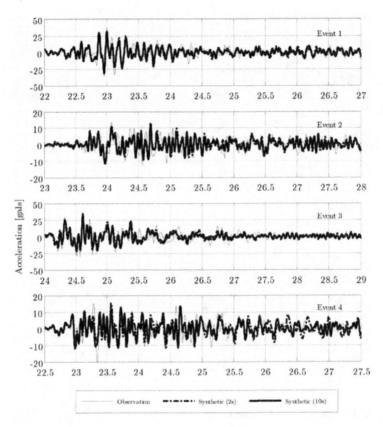

energy partition has larger gradient at early times, indicative of the coherent energy incidence for the horizontally stratified homogeneous medium, as opposed to the heterogeneous real configuration. Even assuming vertical incidence at the borehole level for the latter: (a) the first arrival of seismic energy would be still delayed due to multiple scattering, and (b) this phenomenological energy attenuation would be successively counteracted by nonvertically propagating waves, scattered from adjacent locations within the same profile.

Figure 18. Average coherence function spectrum of surface observations and global optimum synthetics at station iwth04, for events 1-4: Comparison of the results obtained using 10s and 2s time-windows in the objective function definition

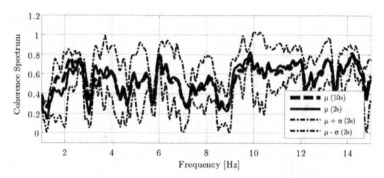

Sensitivity Analysis of the Inverted Attenuation Structure

As illustrated above, the *best-fit* attenuation structure is strongly dependent on the frequency content of the seismogram used in the inversion scheme, stemming from the stochastic nature of the physical process. In particular, uncertainties introduced in the measured process by means of material heterogeneities and incident motion variability, result to observations where the elastic site response is *time-* and *frequency-content* dependent. Based on results illustrated in Section 5, such limitations of the mathematical model are strongly reflected on the inverted attenuation profile, unlike the estimated impedance structure.

To illustrate this concept, Figure 14 depicts the temporal variability of the empirical transfer function recorded at station *iwth*04, during the $M_L = 4.8$ aftershock (06/10/2003 16:24) of the Miyagi earthquake. For this purpose, we apply a 2.2s, 5% tapered sliding window, and compute the surface and borehole frequency spectra and the corresponding empirical transfer functions. The latter is estimated as the amplitude of the complex Fourier spectral ratio of the surface to borehole motion. As can be readily seen, the relatively uniform site response, which corresponds to the first S-wave arrival ($t = 22.80$s), is followed by a highly erratic region. Unlike the former, the response is here governed by late arrivals of scattered energy interacting with waves trapped within the strongly heterogeneous, near surficial layers.

To identify the time window for which the shifted borehole and surface motion show maximum coherency, we compute the cross-correlation coefficient at zero time-lag (i.e., $c_{a_i,a_j}(0)$) using a 2.0s sliding time-window, and plot its temporal variation in Figure 15. The local maximum of the coherence function in the region of the S arrival, corresponds to the time-window that starts exactly at the S-wave arrival, and from this point on, the cross-correlation degrades due to scattering phenomena that *contaminate* the uniform response.

Figure 19. Inverted shear wave velocity structure (left), attenuation (middle) and density (right) at site iwth04 using the average frequency response and travel time as objective functions; also shown are the inverted structures at the same station, obtained by using 2s and 10s windows in the wavelet domain, as well as the reported shear wave velocity at the same station

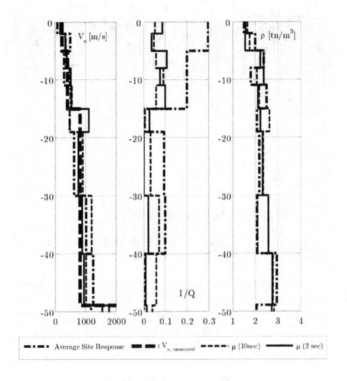

Successively, we investigate the effects of the time-window duration, namely the trade-off between the available information on the process to be simulated, and the uncertainty introduced by nonsimulated phenomena. As shown above, we overcome this uncertainty that manifests in the inverted attenuation structure, by averaging multiple *best-fit* models at the same station, thus obtaining a robust global optimum solution. In the ensuing, we attempt to achieve a more stable local average site response by computing the inverted soil structure at station *iwth*04 using a 10s time-window initiating at the S-wave arrival.

Figure 16 shows the average optimum soil structure, obtained using the 10s time window and events 1-4 in series. Also shown is the standard deviation of the ensemble of inverted profiles, as well as the shear wave velocity structure reported at the site (see Figure 7) and the optimum structure computed using a 2s time-window (see Figure 11). Successively, Figure 17 compares the surface recorded and synthetic time-histories, for the 10s and 2s inversion time-windows. As can be readily seen, results obtained by means of the two inversions are

Figure 20. Comparison of: (a) the optimum synthetic and observation ground surface motion for event 1 at station iwth04, and (b) the theoretical transfer function obtained using the average site response and travel time, the one obtained using a 2s window wavelet-domain inversion, and the average frequency response computed using 10s-windows of 28 events recorded at the site

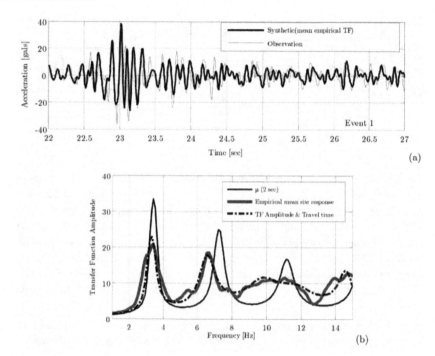

in very good agreement, supported also by the optimum synthetic-observation cross-correlation computed for events 1-4 (see Table 2).

Finally, Figure 18 illustrates the consistency of the two approaches in the frequency domain, by comparing the optimum synthetic-to-observation coherence spectra computed using sliding 5% tapered time-windows with 25% overlap, and duration 2s and 10s correspondingly. In particular, we plot the magnitude-squared coherence estimate $\mathbb{C}_{xy}$ of the observation, x, and synthetic, y, signals using Welch's averaged, modified periodogram method, as follows:

$$\mathbb{C}_{xy} = \frac{\left| P_{xy}\left(f\right) \right|^2}{P_{xx}\left(f\right) \times P_{yy}\left(f\right)} \qquad (31)$$

where $P_{xx}\left(f\right), P_{yy}\left(f\right)$ are the power spectral density functions of x and y correspondingly, and $P_{xy}\left(f\right)$ is the cross-power spectral density function of the two signals. To obtain a global estimate of the signal's similarity, the coherence spectra shown have been also averaged over the individual functions computed for events 1-4.

Figure 21. Empirical transfer functions for sites iwth04 (left) and mygh04 (right), estimated using downhole array seismogram data

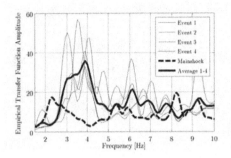 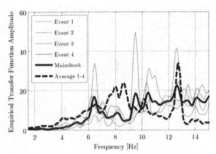

Despite the apparent robustness of the inverted structure to the time-window duration used, the average attenuation of the surficial 30m is $\bar{Q}_{30m}^{10s} = 22$ for the 10s window, whereas the same quantity is estimated $\bar{Q}_{30m}^{2s} = 15$ for the 2s inversion time-window. This can be readily explained if we consider that the additional wavefield information included in the 10s-window case, allows for late arrivals of multiple scattered waves to be included in the simulated system. Since we account both for material and scattering attenuation by means of a common mechanism in our simulations, elongation of the inversion window is interpreted by the algorithm as *less attenuated* energy, hence higher estimated Q in the surficial, strongly heterogeneous layers.

Nonetheless, the deterministic forward operator can approximate the order of magnitude of attenuation in strongly heterogeneous, near-surficial layers with acceptable accuracy for engineering purposes. Furthermore, the average attenuation of the profile at station *iwth04* is estimated as $\bar{Q}_{109m}^{10s} = 19$ for the 10s window, and $\bar{Q}_{109m}^{2s} = 21$ for the 2s window, justifying the agreement of the computed responses in time and frequency. On the other hand, the impedance structure is shown to be rather insensitive to small-scale heterogeneity scattering effects, resulting in weighed average shear wave velocity $\bar{v}_{30m}^{2s}$ 637 m/s $\approx$ 606m/s $= \bar{v}_{30m}^{10s}$ for the top 30m, and $\bar{v}_{109m}^{2s} = 937$m/s ≈ 958m/s $= \bar{v}_{109m}^{10s}$ for the entire profile.

To complete the parametric investigation on the attenuation structure sensitivity to the inversion objective function, we compute the optimum profile at station *iwth04* using (a) the average

Table 3. Optimum synthetic-to-observation normalized cross-correlation at site iwth04 for events 1-4: Comparison of the wavelet domain optimization scheme with 2s and 10s time-windows, and the borehole-to-surface average travel time and frequency site response as objective functions

	Event 1	Event 2	Event 3	Event 4
2s window	0.79516	0.6144	0.78051	0.48173
10s window	0.70232	0.56498	0.69949	0.60906
Average TF	0.75992	0.55174	0.68579	0.51451

Figure 22. Equivalent linear shear wave velocity structure at stations 1-5: Comparison between forwardly estimated configuration upon convergence of the modified iterative scheme, and inversely determined model by means of nonlinear site response optimization

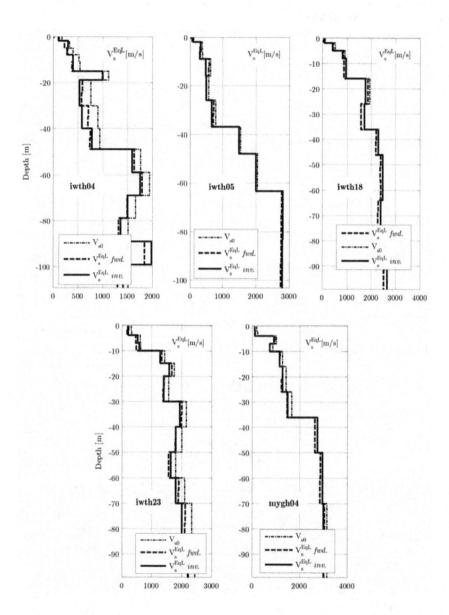

elastic site response in the frequency domain (estimated using 10s time-windows for the ensemble of 28 low-amplitude seismograms recorded at the station) as objective function of the genetic algorithm, and (b) only the average borehole-to-surface travel time as objective function of the local hill-climbing operator. Based on the stability of the estimated travel time illustrated for 4 events in Figure 10, and the erratic behavior of the empirical transfer functions in the high-frequency region, we thus attempt to further minimize the effects of the stochastic nature of the physical process.

Figure 23. Strong motion frequency response at sites 1-5: Comparison between the forward frequency-dependent equivalent linear, the inverse equivalent linear, and the empirical nonlinear site response

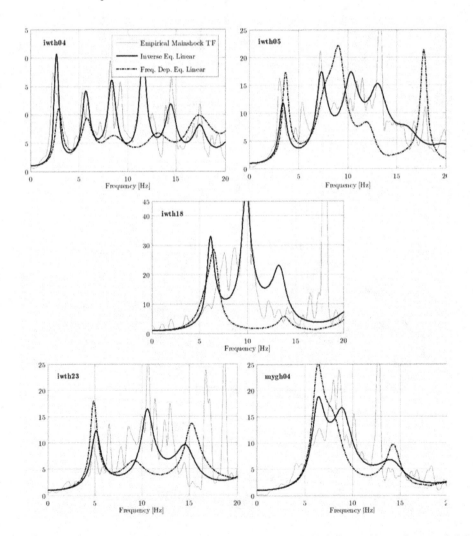

Figure 19 shows the global optimum soil structure obtained, compared to the average inverted structure of the original (2s window) and alternative (10s window) hybrid optimization schemes. As can be readily seen, the inverted velocity and density structures are in good agreement, illustrating the effectiveness of the algorithm to capture the up-going and down-going reflections and refractions of the seismic energy within the multiple horizontally stratified soil layers.

The attenuation structure estimated for the average site response, however, deviates from the two previous approaches, especially in the near-surficial layers. In particular, the average attenuation of the top 30m is now estimated as $\bar{Q}_{30m}^{average} = 6$, substantially larger than the value predicted above. This phenomenon is purely attributed to the averaging process, which smoothes the site response and removes the stochastic nature of measured data, partially reducing the amplitude of the low-frequency components, but mainly broadening the high-frequency region of the spectrum. As a result, accommodation of a less erratic response through an optimization scheme is translated to increased near-surficial attenuation.

Nonetheless, the optimum synthetic-to-observation cross-correlation is comparable to the values obtained above for the 2s- and 10s-window wavelet-domain inversion processes, as shown in Table 3 for the 4 events under investigation. A typical example is shown in Figures 20a and 20b, namely the synthetic surface motion predicted using the best-fit model for event

Figure 24. Cumulative results on the frequency-dependent shear modulus reduction distribution with depth, as estimated at f = 0Hz for stations 1-5 upon convergence of the modified iterative scheme. Note that for comparison, depth is here shown normalized by the total surface-to-borehole depth of the corresponding station.

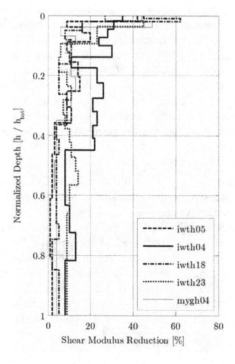

1, and the corresponding theoretical transfer function, compared to the one obtained for the global optimal model of the original hybrid optimization scheme (2s window).

Strong Motion Data Inversion

In this section, the hybrid inversion algorithm is employed to estimate the effects of strong ground motion on the dynamic soil response at sites 1-5, using the mainshock (5/26/2003, 18:24 GMT) downhole array seismogram recordings. It is well known that soil exhibits nonlinear behavior when undergoing time varying deformations caused by intense seismic loading, manifested in site response through shift of the fundamental frequency of the profile and hysteretic material energy absorption. Typical effects of the nonlinear soil behavior under transient loading can be readily observed by comparing the empirical nonlinear site response at stations *iwth*04 and *mygh*04 with the average empirical linear site response computed using the ensemble of available aftershocks at the corresponding locations (Figure 21a and 21b respectively). Also shown are the empirical transfer functions of the individual events. Note that the erratic frequency response of these events illustrates that the averaging process of the elastic site response eliminates the stochastic nature of seismic wave propagation caused by near-surficial heterogeneities.

Successively, the proposed algorithm is applied using the mainshock borehole and ground surface recordings. The nonlinear cyclic material response experienced by the surficial layers at sites 1-5 (e.g., Figure 21) implies that during the M_w 7.0 mainshock, the dynamic soil properties were being continuously adjusted to the instantaneous levels of strain imposed by strong incident motion. Acknowledging that a linear forward operator wouldn't be suitable to invert simultaneously for the amplitude and phase angle of a nonlinear recorded process, the objective functions of the algorithm were modified to estimate the optimum equivalent linear shear wave velocity and hysteretic material damping using the nonlinear site response in the frequency domain as objective function of both the genetic algorithm and the local hill-climbing operator. In this case, convergence of the algorithm was accelerated by seeding the first parental generation with the global optimum elastic structure estimated above. Note that by means of the strong motion site response inversion, we estimate an equivalent linear layered medium whose impedance structure is consistent with the average frequency response computed on the basis of the mainshock recordings.

To assess the validity of the strong motion inversion results, the low-strain inverted soil structures at stations 1-5 were subjected to the mainshock time history (05/26/03 18:24GMT), and the surface ground motion was estimated by means of the modified equivalent linear method by Assimaki and Kausel (2002). By contrast to the original computational scheme proposed by Seed and Idriss (1969), this method accounts for the amplitude-dependent nature of dynamic soil properties by assigning frequency-dependent modulus degradation and damping. For the latter, the dynamic soil properties by Vucetic and Dobry (1992) for cohesive soils with PI = 10-20% were selected, due to lack of detailed geotechnical information at the site. For further details on the forward and inversely estimated equivalent linear structure, the reader is referred to Assimaki and Steidl (2005).

Results are shown in Figure 22, where the inverted shear wave velocity distribution is compared to the strain-compatible profile corresponding to the last iteration of the equivalent linear algorithm. The consistency of the forward and inverse analysis of downhole strong motion recordings, illustrates that inversion of downhole strong motion seismograms can be used to assess the equivalent linear impedance of relatively stiff, cohesive sites, thus inferring the degree of nonlinearity exhibited by the configuration when subjected to high amplitude cyclic shear motion.

It should be noted, however, that the variation of inverted attenuation with depth was found both to deviate substantially from the high-strain hysteretic material damping, and to depend strongly on the bounds assigned to the stochastic search space. This result stems from the multitude of attenuation mechanisms involved in the physical mechanism of nonlinear site response. While the hysteretic low-strain material damping and high-frequency scattering mechanisms already resulted to instabilities in the inverted elastic attenuation structure, the linear forward operator is here forced to account for strain-induced heterogeneities of the surficial layers as well; the distribution of these heterogeneities, which characterize the spatial variability of the nonlinearly responding configuration, is controlled by the spatially varying low-strain impedance structure and its effects on the direction of propagation of incident seismic waves.

Figure 23 compares the frequency response of the inverted equivalent linear impedance structure shown in Figure 22 to (a) the empirical nonlinear site response defined as the surface-to-borehole Fourier amplitude spectral ratio, and (b) the frequency-dependent equivalent linear site response at the last iteration of the algorithm. As can be readily seen, the equivalent linear impedance structure consistency is reflected on the fundamental frequency agreement of the configurations, while the amplitude deviation stems from the hysteretic material damping incompatibility between the forward and inversely estimated equivalent linear attenuation structures.

Based on the aforementioned conclusions, the inverted hysteretic damping profile was not considered reliable for the purpose of this study. Conversely, the equivalent linear inverse impedance was found to be insensitive both to the bandwidth and variation of deterministic search space bounds with depth, illustrating robustness of the estimated structure.

Finally, Figure 24 shows the percentage of shear modulus reduction as a function of the normalized borehole depth h / h_{tot}, estimated by means of the mainshock inversion at sites 1-5. Clearly, the nonlinear cyclic shear response is manifesting in the upper 20% of the total profile thickness, while for the stiff site conditions under investigation, the secant modulus reduction of the surficial layers is quite considerable (by average 30%-40%) and elicits strong nonlinearity of the transient material response, a phenomenon that was consistently approximated by means of the frequency-dependent algorithm and the hybrid optimization scheme employed for the mainshock recordings.

Conclusion

We have presented a hybrid optimization scheme for downhole array seismogram inversion, employed to provide estimates of the impedance and attenuation structures at 5 downhole arrays in Japan using low-amplitude aftershock recordings from the Sanriku-Minami earthquake.

The seismogram inversion algorithm combines a traditional genetic algorithm defined in the wavelet domain, with a local hill-climbing operator, namely a nonlinear least-square optimization, in the frequency domain. As a result, it has the following advantages: (a) using normalized wavelet decomposition of the ground surface seismograms, equal weight is assigned across all frequency bands of interest and high observation-synthetic cross-correlation is achieved throughout the 1-15Hz frequency range; and (b) while locally improving the parental population of each generation in the genetic algorithm, the convergence rate of the inversion scheme is considerably accelerated when compared to traditional genetic algorithms.

To eliminate the observation error propagation and the computational error arising from the fact that the forward model does not account for small-scale material heterogeneities, the inversion process is repeated in series for multiple recordings selected on the basis of their signal quality. Successively, the global optimum solution is defined as the average structure of the individual *best-fit* models. Results show that the hybrid optimization scheme can efficiently provide estimates of the shear wave velocity, attenuation and density profiles of near-surficial soil formations using downhole array recordings, reflected in the illustrated agreement between the synthetic response of the global optimal profiles and the corresponding ground surface observations.

The applicability of the inversion scheme has been also illustrated for strong ground motion. It has been shown that high intensity downhole array seismogram inversion enables the estimation of the nonlinearity extent exhibited by near-surficial formations during strong transient loading. Strong motion results were also compared with the reduced dynamic stiffness, estimated upon convergence of the frequency-dependent equivalent linear algorithm.

Through a comprehensive analysis of weak and strong motion acceleration time-histories, it has been shown that by means of optimization algorithms, downhole array recordings may provide valuable information on the elastic and nonlinear in-situ material response under true seismic loading, which can be used to enhance the understanding of complex site response phenomena, to assess the effectiveness of simplified and elaborate nonlinear site response algorithms, and ultimately to establish a unified methodology that will account for site effects in current state-of-practice based on physical fundamental principles instead of phenomenological, site specific approximations.

Acknowledgments

This research was partially supported by the Southern California Earthquake Center No. 0878; SCEC is funded by NSF Cooperative Agreement EAR-0106924 and USGS Cooperative Agreement 02HQAG0008. Additional support was provided by the Institute for Crustal Studies No. 0691, at the University of California, Santa Barbara, CA 93105. The author would also like to acknowledge the significant contribution of the following individuals: Associate Research Seismologists Dr. J. Steidl and Dr. P. C. Liu, and Prof. R. Archuleta from the Institute of Crustal Studies, U.C. Santa Barbara, CA.

References

Abercrombie, R. E. (1995). Earthquake source scaling relationships from -1 to 5 M_L using seismograms recorded at 2.5km depth. *Journal of Geophysical Research, 100*(24), 15-24, 36.

Aguirre, J., & Irikura, K. (1997). Nonlinearity, liquefaction, and velocity variation of soft soil layers in Port Island, Kobe, during the Hyogoken Nanbu earthquake. *Bulletin of the Seismological Society of America, 87,* 1244–1258.

Assimaki, D., & Steidl, J. H. (2005). Inverse analysis of weak and strong motion downhole array data from the Mw7.0 Sanriku-Minami Earthquake. Soil *Dynamics and Earthquake Engineering,*

Assimaki, D., & Kausel, E. (2002). An equivalent linear algorithm with frequency—and pressure—dependent moduli and damping for the seismic analysis of deep sites. *Soil Dynamics and Earthquake Engineering, 22,* 959-965

Bendat, J. S., & Piersol, A. G. (1996). *Engineering applications of correlation and spectral analysis* (2nd ed.). Wiley.

Bersini, H., & Renders, B. (1994). *Hybridizing genetic algorithms with hill-climbing methods for global optimization: Two possible ways.* Paper presented at IEEE International Symposium Evolutionary Computation, Orlando, FL.

Bonilla, L. F., Steidl J. H., Lindley G. T., Tumarkin A. G., & Archuleta R. J. (1997). Site amplification in the San Fernando Valley, California: Variability of site-effect estimation using the *S*-wave, coda, and H/V methods. *Bulletin of the Seismological Society of America, 87,* 710–730.

Boore, D. M., & Joyner W. B. (1997). Site amplification for generic rock sites. *Bulletin of the Seismological Society of America, 87,* 327–341.

Borcherdt, R. D. (1970). Effects of local geology on ground motion near San Francisco Bay. *Bulletin of the Seismological Society of America, 60,* 29–61.

Brune, J. N. (1970). Tectonic stress and the spectra of seismic shear waves from earthquakes. *Journal of Geophysical Research, 75*(4), 997-5009.

Chui, C. K. (1992). *Wavelets: A tutorial in theory and applications (Wavelet analysis and its applications, vol.2)*. San Diego, CA: Academic.

Daubechies, I. (1988). Orthonormal basis of compactly supported wavelets. In *Proceedings of the CBMS-NSF Regional Conference Series in Applied Mathematics* (Vol. 61, p. 357). Philadelphia: Society for Industrial and Applied Mathematics.

Daubechies, I. (1992). *Ten lectures on wavelets*. Philadelphia: Society for Ind. and Applied Mathematics.

Davis, L. (1991). *The handbook of genetic algorithms*. New York: Van Nostrand Reingold.

Earthquake Spectra. (2000, December). *1999 Kocaeli, Turkey, earthquake reconnaissance report, 16*(Suppl. A).

Field, E. H. (1996). Spectral amplification in a sediment-filled valley exhibiting clear basin-edge induced waves. *Bulletin of the Seismological Society of America, 86,* 991–1005.

Field, E. H., & Jacob, K. H. (1995). A comparison and test of various site response estimation techniques, including three that are non reference-site dependent. *Bulletin of the Seismological Society of America, 85,* 1127–1143.

Foufoula-Georgiou, E., & Kumar, P. (1994). *Wavelets in geophysics*. San Diego, CA: Academic.

Frankel, A., & Wennerberg, L. (1987). Energy-flux model of seismic coda: Separation of scattering and instrinsic attenuation. *Bulletin of the Seismological Society of America, 77,* 1223-1251.

Goldberg, D. (1989). *Genetic algorithms in search, optimization, and machine learning*. Addison Wesley.

Goupillaud, P. A., Grossmann, A., & Morlet, J. (1984). Cycle-octaves and related transforms in seismic signal analysis. *Geoexploration, 23,* 85-102.

Grossmann, A., & Morlet, J. (1984). Decomposition of handy functions into square integrable wavelets of constant shape. *SIAM Journal of Mathematical Analysis, 15,* 723-736.

Gustafsson, F. (1996). Determining the initial states in forward-backward filtering. *IEEE Transactions on Signal Processing, 44,* 988-993.

Hartzell, S. H. (1996). Site response estimation from earthquake data. *Bulletin of the Seismological Society of America, 82,* 2308–2327.

Hartzell, S. H., et al. (1996a). The 1994 Northridge, California, earthquake: Investigation of rupture velocity, rise time and high-frequency radiation. *Journal of Geophysical Research, 101,* 20,091-020,108.

Hartzell, S. H., Leeds, A., Frankel, A., & Michael, J. (1996b). Site response for urban Los Angeles using aftershocks of the Northridge earthquake. *Bulletin of the Seismological Society of America, 86,* 5168–5192.

Holland, J. (1975). *Adaptation in natural and artificial systems*. Ann Arbor: The University of Michigan Press.

Holschneider, M. (1995). *Wavelets: An analysis tool*. New York: Oxford University Press.

Houck, C. R., Joines, J.A., & Kay, M.G. (1996). Comparison of genetic algorithms, random restart and two-opt switching for solving large location-allocation problems. *Computers and Operations Research, 23,* 587-596.

Housner, G. W. (1990). Competing against time. *Report to Governor George Deukmejian from the Governor's Board of Inquiry on the 1989 Loma Prieta Earthquake.*

Iai, S., Morita, T., Kameoka, T., Matsunaga, Y., & Abiko, K. (1995). Response of a dense sand deposit during 1993 Kushiro-Oki earthquake. *Soils and Foundations, 35,* 115–131.

Kato, K., Aki, K., & Takemura, M. (1995). Site amplification from coda waves: Validation and application to S-wave site response. *Bulletin of the Seismological Society of America, 85,* 467–477.

Kinoshita, S. (1998). Kyoshin Net (K-Net). *Seismological Research Letters, 69,* 309–332.

Kokusho, T., Sato, K., & Matsumoto, M. (1996). Nonlinear dynamic soil properties back-calculated from strong motions during Hyogoken-Nambu earthquake. In *Proceedings of the 11th World Conference on Earthquake Engineering*, Acapulco, Mexico (Paper No. 2080).

Korn, M. (1997). Modelling of teleseismic P coda envelope: Depth-dependent scattering and deterministic structure. *Physics of Earth and Planetary interior, 104,* 23-36.

Madariaga, R. (1976). Dynamics of an expanding circular fault. *Bulletin of the Seismological Society of America, 66,* 639-667.

Mallat, S. (1989a). A theory for multiresolution signal decomposition: The wavelet representation. *IEEE Transactions on Pattern Analysis and Machine Intelligence, 11,* 674-693.

Mallat, S. (1989b). Multifrequency channel decomposition of images and wavelet models. *IEEE Transactions on Acoustic Speech and Signal Analysis, 37,* 2091-2110.

Margheriti, L., Wennerberg, L., & Boatwright, J. (1994). A comparison of coda and S-wave spectral ratio estimates of site response in the southern San Francisco Bay area. *Bulletin of the Seismological Society of America, 84,* 1815–1830.

Meyer, Y. (1992). *Wavelets and operators.* New York: Cambridge University Press.

Michalewicz, Z. (1994). *Genetic algorithms + data structures = evolution programs.* New York: Springer-Verlag.

Morlet, J., Arens, G., Fourgeau, E., & Giard, D. (1982a). Wave propagation and sampling theory: I. Complex signal and scattering in multilayered media. *Geophysics, 47,* 203-221.

Morlet, J., Arens, G., Fourgeau, E., & Giard, D. (1982b). Wave propagation and sampling theory: II. Sampling theory and complex waves. *Geophysics, 47,* 222-236.

Ohsaki, Y. (1969). *The effects of local soil conditions upon earthquake damage.* Proceedings of the Seventh ICSMFE Specialty Session on Soil Dynamics.

Rosenblueth, E. (1960). Earthquake of 28 July 1957 in Mexico City. *Proceedings Second WCEE, 1,* 359-379.

Sato, K., Kokusho, T., Matsumoto, M., & Yamada, E. (1996). Nonlinear seismic response and soil property during strong motion. In *Special issue of soils and foundations in geotechnical aspects of the January 17, 1995, Hyogoken-Nambu earthquakes* (pp. 41–52). Journal Soils and Foundations Issued by Japanese Geotechnical Society.

Satoh, T., Fushimi, M., & Tatsumi, Y. (2001). Inversion of strain-dependent nonlinear characteristics of soils using weak and strong motions observed by borehole sites in Japan. *Bulletin of the Seismological Society of America, 91*(2), 365-380.

Satoh, T., Kawase, H. & Sato, T. (1995). Evaluation of local site effects and their removal from borehole records observed in the Sendai region, Japan. *Bulletin of the Seismological Society of America, 85,* 1770–1789.

Seed, H. B. & Idriss, I. M. (1969). Influence of soil conditions on ground motions during earthquakes. *Journal of Soil Mechanics and Foundations Division, ASCE, 95*(1), 99–137.

Seed, H. B., & Idriss, I. M. (1970). Analyses of ground motions at Union Bay, Seattle, during earthquakes and distant nuclear blasts. *Bulletin of the Seismological Society of America, 60,* 125–136.

Seed, H. B., & Romo, M. P.(1987). *Relationships between soil conditions and earthquake ground motions in Mexico City in the earthquake of September 19, 1985* (Rep. No. EERC-87-15). Berkeley: University of California.

Seed, H. B., Whitman, R. V., Dezfulian, H., Dobry, R., & Idriss, I. M. (1972). Soil conditions and damage in the 1967 Caracas earthquake. *Journal of Soil Mechanics and Foundations, ASCE, 98,* 787-806

Seed, H. B., Whitman, R.V., Dezfulian, H., Dobry, R., & Idriss, I. M. (1972). Soil conditions and damage in the 1967 Caracas earthquake. *Journal of Soil Mechanics and Foundations, ASCE, 98,* 787-806.

Sivaji, C., Nidhizawa, O., Kitagawa, G., & Fukushima, Y. (2002). A physical-model study of the statistics of seismic waveform fluctuations in random heterogeneous media. *Geophysical Journal International, 148,* 575-595.

Snieder, R., & Trampert, J. (1999). Inverse problems in geophysics. In A. Wirgin (Ed.), *Wavefield inversion* (pp. 119-190). New York: Springer-Verlag.

Soils and Foundations. (1996, January). *Special issue on geotechnical aspects of the January 17, 1995, Hyogoken-Nambu earthquake.*

Steidl, J. H., Tumarkin, A. G., & Archuleta, R. J. (1996). What is a reference site? *Bulletin of the Seismological Society of America, 86,* 1733–1748.

Stoffa, P. L., & Sen, M. K. (1991). Nonlinear multiparameter inversion using genetic algorithms: Inversion of plane-wave seismograms. *Geophysics, 56,* 1794-1810.

Su, F., Anderson, J. G., Brune, J. N., & Zeng, Y. (1996). A comparison of direct S-wave and coda wave site amplification determined from aftershocks of Little Skull Mountain earthquake. *Bulletin of the Seismology Society of America, 86,* 1006–1018.

Tezcan, S. S., Yerlici, V., & Durgunoglou, H. T. (1979). A reconnaissance report for the Romanian earthquake of 4 March 1977. *Engineering and Structural Dynamics, 6,* 397-421

Trampert, J. (1998). Global seismic tomography: The inverse problem and beyond. *Inverse Problems, 14,* 371-385.

Vucetic, M., & Dobry, R. (1991). Effect of soil plasticity on cyclic response. *Journal of Geotechnical Engineering, ASCE, 117,* 89-107.

Wen, K., Beresnev, I., & Yeh, Y. T. (1994). Nonlinear soil amplification inferred from downhole strong seismic motion data. *Geophysical Research Letter, 21,* 2625-2628.

Wornell, G. W. (1995). *Signal processing with fractals: A wavelet-based approach.* Englewood Cliffs, NJ: Prentice Hall.

Zeghal, M., & Elgamal, A. W. (1994). Analysis of site liquefaction using earthquake records. *Journal of Geotechnical Engineering, 120*(6), 996–1017.

Chapter XIII

Genetic Algorithms in Structural Identification and Damage Detection

Chan Ghee Koh, National University of Singapore, Singapore

Michael John Perry, National University of Singapore, Singapore

Abstract

Genetic algorithms (GA) have proved to be a robust, efficient search technique for many problems. In this chapter, the latest developments by the authors in the area of structural identification and structural damage detection using genetic algorithms are presented. A GA strategy involving a search space reduction method (SSRM) using a modified genetic algorithm based on migration and artificial selection (MGAMAS) is first used to identify structural properties in multiple degree-of-freedom systems. The SSRM is then incorporated in a structural damage detection strategy using response measurements both before and after damage has taken place. Numerical studies on 10 and 20 degree-of-freedom systems show that a small damage of only 2.5% can be accurately and consistently identified from incomplete acceleration measurements in the presence of 5% input and output noise.

Introduction

Engineering analysis can be broadly categorized as direct analysis and inverse analysis. Direct analysis for structural systems aims to predict structural response (output) for given excitation (input) and known system parameters, whereas inverse analysis deals with identification of structural parameters based on given input and output (I/O) information. The latter may be termed as "structural identification" and falls within the larger domain of system identification. Structural identification can be applied to update or calibrate structural models so as to better predict response and achieve more cost-effective designs. It can also be used for structural health monitoring and damage assessment in a nondestructive way by tracking changes in pertinent structural parameters. This is especially useful for identifying structural damage caused by natural actions such as earthquakes. For structural control applications, identification of actual parameters is essential for effective control.

From computational point of view, structural identification presents a very challenging problem particularly when the system involves a large number of unknown parameters. Besides accuracy and efficiency, robustness is an important issue for selecting the identification strategy. Presently the main hurdle is the lack of a robust and intelligent computational strategy to identify parameters, given limited number of sensors and inevitable noise in reality. Many studies on structural identification have adopted classical methods such as extended Kalman filter (e.g., Hoshiya & Sutoh, 1993), least squares (e.g., Caravani, Watson, & Thomson, 1977) and maximum likelihood methods. These methods are typically gradient based and point-to-point search. The solutions may converge falsely to a local optimal point rather than the global optimal point, depending largely on the initial guess. On the other end of spectrum, exploration methods such as random search may be used to increase the chance of global convergence but are obviously very time consuming for large systems due to the huge combinatorial possibilities.

A soft-computing approach based on genetic algorithms (GA) is employed in this chapter as the main search engine. Providing a remarkable balance between exploitation of good candidates and exploration by random chances, this method has been shown to possess several crucial advantages over classical methods in the context of structural health monitoring and damage identification. The advantages include enhancement of global convergence by conducting population-to-population search, no requirement of gradient information, relative ease of implementation, convenient use of any measured response in defining the fitness function, and robust self-start feature with random initial guess in a specified search range. Besides, it has a high level of concurrency and is thus suitable for distributed computing. Nevertheless GA cannot be treated as a black box, lest the computational time would be too prohibitive for real problems. Much understanding and additional treatments are needed to make the GA approach work effectively. It is the aim of this chapter to provide an overview of the state-of-the-art development of GA-based structural identification.

The chapter will begin with some background on the development of genetic algorithms and then describe how genetic algorithms can be utilized for the identification of structures. Identifying structural systems poses a difficult problem due to the large number and interdependence of structural parameters, particularly when the mass of the structure is unknown. In addition it is generally not practical to obtain measurements of all degrees of freedom in a structure and any measurement that is obtained will inevitably be contaminated with noise.

The challenge of identifying mass, stiffness and damping properties from incomplete, noisy measurements is thus the focus of this chapter. In recent years efforts have been made to alter the operators and the overall architecture of GA in order to achieve more accurate and robust identification strategies for these problems. One of these identification strategies is presented in this chapter. The structural identification strategy presented works on the whole structure and achieves very good accuracy by progressively reducing the search limits of the structural parameters as the identification proceeds. This search space reduction method, or SSRM (Perry, Koh, & Choo, 2006), uses a modified genetic algorithm as the search engine and has proved very reliable for identification of structural systems. Structural damage detection is a natural extension from structural identification. There are two possibilities when it comes to damage detection. Firstly, damage can be identified with no prior measurement of the undamaged structure, or secondly damage can be identified where measurement of the structure both before and after damage is available. For the first case, we have no choice but to identify the structural properties and compare these to some theoretical values in order to identify the magnitude and location of damage. In this case the SSRM method above can be utilized directly and no additional development is required. For the second case, however, the additional information of the undamaged structure can be utilized in developing an improved strategy. The final section in this chapter therefore describes a damage detection strategy (Koh & Perry, 2005) for the case where we have measurements of the structure both before and after damage has taken place. The strategy assumes that the structural mass, stiffness and damping are unknown, and the damage can be quantified and detected as a change in the stiffness of the damaged member. The strategy uses the SSRM to carry out the identification of the structure and uses the parameters identified for the undamaged structure to guide the identification of the structure after damage has occurred.

Genetic Algorithms

Genetic algorithms are developed based on Darwin's theory of natural selection and survival of the fittest. Darwin observed that individuals with characteristics better suited for survival in their given environment would be more likely to survive to reproduce and have their genes passed on to the next generations. Through mutations, natural selection, and reproduction, the population could evolve and adapt to changes in the environment. The first model of adaptation-processes based on GA was by Holland (1975). Adaptation is regarded a process of progressive variation of structures, leading to an improved performance. He recognized the similarities between natural and artificial systems and sought ways in which the operators acting to shape the development of natural systems could be modelled mathematically. He recognized that operators such as crossing over and mutation that act in natural systems were also present in many artificial systems and proposed that computers could be programmed by specifying what has to be done rather than how to do it. Over the past 3 decades, the area of GA has been widely developed and applied. A basic coding using binary representation and mutation, crossover and reproduction operators formed the early basis for application into mathematical problems. Later as application moved into more complex areas, new coding schemes and operators were developed to adapt to the problems under study. In recent years, efforts have also been made to alter the architecture of GA and to incorporate local search algorithms (Koh, Chen, & Liaw, 2002) to further improve the performance and to help reduce the problems associated with standard GA where a trade off exists between exploration and exploitation of the possible solutions. Many have argued that these new methods deviate from classical GA and as such have come up with new names such as evolution programs and so on in order to acknowledge the deviation from binary encoding and operators. In this chapter, however, the term GA is used as the underlying principle is still the same, although the coding and architecture may not exactly resemble a classical GA.

Unlike classical search engines which develop a single solution, genetic algorithms work on a group of candidate solutions, commonly referred to as a population of individuals. Each individual represents a possible solution to the given problem and the quality of the solution is quantified by use of a fitness function. In a classical GA the individuals are represented as a binary string representing the parameters of interest. Through crossover, mutation and reproduction the overall fitness of the population can improve over time leading to better solutions. In keeping with the evolutionary idea, each cycle of operations is called a generation. In order to introduce the genetic operators and gain understanding of GA a basic example involving the maximization of a mathematical function is described below. Those wishing to read further into simple GAs and how they work are referred to part 1 of the excellent book, *Genetic Algorithms + Data Structures = Evolution Programs*, by Michalewicz (1994).

Consider the problem of maximizing the function $f(x)$ as given in equation (1) over the range of $-50.0 \leq x \leq 50.0$. This function, shown in Figure 1, contains a global maxima at $x = 0$ and would be difficult to solve using classical optimization methods due to the many local maxima about the global solution.

$$f(x) = 0.5 - \frac{\sin^2 x - 0.5}{1 + 0.01x^2} \tag{1}$$

The layout of a simple GA that may be used to maximize this function is shown in Figure 2. In this example a basic binary encoding is used. As the search range is $-50.0 \leq x \leq 50.0$ and assuming we want an accuracy of at least two decimal places a binary string of length 14 is required. This binary number can represent integers from 0 to $2^{14}-1$ and so the binary

Figure 1. Function to be optimised

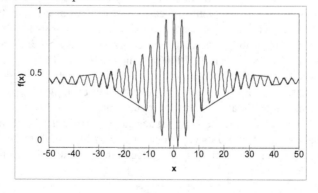

Figure 2. A simple GA

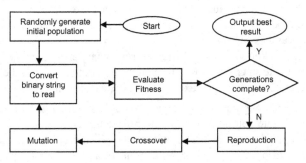

to real conversion is made as shown in equation (2), where I is the integer represented by the binary digits (For example the binary string 10011001010001, represents the integer $2^{13} + 2^{10} + 2^9 + 2^6 + 2^4 + 2^0 = 9809$, and is converted to the real number $x = 9.87$).

$$x = -50 + 100 \times \frac{I}{2^{14} - 1} \qquad (2)$$

The GA begins by randomly generating a set of initial candidate solutions. This is done by randomly assigning either a 0 or a 1 to each bit of each individual within the initial population. The binary strings are converted to real numbers using equation (2) and then the fitness of each solution is calculated. The fitness is a measure of the quality of a given individual. As the objective in this case is to maximize $f(x)$, and $f(x)$ is greater than 0 for all values of x, the function value gives an indication of the quality of the solution and can be used directly as the fitness function. Reproduction is designed to give fitter individuals, more representation in future generations. One simple way to carry out reproduction is using the so called roulette wheel method. Each individual is assigned a selection probability proportional to its fitness and selection is made with replacement until the new population is full. This method encourages multiple selections of fitter individuals and filters out the weakest individuals. Crossover and mutation allow the GA to discover new solutions. In this example a simple crossover is used. A crossover rate determines the chance of an individual being involved in a crossover and once selected two individuals (parents) are paired up for the crossover to take place. The crossover point is randomly selected and the ends of the parents switched to form two new individuals (offspring). For example if the parent strings 11100011100111 and **10001110000101**, representing the values $x = 38.92$ and $x = 5.50$, are crossed after the 4th bit the offspring created are 11101110000101 and **10000011100111**, $x = 43.01$ and $x = 1.41$. The crossover operator simply recombines information, which already exists but is unable to explore areas not included in the population. For example, the parents above both contain a zero at position 4 and no crossover can change this value to a 1. Mutation is therefore needed to ensure the whole search space can be explored. Mutation works by changing individual bits from 1 to 0 or vice versa. The chance of a bit being mutated is determined by the mutation rate and all bits are treated in the same way. For example if the

Figure 3. Function maximisation example

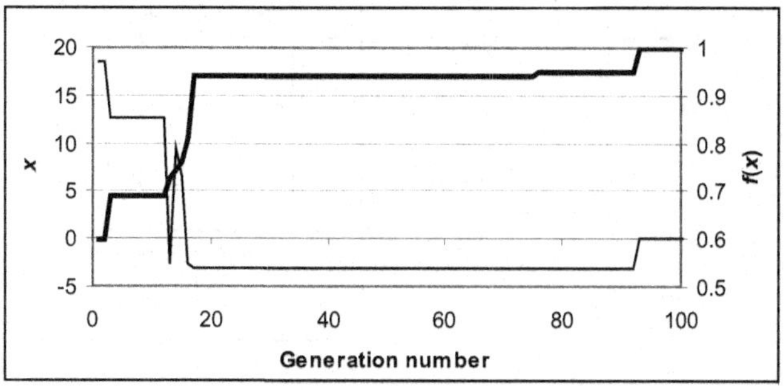

Note: Normal line = x value, Bold line = function value

second and seventh bits of the individual 11100011100111 undergo mutation it will become 10100011110111. The whole process of fitness evaluation, reproduction, crossover and mutation is repeated for a given number of generations and the best solution obtained is output. As an example the GA is applied to the given maximization problem using a population of 20, crossover rate of 0.8, mutation rate of 0.05 and 100 generations. The best solution at the end of each generation is plotted in Figure 3 to illustrate how the GA evolves the solution over time. In the figure it is seen that the solution quickly converges to a local maxima of 0.943 at $x = -3.012$ which is close to the global maxima of 1.000 at $x = 0$. It is also observed that the solution is able to escape the local maxima in the 92nd generation, giving a value of $x = 0.021$ as the final result. This example highlights an important feature of GA. A major strength of GA is the ability to escape form local optima to find the global optima solution. While in this case the solution found the global maxima, it may not always be the case. In developing a GA, the reliability and robustness of the solution is therefore very important. It is possible to influence the search by selecting appropriate mutation and crossover rates, but in general there will be a trade-off between exploration (broad search) and exploitation (local search). For example a small mutation rate will help us to explore the spaces around the current solutions but will make it difficult to jump to completely new areas, whereas a large rate will help cover more ground, but at the expense that the solutions will find it harder to converge. This trade-off between exploration and exploitation of solutions has long been an issue with simple genetic algorithms and is one of the key motivations behind the strategy developed in the following sections.

Structural Identification

In structural engineering, system identification can be applied to determine unknown parameters of a structure. Termed as structural identification, this approach can be utilised for the nondestructive assessment and monitoring of structures.

Figure 4. Effect of noise on structural identification

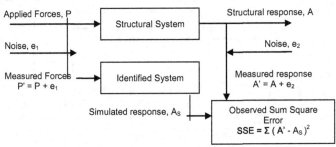

$$\mathbf{M\ddot{x} + C\dot{x} + Kx = P} \tag{3}$$

The dynamic equation of motion of a structural system can be written as shown in equation (3), where $\mathbf{M}$, $\mathbf{C}$, and $\mathbf{K}$ are the mass, damping and stiffness matrices, $\mathbf{x}$ is the displacement vector, and $\mathbf{P}$ is the input force vector. GA methods can be used for the identification of such systems in the time domain as shown in Figure 4, where noise is present on both the measured forces and accelerations. The forward analysis is carried out by numerically solving the dynamic equation using the trial parameter values of the identified system and the measured force input. The dynamic time-history generated is compared with the measured time-history and the GA works to minimise the sum of square error by altering the parameters of the identified system.

For practical reasons, the following points have to be considered:

- The method should not require an unreasonably good initial guess of the parameters in order to converge.

- Real I/O measurements contain noise and the method should be tested in the presence of I/O noise.

- The method should operate on incomplete measurements as it is not practical to have measurements at all DOFs in a structure.

- Dynamic measurements are usually obtained using accelerometers and numerical error is inevitable in integration of acceleration to compute velocity and displacement. It is therefore preferable to utilise accelerations directly in the identification procedure.

An identification strategy involving a search space reduction method (SSRM) and a modified GA based on migration and artificial selection, (MGAMAS) is presented in this section. The motivation behind the development of the SSRM comes from the fact that for GA the convergence rate and accuracy of identification are highly dependent on the size of the search space. By adaptively reducing the limits of the search a more accurate and reliable identification is possible. The heart of the method is the MGAMAS. This algorithm, based on the GAMAS (Potts, Giddens, & Yadav, 1994), has been developed in order to provide a good identification technique that can simultaneously explore the search space and focus on promising individuals. The MGAMAS includes a reduced input data procedure and other special features that are shown to greatly reduce the computational time and to increase the accuracy of identified parameters.

Search Space Reduction Method

The proposed search space reduction method (SSRM) aims to increase the accuracy and reliability of identification by reducing the search space. The SSRM is schematically shown in Figure 5 and the MGAMAS is further explained in the next subsection. The basic idea is simple. Let the search space reduce for those parameters that converge quickly in order to reduce the time wasted looking far outside the area where the optimal solution lies. This is achieved by carrying out several runs of the MGAMAS, following which the mean and standard deviation of the identified parameters are computed. The standard deviation gives us an indication of the certainty to which the parameter has been identified and the search space can be reduced accordingly. If the standard deviation is small it is very likely that the mean is close to the optimal parameter value and the search limits can be reduced. Conversely

Figure 5. Block diagram of SSRM

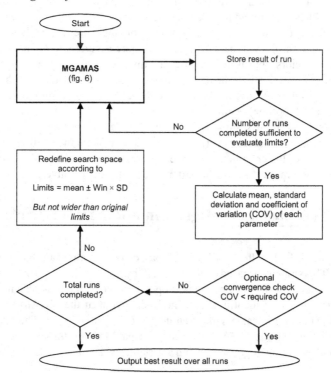

if the standard deviation is large we should continue to search broadly for that parameter. Eventually as some parameters converge almost exactly, the SSRM effectively reduces the number of unknown parameters and those remaining can be identified more easily. The final result output is the best result over all of the runs conducted.

The basic parameters that define the SSRM are the number of runs to be used for evaluation of the search space, the width of the reduced search space window and the total number of runs to be carried out. Here a "run" refers to an identification run using the MGAMAS. In addition a convergence exit criteria may be included to exit from the system early if prescribed convergence is achieved. The number of runs to be used for evaluation of the search space should be selected such that it is sufficient to get a good estimation of the mean of the parameters but not so large that it includes too many old results that will slow the convergence. More runs will make the system more robust but at the price of increase in total computational time. In general the number of runs should be chosen such that a "bad" run will not overly affect the mean parameter values obtained. In practice it is found that four or five runs work well. The evaluation of the search space is conduced after each run using the most recent runs. For example, if four runs are used, after the eighth run the search space is evaluated using runs five to eight. The "width of window" parameter defines how quickly the search space is reduced according to search space = Mean ± Window × standard deviation, but not wider than the original limits. In the SSRM the mean values of parameters are calculated using weighted results whereby the more recent runs are given a higher weighting. This is to recognise that the results should improve as the search space is reduced. It is important to choose a window that is small enough to achieve convergence but wide enough that the actual solution will be contained within the new search space. In practice a value of window width of about 4 has been found to give good performance. The total number of runs to be used depends mostly on the accuracy required. In theory the search space should reduce after each run and so the results will become more and more accurate. In general accuracy will be limited due to factors such as noise and after a time no further improvement in accuracy is possible. Other factors such as the number of runs, the window width and the MGAMAS parameters also need to be carefully considered to ensure the algorithm does not converge prematurely to a sub optimal result. If desired, a convergence criterion can be prescribed so as not to waste time if the results converge quickly. In this case the ratio of the standard deviation and mean (coefficient of variation) is used. In general the coefficient of variation gives an indication of the error in the parameter value and is therefore also useful to check at the end of the program to see to what extent the results have converged.

Modified GA Based on Migration and Artificial Selection

The heart of the SSRM is the modified GA based on migration and artificial selection (MGAMAS). This strategy is based on the GAMAS by Potts et al. (1994) but uses a floating-point representation and includes several new operators and techniques designed to increase the speed and accuracy of identification. The basic layout of the MGAMAS is shown in Figure 6 and the important features of the strategy are discussed below. The most important features that distinguish the MGAMAS from "normal" GA are the inclusion of multiple species, artificial selection, regeneration and a variable data length procedure. In addition to

Figure 6. Block diagram of MGAMAS

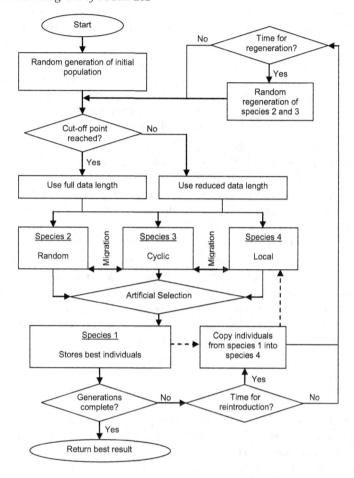

these important points the strategy also includes a rank selection procedure, new mutation operators and a tagging procedure to help maintain diversity in the best solutions.

Individuals are represented using floating-point numbers in vector form. Here each parameter is represented by a single value and the vector of all parameters makes up an individual. The floating-point representation is more natural than the binary encoding traditionally used in GA. In addition the floating-point representation allows for new operators, such as nonlinear cyclic mutation, that would not have been possible in a binary system.

The real power of the original GAMAS and this MGAMAS strategy lies in the division of the population into species. One of the problems with GA has always been the trade off between exploration and exploitation. That is, it is difficult to set the GA parameters in such a way that we can focus in on promising areas while still maintaining a broad search. With multiple species this problem is greatly reduced. As one species searches broadly, another can be designed to search locally around the best solutions already found. In the

MGAMAS, four species have been adopted. Species 1 is used to store the best results, while species 2-4 conduct searches increasing in focus from a very broad random search to a more refined local search. The operators used to achieve this are discussed further in the relevant sections below.

To ensure that species 4 operates on a set of good solutions a reintroduction is required. This involves inserting individuals from species 1 into species 4 at a prescribed interval. The number of times that reintroduction should be done must consider that the best solutions need to be improved, and that species 4 needs some time to develop new solutions. Nevertheless it is generally found that a large number of reintroductions gives the best results.

A well-known problem with GA is that the solutions may converge to local optima and find it difficult to escape to find the global optimum solution. Regeneration involves the complete random replacement of a species. In this way the process is effectively restarted and new optimum may be found. In the MGAMAS scheme developed, only species 2 and 3 are regenerated. This allows species 4 to focus on refining the previously generated solutions while species 2 and 3 search for new possibilities. The number of times regeneration is carried out is generally small to allow for sufficient time for good solutions to develop.

Migration allows for exchange of information between species. Just as human movements between cities or companies can help transfer knowledge and ideas, the migration of individuals between species can help share important information. The migration operation involves swapping randomly selected individuals between species 2 and 3 and also between species 3 and 4. The number of individuals moving is controlled by the migration rate. Generally a rate of movement of about 5% per generation works well.

Mutation Operators

One of the benefits of multiple species and floating-point representation is that many different mutation operators are possible. Three different mutation operators and rates of mutation are used for the species in the MGAMAS. The mutation operators are designed to give each species a different strength so that the whole system can be effective. In each case mutation is carried out on a single value of the individual and the mutation rate determines the probability of an individual parameter being mutated. A random number generator then determines the magnitude of the mutation to be carried out. A graphical representation of the average mutation provided by the operators for species 3 and 4 is shown in Figure 7 for a case where three regeneration cycles are used and the random number generated is 0.5. Note that the completely random mutation of species 2 would have a value of 0.25 in this case and would be independent of the generation number. The random mutation used in species 2 simply involves random regeneration of a single value. That means that the selected parameter within the individual will be assigned a value randomly distributed within the parameter limits. The Nonuniform mutation operator reduces the average magnitude of mutations as the analysis proceeds and has been shown to help increase the accuracy and convergence rate in mathematical optimisation problems (Michalewicz, 1994). The cyclic nonuniform mutation operator used in species 3 is based on this operator but was designed with the regeneration operation in mind. The idea is to allow for larger mutations after regeneration has taken place and then to gradually reduce the size of the mutations as

Figure 7. Average magnitude of mutations for species 3 and 4

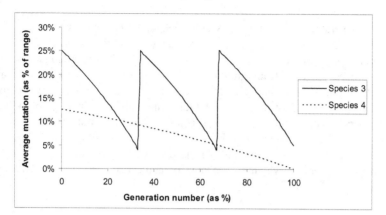

the solutions develop. This means the average size of the mutations will decrease gradually within each regeneration cycle and then increase again after the regeneration, resulting in a cyclic mutation. To achieve this objective the following operator is used;

$$x_i = x_i + (UL_i - x_i) \times \left(1 - r_1^{\left(1 - \frac{0.9\,MOD(g,R)}{R}\right)}\right); \quad for \quad r_2 = 0$$

$$= x_i + (LL_i - x_i) \times \left(1 - r_1^{\left(1 - \frac{0.9\,MOD(g,R)}{R}\right)}\right); \quad for \quad r_2 = 1 \tag{4}$$

UL_i and LL_i are the upper and lower limits of the search space for the *i-th* parameter x_i. r_1 is a random number in the range [0 1] and r_2 is randomly selected as either 0 or 1. $MOD(g,R)$ is the remainder when the generation number g is divided by the number of generations between regenerations R. The factor of 0.9 is to ensure that the size of the mutation is not too small as the generations approach a regeneration point. As species 4 is designed to refine the best solutions, small mutations are preferred. A local nonuniform mutation method is used, whereby the size of mutations is gradually reduced as the analysis proceeds. The following operator achieves this mutation:

$$x_i = x_i + 0.5 \times (UL_i - x_i) \times \left(1 - r_1^{(1 - g/G)}\right); \quad for \quad r_2 = 0$$

$$= x_i + 0.5 \times (LL_i - x_i) \times \left(1 - r_1^{(1 - g/G)}\right); \quad for \quad r_2 = 1 \tag{5}$$

Here G is the total number of generations to be run and the multiplier (0.5) encourages smaller mutations as can be seen in Figure 7.

Crossover Operators

Two crossover operators are used in the MGAMAS, namely a simple crossover and multipoint crossover. The simple crossover is similar to the crossover performed in binary GA as the top part of one individual is switched with the bottom of another and vice versa. The switching position is calculated randomly. The multipoint crossover allows crossover at each parameter. Pairs of individuals are selected for crossover and then crossover performed on each value in the individual by generation of a random number. The number of individuals involved in crossover for a given generation is controlled by the crossover rate. Where both simple and multipoint crossovers are to be used the total crossover rate should be considered. For example, if a crossover rate of 0.4 is used for each of the forms of crossover, the effective total crossover rate, which is the chance of an individual being involved in at least one crossover is $1-0.6^2 = 0.64$.

Fitness Evaluation and Reproduction Procedure

In the MGAMAS fitness is evaluated from the total sum of square error between the simulated and measured accelerations as was shown in Figure 4. Accelerations are calculated using the unconditionally stable Newmark constant average acceleration method. The method is based on an assumption of constant acceleration during each time step and the system is solved for incremental values at each step. The fitness is evaluated as the inverse of the sum of square error between the simulated and measured data. Generally selection would then be carried out by allocating a selection probability to each individual based on its fitness. It is noted however that, as the identification proceeds, a lot of individuals will have very similar fitness values and the selection procedure will become almost random. To avoid this problem a ranking procedure is used to determine the selection probabilities. Within each species the individuals are ranked, the worst individual assigned a rank of 1 and the best a rank equal to the population size. Reproduction is then carried out by the commonly used roulette wheel method whereby an individual's chance of selection is proportional to its rank as shown in equation (6).

$$P_{selection} = \frac{R_i}{\sum R} \tag{6}$$

This procedure ensures that the fittest individual will always have twice the chance of selection of an average individual. It must be noted here, that the original fitness values must still be used for the artificial selection. This is very important as a common measure of fitness must be used for the comparison between individuals in different species. As the name of the method implies, artificial selection is crucial to the functioning of the MGAMAS. Artificial selection involves ensuring that the fittest individuals generated by any of the species are stored in species 1 for future refinement by species 4. This is a simple procedure involving the comparison of the fitness value of the weakest individual in species 1 with those in species 2, 3 and 4. If any individuals are better, then they replace the worst individuals in species 1 so that species 1 always contains the best solutions. The problem with artificial selection is that the same individual could be selected many times, saturating species 1. To eliminate this possibility a new idea of tagging is proposed. It is desirable to have as many good solutions as possible retained to avoid excessive focus on one, possibly suboptimal solution. The tagging procedure is achieved as follows. All individuals are initially assigned a 0 tag. If an individual is selected for species 1 its tag is changed to 1. The tag follows the individual wherever it goes, through migration, selection and reintroduction. If an individual is altered in any way through mutation, crossover or regeneration it no longer represents the same individual and its tag is changed back to 0 making it available again for selection to species 1.

Reduced Data Length Procedure

In the identification procedure the simulated time-history response of the system must be calculated for comparison with the measured values. This is of course the most computation-ally expensive part of the whole process and is responsible for most of the time used. To improve computational efficiency, a variable data length procedure is proposed. The idea is to use a small portion of the total available time-history response data to roughly identify the parameters before increasing to the full data later in the process. In the MGAMAS, this is achieved by specifying a cut-off point where the evaluation switches from reduced data to full data. The cut off point is specified as a fraction of the total generations to be run. The cut-off point and the length of the reduced data to use again depend on the problem but an indication is given in the numerical study. In general if noise is present a longer reduced data sequence will be required to help average out the effect of the noise. The time savings using this procedure can be very significant. For example if 20% of the full data is used for 75% of the generations, the time saved is 60%.

Numerical Studies

In order to demonstrate the effectiveness of the SSRM method, numerical simulations of multiple DOF shear buildings as shown in Figure 8 are considered. The structures are char-acterised by flexible columns, rigid beams and lumped mass at each floor level, effectively reducing the response to a single translation at each level. Rayleigh proportional damping, as shown in equation (7), is applied in all cases allowing for mass, stiffness and damping matrices to remain banded and constant over time. The damping assumes a proportion of critical damping of 5% in the first two modes of vibration in order to determine the damping parameters α and β. In the identification, damping parameters α and β are assumed unknown resulting in $N + 2$ unknown parameters for the known mass case and $2N + 2$ parameters for the unknown mass case, for an N-DOF system. A numerical integration scheme utilis-ing the unconditionally stable Newmark constant average acceleration method and an LU factorisation procedure can then be used to efficiently generate the time history responses required by the GA strategies.

Figure 8. n-DOF structure for numerical study

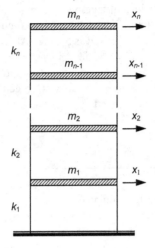

Table 1. Structural properties

10-DOF	
Stiffness (kN/m)	
Levels 1-4	5000
Levels 5-8	4000
Levels 9-10	3000
Mass (kg)	
Levels 1-5	6000
Levels 6-10	4200
Natural Period of Vibration (s)	
First mode	1.321
Second Mode	0.505
20-DOF	
Stiffness (kN/m)	
Levels 1-10	5000
Levels 11-15	4000
Levels 16-20	3500
Mass (kg)	
Levels 1-10	4000
Levels 11-20	3000
Natural Period of Vibration (s)	
First mode	2.123
Second Mode	0.797

Table 2. Location of acceleration measurements

System		Floor Levels
Known Mass	10-DOF	2, 4, 7, 10
	20-DOF	2, 4, 7, 10, 12, 14, 17, 20
Unknown Mass	10-DOF	1, 2, 4, 6, 8, 10
	20-DOF	1, 2, 3, 4, 6, 8, 10, 12, 14, 16, 18, 20

$$\mathbf{C} = \alpha\mathbf{M} + \beta\mathbf{K} \quad ; \quad \varsigma_r = \frac{\alpha}{2\omega_r} + \frac{\beta\omega_r}{2} \tag{7}$$

The structural properties of the systems considered given in Table 1. Input forces are applied at every fifth level of the structures as random white Gaussian noise with the RMS of the force scaled to $1,000N$. Acceleration measurements are obtained at 40% of the levels for the known mass cases and at 60% of the floor levels for the more difficult unknown mass cases as given in Table 2. The search limits are taken as 0.5 to 2.0 times the exact values.

It is important to note that for the programs developed all mass, stiffness and damping variables are assumed unknown. For the cases where masses are known, the same program is used by imposing very narrow limits on the (known) masses. The effect of this procedure must be considered, as it is different than the case where masses are input directly into the program and not treated as unknowns. The most obvious effect is that the effective crossover rate is reduced for the simple crossover as a crossover point anywhere within the mass variables will have the same effect. The effect of this is seen in the results given where a higher crossover rate is needed for the known mass problems.

The numerical study is presented in two stages. First, to develop an understanding of the GA parameters required and to illustrate the effectiveness of the strategy, comparison is made with a simple GA (SGA) with a single population and simple random mutation. 10 and 20-DOF known mass systems are considered, and trials repeated for different combinations of GA parameters. Next, the SSRM is applied to the more difficult, unknown mass systems. The robustness of the strategy is demonstrated in the presence of 5% and 10% I/O noise and the reduced data length procedure is used to reduce the computational time. The identification results presented in Tables 3 and 4 are average results over 25 runs. The input forces and noise pattern was freshly generated for each run to avoid any bias that might result from using the same input for all of the 25 runs.

Known Mass Systems: Comparison with a Simple GA

When conducting tests to compare identification results two approaches are possible. The first is to compare the time taken in achieving a given accuracy and the second is to compare the accuracy that can be achieved in a given time. In this study the latter method is used whereby the total number of evaluations is fixed for each system. The total evaluations refers to the number of times the time history simulation is carried out and is set at

Table 3. Comparison of results for known mass system

GA Parameters	10-DOF			20-DOF		
	SGA	MGAMAS	SSRM	SGA	MGAMAS	SSRM
Runs	-	-	4	-	-	4
Total Runs	-	-	9	-	-	9
Window	-	-	4.0	-	-	4.0
Population	113	9 x 3*	19 x 3*	226	9 x 3*	19 x 3*
Generations	176	741	82	354	1404	156
Data length	200	200	200	200	200	200
Time step	0.01	0.01	0.01	0.01	0.01	0.01
Regeneration	-	2	2	-	3	3
Reintroduction	-	30	30	-	42	42
Migration	0.05	0.05	0.05	0.05	0.05	0.05
Crossover $^{\psi}$	0.96	0.80	0.80	0.96	0.80	0.80
Mutation $^{\Omega}$	0.05	0.05	0.20	0.05	0.05	0.10
Results						
Time taken	12s	12s	12s	100s	100s	100s
Mean error – k	4.22%	1.36%	0.43%	8.33%	2.87%	0.52%
Max error – k	12.36%	4.22%	1.21%	31.28%	9.28%	1.60%
Mean error – c	12.33%	5.68%	1.56%	15.81%	4.03%	0.64%
Max error – c	20.76%	10.10%	2.76%	28.97%	7.57%	1.21%

Note:

* *Same population size used for each species*

$^{\psi}$ *Same crossover rate applied for simple and multi point. For SGA only simple crossover used.*

$^{\Omega}$ *Same mutation rates applied for all species*

20,000 and 80,000 for the 10 and 20 DOF systems respectively. The computational times are approximately 12s, and 100s for analysis conducted on a standard Pentium 4, 3-GHz PC. For each parameter of interest two or three different values are tested resulting in a large number of possible combinations. The testing of parameters is spilt into two sections. First, all possible combinations of the main parameters are trialed. The main parameters considered are population size, number of runs, total runs, crossover rate, mutation rate, number of regenerations and number of reintroductions. Following this, additional tests are conducted by varying the other parameters about the optimum values identified in main tests. This procedure is used as including all parameters in the main tests would result in an unreasonable number of tests to be carried out. With the GA parameter variations selected there are 216, 72, and 18 combinations for each system for the main tests of SSRM, MGAMAS, and SGA respectively.

Results shown in Table 3 are for identification carried out using the best GA parameter combinations trialed. The average and maximum error of stiffness and damping properties is presented. Due to the fact that the damping parameter α has only a small contribution to the overall response, its value is generally poorly estimated. This however is not really of concern as the damping is dominated by the stiffness proportional part and so errors in the mass proportional part, α will have little effect on the total damping present. Of most importance from a practical point of view are the errors in the estimated stiffness values and the computational time used.

The results show that, given a fixed time, the MGAMAS can give far better results than the SGA. It is noted however that there is still room for improvement. While the average error achieved using the MGAMAS is good the maximum error is reasonably high. Carrying out the identification procedure for a larger number of generations can help to reduce the errors but parameters with significantly large identification errors (outliers) are still a major problem. In a real case it is difficult to know simply by looking at the results, which results are accurate and which are not. Thus it is important to improve the reliability of results by reducing the errors of outliers. This issue is addressed by the SSRM where utilising mean and standard deviation of the results over several runs enables the procedure to significantly reduce outliers. This is seen in the results where the maximum error in stiffness is only 1.21% and 1.60% for the 10 and 20-DOF systems respectively.

Unknown Mass Systems with Noise: Illustration of SSRM

In order to demonstrate the effectiveness of the SSRM on more difficult, unknown mass systems, the same 10-DOF and 20-DOF systems are considered. Comparison with the SGA is not shown here as the SGA is unable to identify the systems and the results are not worth comparing. As with the known mass cases, several trials were conducted in order to establish reasonable GA parameters. The GA parameters used and results obtained are presented in Table 4. The estimations of mass and stiffness parameters are very good, even in the presence of large I/O noise. To be able to identify 42 unknowns for a 20-DOF system when 5% or 10% noise is present in both the input forces and measured accelerations is very pleasing. The maximum error in stiffness of only 3.8% under 5% noise is very good and allows the results to be considered with some confidence. The computational time of 42mins for the 20-DOF system is very good however it is expected that for larger and more realistic systems this time could become very large. The option of the use of distributed or parallel computing (Koh, Wu, & Liaw, 2002) could then be very relevant. It is also noted that, while the unknown mass systems present a far greater challenge as compared to systems where the mass is known, for most practical applications the mass should be at least approximately known, and thus a smaller search space for mass may be adopted, resulting in even better and faster identification.

Table 4. Results for unknown mass system based on SSRM

GA Parameters	10-DOF		20-DOF	
	5% noise	10% Noise	5% noise	10% Noise
Runs	5		5	
Total Runs	15		15	
Window	4.0		4.0	
Population	65 x 3*		90 x 3*	
Generations	342		494	
Data length	500/200/50 ⁹		500/200/50 ⁹	
Time step	0.01s		0.01s	
Regeneration	3		3	
Reintroduction	120		200	
Migration	0.05		0.05	
Crossover ᵛ	0.40		0.40	
Mutation Ω	0.20		0.10	
Results				
Time taken	10m 30s		42m	
Mean error – k	1.60%	2.98%	1.38%	2.78%
Max error – k	3.62%	6.62%	3.83%	8.64%
Mean error – m	1.50%	3.00%	1.51%	3.00%
Max error – m	3.79%	6.81%	4.02%	10.40%
Mean error – c	6.29%	8.41%	6.70%	14.69%
Max error – c	11.43%	15.29%	12.90%	20.36%

Note:

* Same population size used for each species

ᵛ Same crossover rate applied for simple and multi point. For SGA only simple crossover used.

Ω Same mutation rates applied for all species

⁹ Full data length is 500. Reduced length of 200 is used for 50% of the generations.

Structural Damage Detection

There are two possibilities when it comes to damage detection: (a) Damage can be identified with no prior measurement of the undamaged structure, and (b) damage can be identified utilizing previous measurement of the undamaged structure. For the first case we have no choice but to identify the structural properties and compare these to some theoretical values in order to identify the magnitude and location of damage. In this case, the SSRM developed

Figure 9. Block diagram of damage detection strategy

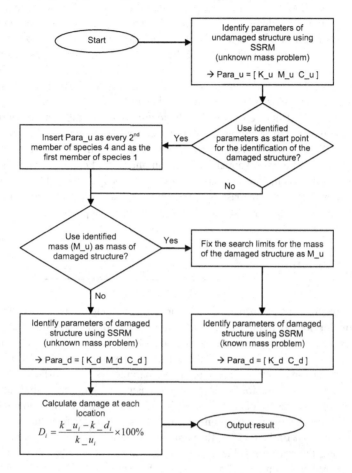

in the previous section can be utilized directly and no additional development is required. For the second case however, the additional information of the undamaged structure can be utilized in developing an improved strategy. This section therefore deals with the case where we have measurements of the structure both before and after damage has taken place. The strategy assumes that the structural mass, stiffness and damping are unknown, and the damage can be quantified and detected as a change in the stiffness of the damaged member. The aim is to detect the magnitude and location of the damage from the measured response of the structure before and after the damage takes place.

The damage detection strategy, shown in Figure 9, uses the SSRM to identify the structure before and after damage has occurred. It is seen that the strategy contains some options for the analysis to be conducted on the damaged structure. First, it is possible to use the parameters identified for the undamaged structure as a starting point for the identification of the damaged structure. This has obvious benefit in that only the changes (damaged members) need to be identified, resulting in a more accurate identification. This option is implemented by setting half of the individuals in species 4 to the values identified for the undamaged structure. The other half of the species as well as species 2 and 3 are initialized randomly. The value of half is chosen such that a lot of the values will be present but some random results exist so as to ensure good performance of the crossover operation in the early stages. Second, there is an option of fixing the mass based on the mass of the undamaged structure. This option is useful if we are sure that the mass has not been altered since the measurement of the undamaged structure was made. The identification of the damaged structure is reduced from an unknown mass problem, to a much easier known mass problem, which can be identified with better speed and accuracy. There is also an added benefit that changes in stiffness will not be masked by an apparent change in the mass. For cases where significant changes in structural mass may have occurred however, this option should not be used and a full identification including the mass of the damaged structure must be done. Finally the damage is calculated as the loss in stiffness of the structure as a percentage of the original undamaged stiffness.

Numerical Studies

In order to demonstrate the performance of the proposed strategy and to study the effect of the identification options available, numerical trials are carried out. The same multiple DOF shear buildings used in the previous section are considered. It is assumed that, prior to the identification of the undamaged structure, the structural mass, stiffness and damping are all unknown. This results in a total of $2n + 2$ unknown parameters for an n-DOF system. The structural response is simulated for 500 steps at 0.01s using random force inputs at every 5th level and acceleration measurement at 60% of the floors as given for the undamaged case in Table 2. White Gaussian noise is again added to the simulated forces and accelerations such that the noise is 5% of the RMS of the signal. In order to simulate the response for the damaged structure, the stiffness value at the fourth-floor level is reduced. Three damage cases are considered representing a 2.5%, 5%, and 10% reduction in story stiffness. Each system was simulated 25 times with fresh input forces and noise in each case and the results presented are the average over these 25 runs. For all trials the parameter search limits are set as half to double the actual parameter values of the undamaged structure.

The trials presented here are designed in order to illustrate the effect that fixing the mass and using the undamaged parameters as a starting point has on the quality of the damage detection. Therefore, for each system and damage level there are four identification options to be trialled. Only the identification of the damaged structure is affected by these options. In all cases the undamaged structure is first identified as an unknown mass problem using the GA parameters given for unknown mass systems in Table 6. If the mass is not fixed based on this result, the same GA parameters are then used to identify the damaged structure. If the option to fix the mass is used however, the GA parameters for the known mass system can be used and the computational time greatly reduced. In all cases the reduced data length procedure is used with a reduced length of 200 being used for 50% of the generations. The resulting computational times are indicated in Table 5 for analysis conducted on a Pentium 4, 3-GHz PC. The total analysis time is the sum of the analysis for the undamaged and dam-

Figure 10. Illustration of damage detection error for 10-DOF structure

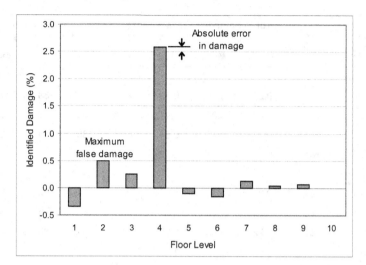

aged structures depending on the option chosen. For example, when the mass is fixed based on the undamaged parameters identified, the total time is 11m 10s and 48m 40 sec for the 10 and 20-DOF systems, respectively.

A summary of the identification results is presented in Tables 6 and 7. There are three components to the results presented. These considerations can be understood by viewing the typical plot of damage results shown in Figure 10 where the actual damage simulated was

Table 5. GA parameters used in damage detection

	Known Mass		Unknown Mass	
	10-DOF	20-DOF	10-DOF	20-DOF
Time	30s	4min 10s	10min 40s	44min 30s
Population sizes	9 x 3	19 x 3	65 x 3	90 x 3
Runs / Total runs	4 / 10	4 / 10	5 / 15	5 / 15
Generations	100	200	200	300
Crossover	0.8	0.8	0.4	0.4
Mutation	0.2	0.1	0.2	0.1
Window	4.0	4.0	4.0	4.0
Migration	0.05	0.05	0.05	0.05
Regeneration	2	3	3	3
Reintroduction	30	50	100	150

Table 6. Damage detection results for 10-DOF structure

Use para_u	Fix mass	Error in damage (%)		Max false damage (%)		Identification Success (%)		
		Mean	Max	Mean	Max	1×	2×	4×
2.5% damage								
Yes	Yes	0.22	0.85	0.49	2.31	96	88	80
Yes	No	0.65	3.04	1.41	7.85	92	72	48
No	Yes	0.24	1.64	0.84	4.67	88	80	68
No	No	0.99	3.67	2.19	7.34	56	44	40
5% damage								
Yes	Yes	0.21	1.15	0.63	2.88	100	96	80
Yes	No	0.55	1.92	1.91	7.36	96	76	60
No	Yes	0.22	0.64	0.90	4.20	100	88	76
No	No	1.00	5.02	1.19	6.28	100	84	68
10% damage								
Yes	Yes	0.13	0.39	0.45	3.89	100	100	96
Yes	No	0.68	2.20	1.35	6.18	100	96	92
No	Yes	0.26	1.23	0.67	1.96	100	100	100
No	No	1.18	5.10	1.84	6.93	100	92	76

2.5% in the 4th story. The first consideration is the absolute error in the damage identified at the fourth (damaged) level. Just as important however is ensuring that damage is not falsely reported at other floors, and in considering this, the result of the maximum damage identified on the undamaged floors is also presented. For both of these considerations both the mean and maximum values over the 25 runs are presented. Finally, to be of practical use the damage identified should exceed any false damage by a reasonable margin. The success rate of the identification is therefore given in terms of the percentage of the 25 trials conducted where the damage exceeded the false damage by a given factor. In the table the success rates are given such that the damage is one, two, or four times greater than the largest false damage reported.

The numerical results demonstrate that the strategy is able to accurately and consistently identify even small levels of damage corresponding to a change in story stiffness of only 2.5%. The performance of the strategy is improved significantly by using the option of fixing the mass based on the result of the undamaged structure. The results also improve, although to a lesser extent, when the undamaged parameters identified are used as a starting point for the identification of the damaged structure. In addition to improved accuracy, fixing the mass based on the undamaged result reduces the computational time significantly. The results also highlight an important fact that although the identification is very good, it is not perfect, and in some cases the identification may fail. This is seen for example in

Table 7. Damage detection results for 20-DOF structure

Use para_u	Fix mass	Error in damage (%)		Max false damage (%)		Identification Success		
		Mean	Max	Mean	Max	1×	2×	4×
2.5% damage								
Yes	Yes	0.11	0.39	0.55	4.04	96	88	76
Yes	No	0.28	0.95	0.78	2.66	100	84	56
No	Yes	0.24	2.23	0.58	2.96	96	92	80
No	No	0.63	2.70	1.34	5.44	76	72	40
5% damage								
Yes	Yes	0.14	0.82	0.57	2.93	100	92	84
Yes	No	0.34	2.26	1.04	5.24	96	92	76
No	Yes	0.21	0.93	0.74	4.97	100	92	88
No	No	0.58	1.97	1.32	5.74	96	88	76
10% damage								
Yes	Yes	0.14	0.79	0.42	1.96	100	100	100
Yes	No	0.68	6.47	1.20	3.59	100	100	96
No	Yes	0.15	0.46	0.74	2.96	100	100	96
No	No	0.61	5.36	1.24	4.51	100	100	88

the results for the 10-DOF system with 2.5% damage, whereby there were 3 cases (12%) of the 25 trials conducted where the damage identified at the fourth floor is not more than twice that of any other floor in the structure. In one of those cases the damage identified at an undamaged floor actually exceeded that identified at the damaged (fourth) level. While this failure only occurs once in 25 trials conducted it must be considered. In application to a real structure it would therefore be recommended that the identification be carried out more than once to ensure consistency and validity of the result. This would almost guarantee good results as it would be highly unlikely that the same false result would be identified more than once based on different tests.

A Note on Experimental Tests and Extension to Realistic Problems

The numerical examples above clearly indicate the potential of the strategy of identifying even small amounts of damage in simulated structures. However, as with any method, it is always more convincing however when the results can be verified on real systems where the numerical models cannot exactly match the behavior of the system under study. With this in mind, tests on a seven-story steel shear building model have been carried out at the National University of Singapore. Early results of these tests indicate that damage representing as little as a 4% change in stiffness is accurately and consistently identified using as few as two acceleration measurements. These results are very encouraging for future application to larger and more realistic systems. It is expected, however, as the systems become larger separation of true damage from reported false damage in a reasonable computation time may become difficult. The use of substructuring and/or parallel computing techniques (Koh, Hong, & Liaw, 2003; Koh, Wu, & Liaw, 2002) may be necessary in achieving the desired results.

Conclusion

In recent years, genetic algorithms have been successfully applied to a wide range of search and optimisation problems. Classical genetic algorithms however may encounter problems in solving the complex systems encountered in structural engineering. In this chapter a modified genetic algorithm strategy has been presented. This strategy uses the ideas of evolution and survival of the fittest from genetic algorithms and introduces a new architecture and operators designed to better develop solutions to structural problems. The novel strategy works on multiple populations or 'species' and balances the search with both broad and local search capability. The search space reduction method (SSRM) further improves the accuracy of results by progressively reducing the limits of the search space as the identification proceeds. Finally the SSRM is incorporated into a damage detection strategy, whereby the identification of the structure both before and after damage has taken place allows for accurate identification of small levels of damage, even when measurement is incomplete and contaminated with noise. The numerical studies presented demonstrate the effectiveness of the strategy on 10 and 20-DOF structures, where a damage representing a 2.5% change in stiffness is accurately and consistently identified in the presence of 5% noise.

References

Caravani, P., Watson, M. L., & Thomson, W. T. (1977). Recursive least-squares time domain estimation of structural parameters. *Journal of Applied Mechanics,* 135-140.

Holland, J. H. (1975). *Adaptation in natural and artificial systems.* Ann Arbour: University of Michigan Press.

Hoshiya, M., & Sutoh, A. (1993). Kalman filter–Finite element method in identification. *Journal of Engineering Mechanics, 119*(2), 197-210.

Koh, C. G., Chen, Y. F., & Liaw, C. Y. (2002). A hybrid computational strategy for identification of structural parameters. *Computers and Structures, 81*, 107-117.

Koh, C. G., Hong, B., & Liaw, C. Y. (2003). Substructural and progressive structural identification methods. *Engineering Structures, 25*, 1551-1563.

Koh C. G., & Perry, M. J. (2005, October 19-21). *Damage detection of structures using a modified genetic algorithm.* Keynote paper presented at the Ninth International Conference on Inspection, Appraisal, Repairs and Maintenance of Structures, Fuzhou, China. ·

Koh, C. G., Wu, L. P., & Liaw, C. Y. (2002, August 23-24). *Distributed computing strategy for structural monitoring and diagnostics.* Paper presented at US-Korea Workshop on Smart Structural Systems, Busan, South Korea.

Michalewicz, Z. (1994). *Genetic algorithms + data structures = evolution programs* (2nd ed.). Berlin: Springer-Verlag.

Perry, M. J., Koh, C. G., & Choo, Y. S. (2006). Modified genetic algorithm strategy for structual identification. *Computers and Structures 84,* (pp. 529-540).

Potts, J. C., Giddens, T. J., & Yadav, S. B. (1994). The development and evaluation of an improved genetic algorithm based on migration and artificial selection. *IEEE Trans. on Systems, Man and Cybernetics, 24*(1), 73-86.

Chapter XIV

Neural Network-Based Identification of Structural Parameters in Multistory Buildings

Snehashish Chakraverty, Central Building Research Institute, India

Abstract

A detailed study of the capabilities and powerfulness of soft computing techniques such as artificial neural network with respect to the identification of structural parameters and structural responses are presented. This chapter includes the definition of neural architectures and system identification of multistory structure. An efficient identification algorithm for the multistory structure subject to initial condition and ground displacement is presented. Response identification subject to real earthquake data has also been discussed. Several example problems are incorporated to show the efficiency and reliability of the proposed algorithm.

Introduction

System identification methods in structural dynamics, in general are formulated as inverse vibration problems to identify properties of a structure from measured data. The dynamic behaviour of complicated systems often needs to be investigated by system identification, since it usually has to meet certain requirements. The use of computers and efficient mathematical tools allow an identification of the process dynamics by evaluating the input and output signals of the system. The result of such process identification is usually a mathematical model, by which the dynamic behaviour can be estimated or predicted.

As regards the publications by Natke (1982), Masri, Bekey, Sassi, and Caughey (1982), Masri, Sassi, and Caughey (1982), and Schoukens and Pintelon (1991) presented various methodologies for different type of problems in system identification. Ibanez (1979) has reviewed various techniques for improving structural dynamic models and Datta, Shrikhande, and Paul (1998) reviewed problems related to system identification of buildings done until that date. Some recent related publications may be mentioned as those of Loh and Tou (1995); Yuan, Wu, and Ma (1998); Quek (1999); Sanayei, McClain, Wadia-Fascetti, and Santini (1999); Lus, Betti, and Longman (1999); Huang (2001); Brownjohn (2003); Wrobleski and Yang (2003); Yang, Lei, Pan, & Huang (2003); and Chakraverty (2004a, 2005a).

It is known that the systems, which may be modeled as linear, the identification problem often turns into a nonlinear optimization problem. This requires an intelligent iterative scheme to get the required solution. There exist various online and off-line methods, namely the Gauss-Newton, Kalman filtering and probabilistic methods, such as maximum likelihood estimation. However, the following two basic difficulties are faced often for the identification problem with a large number of parameters:

1. The objective function surface may have multiple maxima and minima and the convergence to the correct parameters is possible only if the initial guess is considered as close to the parameters to be identified.

2. The inverse problem in general gives nonunique parameter estimates.

To overcome these difficulties, the present chapter introduces an identification methodology for the structural parameters and responses of multistory structures by the use of powerful technique of artificial neural network (ANN). However, recently number of studies viz. Masri et al. (2000); Chassiakos and Masri (1996); Narendra and Parthasarathy (1990); Bani-Han, Ghaboussi, and Schneider (1999); Huang, Hung, Wen, and Tu (2003); Chakraverty, Sharma, and Singh (2003); Chakraverty (2004b, 2005b); and the references mentioned there in used ANN for the structural identification problems.

Here, for given input to the system, rather than solving the inverse vibration problem, the forward problem for each time step has been solved as usual to generate the solution vector. First the initial (prior) values of the physical parameters (stiffness, etc.) of the system are randomized for the numerical experiment and then using these set of physical parameters the responses have been obtained. The responses and the corresponding parameters are used as the input/output in the neural net. An iterative scheme is proposed to train the neural network. When the iterative training of the network is done for an acceptable accuracy the final converged weight matrix is obtained. Then the physical parameters may be identified if new response data is supplied as input to the net. The procedure has been demonstrated here for multistory structure and the structural parameters are identified using the response of the structure subject to initial condition and horizontal (ground) displacement as examples. The model has been tested for the identification of the stiffness parameters of multistory structure using the prior values of the design parameters and the results are found to be reliable and comparable.

Next, real earthquake data have also been simulated using ANN to identify the structural response in terms of time series for multistory buildings. In this connection an excellent work is by Qi, Yang, and Amini (1997), who have used ANN for identification and control of civil structures subject to earthquake motions. A detailed survey of literatures may be found in that study related to this problem and all those references are not repeated here. Recently, Mathur, Chakraverty, and Pallavi (2004) and Chakraverty, Marwala, and Pallavi (2006) also used ANN for response prediction of single degree of freedom system subject to earthquakes that were triggered in India. Earthquake ground motion at a particular building site is very complicated. Due to the complexity of civil engineering structures and the uncertain nature of their model, it is difficult to formulate the mathematical expressions. For example at present, the models which can accurately describe the interaction between the soil and the structure are not available. The earthquake ground motion, when it is strong enough; it sets the building in motion, starting with the foundation and transfers the motion throughout the rest of the building in a very complex way. Here, again the powerful technique of ANN has been used to model the problem for multistory structure. The results obtained from the converged ANN model and the results from usual structural analysis for multistory structure are found to be in good agreement for a given earthquake acceleration data that triggered in India (Uttarkashi).

Understanding Artificial Neural Network for System Identification

Recently, soft-computing techniques are being widely used in the field of applied science and engineering. These techniques provide efficient solution of a computationally complex and mathematically intractable problem. Inspired by the functioning of human brain, these techniques combine the dynamics of natural systems with computers. One tries to simulate an environment where computer works as an intelligent machine by acquiring knowledge from massively parallel, complex and adaptive architecture of nodes. Artificial neural network is one of the popular soft-computing techniques where the complex problems may be solved based on certain arithmetic operations only and it is not required to employ any analytical method. As such, neural networks have an inherent edge over analytical methods Zurada (1994), Narendra and Parthasarathy (1990), Chakraverty, Singh, and Sharma (2006) and Marwala and Chakraverty (2006).

In various fields of study such as structural dynamics, control, pattern recognition, and so forth, one has to estimate the complete behaviour of the system with a little knowledge about the dynamics of the system. The systems under these fields of study are described with the help of multiple variables and these are sometimes subjected to disturbances. These disturbances in the parameters, generally, lead to the complexities in the modeling process of systems. As such, it becomes an important problem to estimate the behaviour of these systems.

Mathematical model of a dynamic system, which is based on measured/empirical data, is known as system identification. In the system identification problems, a set of inputs and resulting outputs for a system is known and we desire to find a mathematical description or model of the system. The problems in system identification are of two types, direct and indirect problems. Direct problems are defined by the equations governing the system and the parameters of the system are known. These parameters are used to find the response of the system to a specific input. Indirect problems or inverse problems are defined by the output response to a given input, which is known, but either the governing equation or some of the parameters of the physical process are unknown. In this regard, Wang and Lin (1998) used Runge Kutta Neural network for the identification problems. Identification of nonlinear dynamic systems using neural network have been studied by Masri, Chassiakos, and Caughey (1992, 1993). Other interesting papers are by Chassiakos and Masri (1991, 1996), Chakraverty et al. (2003), and Chakraverty (2004b, 2005b), who have also used artificial neural network to the problem of identification. The present chapter explores Neural Network techniques for solving system identification problems of dynamical systems. The fundamental equation of vibration has been used to explain the approach.

Artificial Neural Networks

Artificial neural networks are the outcome of the efforts in the direction of simulating a model for the human brain and its functions. These networks are massive parallel models that imitate human brain. It has also been observed that this technique is useful in solving many other scientific and engineering problems such as identification of systems in structural dynamics. Various neural networks that may be used in the system identification are back propagation networks, Hopfield networks, and Kohonen networks. Here, the most powerful technique such as the back propagation artificial neural network has been used.

As such, artificial neural networks become a good choice to model the problems in which real-time adaptation and fast processing of large amounts of data are mandatory requirements. This has made the use of artificial neural networks possible in various area of high performance such as speech or image recognition, robotics, control and system identification.

Thus, a neural network essentially comprises of a number of nonlinear processing elements referred to as neurons or nodes, arranged in several layers including an input layer, an output layer, and one or more hidden layer(s) in between. Each of the layers consists of one or more neurons and output of every neuron is fed to neurons in the next layer. Signals propagate through the connections. The strength of the transmitted signal depends on the numerical weights associated with each connection in the network. Each neuron receives signals along the incoming connection, performs some simple operations, such as computing the sum of the product of inputted signal and connecting weights and then computing the output signal corresponding to each of the outgoing connection. The connecting weights represent the knowledge stored in the neural network. The neural network can be trained using various training algorithms such as error back-propagation, which is a gradient descent learning algorithm. This algorithm is one of the most popular supervised learning algorithms. This algorithm back propagates the error in the output signal until an acceptable level of error is achieved. Once acceptable level of error in the output is achieved, we freeze the weights and say that network has acquired sufficient knowledge about the system. These weights are then used to compute output signal values for new/testing input signal values.

Analysis and Modelling for the System Identification

The system of differential equation of motion for n-story (supposed as n degrees of freedom, Figure 1) structure without damping is:

$$[M]\{\ddot{x}\} + [K]\{x\} = \{F(t)\} \tag{1}$$

where $[M]$ and $[K]$ are mass and stiffness matrices of the system and $\{F(t)\}$ is the ground acceleration and it is zero if subject to ambient vibration.

Let us consider that the initial conditions are given by:

$$\{x(0)\} = \{x_1(0) \quad x_2(0) \quad \cdots \quad x_n(0)\}^T \tag{2}$$

$$\{\dot{x}(0)\} = \{\dot{x}_1(0) \quad \dot{x}_2(0) \quad \cdots \quad \dot{x}_n(0)\}^T \tag{3}$$

Solution of equation (1) for free vibration with given values of mass and stiffness gives the corresponding eigenvalues and eigenvectors. These are denoted respectively by λ_i and $\{A\}_i$, $i=1,..,n$, where ω^2_i $(=\lambda_i)$ are the system's natural frequencies. Then the modal matrix $[A]$

Figure 1. Multistory building with n levels

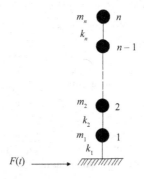

may be constructed which is a partitioned matrix made up of the modal columns or eigen vectors, placed side by side such that:

$$[A] = [\{A\}_1 \quad \{A\}_2 \quad \cdots \quad \{A\}_n]$$

(4)

Also, we denote the diagonal matrix made up of the eigenvalues λ_i, as $[\lambda]_{nxn}$. Next, a new set of coordinates $\{y\}$ related to the co-ordinates $\{x\}$ is introduced by the transformation (Biggs, 1964; Chopra, 1981).

$$\{x\} = [A]\{y\}.$$

(5)

If the system of equation (1) is subjected to an initial velocity only then we have the initial conditions as $\{x(0)\}=0$ with the non-zero values in equation (3). Substituting equation (5) in equation (1) for ambient vibration only with the mentioned initial condition and after the simple analysis one may arrive at the following equation:

$$\{x\} = [A][D][\omega]^{-1}[A]^{-1}\{\dot{x}(0)\}$$

(6)

where:

$$[D] = \begin{bmatrix} \sin(\omega_1 t) & 0 & \cdots \\ 0 & \sin(\omega_2 t) & \cdots \\ \vdots & \vdots & \ddots \end{bmatrix}$$

(7)

where as for the horizontal displacement one may obtain the equations:

$$\{\ddot{y}\} + [\lambda]\{y\} = [P]^{-1}[A]^T\{F(t)\}$$

(8)

where:

$$[P] = [A]^T[M][A]$$

(9)

The final response for this case may be expressed in terms of the original coordinates $\{x\}$ after solving equation (8) for y and then putting in equation (5).

So, response for ambient vibration is given by equation (6) and for the case of horizontal displacement these may be obtained when the right hand side of equation (8) is known. As mentioned earlier the initial design values of the physical parameters of the system are used for the generation of sets of parameters by randomizing the initial design parameters such as the stiffness in the present case. The sets of parameters are generated with the help of uniformly distributed random numbers in [0,1]. The training patterns are the responses of the system and the corresponding sets of parameters for which the responses were obtained. These patterns are now trained using error back propagation training algorithm (EBPTA) of generalized delta learning rule (Rumelhart, Hinton, & Williams, 1986; Schalkoff, 1994).

Error Back Propagation Training Algorithm of ANN

As mentioned before, the first layer is considered to be input layer and the last layer is the output layer in an ANN. Between the input and output layer, there may be more than one hidden layer. Each layer will contain number of neurons or nodes (processing elements) depending upon the problem. These processing elements operate in parallel and are arranged in patterns similar to the patterns found in biological neural nets. The processing elements are connected to each other by adjustable weights.

The input/output behaviour of the network changes if the weights are changed. So, the weights of the net may be chosen in such a way so as to achieve a desired output. To satisfy this goal, systematic ways of adjusting the weights have to be developed, which are known as training or learning algorithm. Neural network (Schalkoff, 1994) basically depends upon the type of processing elements or nodes, the network topology and the learning algorithm. In this investigation, error back propagation training algorithm (EBPTA) with feed forward recall has been used and the typical network is shown in Figure 2a.

In this figure, Z_i, P_j and O_k are input, hidden and output layer respectively. The weights between input and hidden layers are denoted by v_{ji} and the weights between hidden and output layers are denoted by W_{kj}. The procedure may easily be written down for the processing of this algorithm.

Given R training pairs:

$$\{Z_1, d_1; Z_2, d_2; \ldots\ldots\ldots Z_R, d_R\}$$

where Z_i ($I \times 1$) are input and d_i ($K \times 1$) are desired values for the given inputs. Here, the error value is computed as:

$$E = \frac{1}{2}(d_k - O_k)^2, \quad k = 1, 2, \ldots . K \tag{10}$$

The error signal terms of the output (δ_{Ok}) and hidden layers (δ_{Pj}) are written respectively as Zurada (1994).

$$\delta_{Ok} = 0.5 * (d_k - O_k)(1 - O_k^2), \, k = 1, 2, \ldots\ldots . K \tag{11}$$

Figure 2. (a) Layered feedforward neural network, (b) a schematic diagram indicating the implementation of the proposed identification procedure

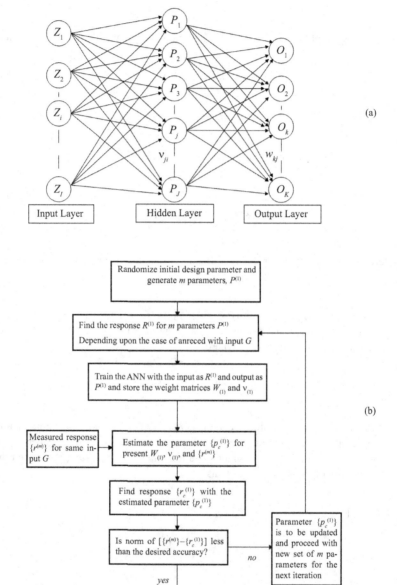

(a)

(b)

$$\delta_{Pj} = 0.5*(1-P_j^2)\sum_{k=1}^{K}\delta_{Ok}W_{Pj}, j=1,2,....J \tag{12}$$

Consequently, output layer weights (W_{kj}) and hidden layer weights (v_{ji}) are adjusted as:

$$W_{kj}^{(New)} = W_{kj}^{(Old)} + \eta\delta_{Ok}P_j, k=1,2.....K \text{ and } j=1,2,.....J \tag{13}$$

$$v_{ji}^{(New)} = v_{ji}^{(Old)} + \eta\delta_{Pj}Z_i, j=1,2.......J \text{ and } i=1,2,.......I \tag{14}$$

where, η is the learning constant.

Identification Algorithm

An iterative methodology for the identification procedure is proposed here which may be written by the following steps:

1. The initial design (structural) parameters are first randomized using the uniformly distributed random numbers in [0,1] and the initial set of the m parameters (say) so generated are denoted as $P^{(1)}$ where:

    $$P^{(1)} = [\{p_1^{(1)}\} \quad \{p_2^{(1)}\} \quad \cdots \quad \{p_m^{(1)}\}].$$

2. When the structure is subjected to the given input G with the above initial parameters, the set of the corresponding responses $R^{(1)}$ in the first iteration are generated with the help of equation (6) for ambient vibration and from equation (5) for the other case (after solving equation (8) for y) as:

    $$R^{(1)} = [\{r_1^{(1)}\} \quad \{r_2^{(1)}\} \quad \cdots \quad \{r_m^{(1)}\}]$$

    where:

    $$\{r_i^{(1)}\} = f(\{p_i^{(1)}\}, G).$$

3. Next the mentioned neural net is trained with the initial responses that are generated from the initial structural parameters as the given input. When the neural net converges, store the converged weight matrices $[W_{(1)}]$ and $[v_{(1)}]$ of equations (13) and (14).

4. Using the measured responses $\{r^{(m)}\}$ of the structure when it is subjected to the same input G, an initial estimate of the parameters are determined as:

$$\{p_c^{(1)}\} = f[(W_{(1)}; v_{(1)}), \{r^{(m)}\}],$$

by direct use of the weight matrices of the trained neural net of step (3).

5. In this step the responses $\{r_c^{(1)}\}$ are obtained by using equation (6) for the first case and from equations (5) and (8) for the other with the obtained estimates of the parameters, $\{p_c^{(1)}\}$ along with the given input G. Then if the norm viz. $\|\{r^{(m)}\} - \{r_c^{(1)}\}\|$ is less than desired accuracy print the current parameters as the final output. Otherwise stiffness patterns are updated suitably utilizing the above estimated parameters $\{p_c^{(1)}\}$ to re-train the neural network and go to the next step.

6. The updated parameters are then used to find the responses again by utilizing equation (6) for ambient vibration and from equations (5) and (8) for the forced case with the given input G.

7. Train the neural network for generating the converged weight matrices $[W_{(2)}]$ and $[v_{(2)}]$ as in step (3).

8. Go to step (4) with converged weight matrices $[W_{(2)}]$ and $[v_{(2)}]$ for obtaining next estimate of the parameters as $\{p_c^{(2)}\}$. Then again the norm, $\|\{r^{(m)}\} - \{r_c^{(2)}\}\|$ is checked for the desired accuracy where $\{r_c^{(2)}\}$ is the response vector obtained by the same way as before for the given input G and using the currently obtained estimates of the parameters $\{p_c^{(2)}\}$.

The above step by step procedure is shown in Figure 2b which clearly indicates the implementation of the proposed identification algorithm.

Results and Discussion

Numerical experiment has been demonstrated for structural system using EBPTA algorithm of ANN with the proposed iterative steps to identify physical properties of the structure. In particular, here, numerical experiment has been shown for two-story lumped mass structure to identify stiffness parameters.

The initial stiffness parameters of the system have been used to generate sets of parameters. Then the corresponding responses are computed from equations (6) and (5) (after solving equation (8) for y) and the ANN has been trained in an iterative cycle using the proposed methodology as above. The procedure has been analyzed by considering different data sets for its reliability. Here, the methodology has been discussed first by giving the results for

Figure 3. Testing of the neural weights using the input pattern (Ambient vibration, Initial parameter, K_1=2500, K_2=2400)

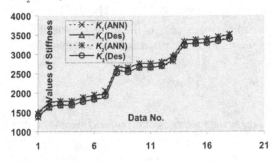

Figure 4. Testing of the neural weights using the input pattern (Ambient vibration, Initial parameter, K_1=1000, K_2=1000)

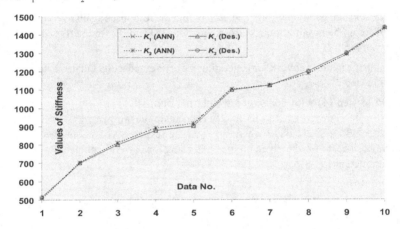

two cases viz. case (1) initial stiffness parameters as (a) K_1 = 2500, K_2 =2400 and (b) K_1 = K_2 =1000 for the first problem and case (2) (a) K_1 = 2200, K_2 =2100 and (b) K_1 = K_2 =1000 only for the second problem with the mentioned structural system. All the parameters are taken to be in consistent units. The system is subjected first to initial condition for case (1) (a) and (b) expressed by the vector (with zero displacement):

$$\{\dot{x}(0)\} = \{8 \quad -8\}^T \text{ and}$$

$$\{\dot{x}(0)\} = \{10 \quad -10\}^T, \text{ respectively,}$$

and then the forcing function is taken for case (2) with zero initial condition as:

Table 1. Effect of number of neurons on the identified parameters (with initial parameter $K_1 = 1000$, $K_2 = 1000$) for Example (1) (b)

Neural Architecture	K_1		K_2	
	Desired	Neural	Desired	Neural
3-10-2	1500	1507.93	1500	1507.92
3-14-2		1496.98		1497.15
3-16-2		1496.63		1497.42
3-20-2		1497.67		1493.82
3-10-2	900	912.78	900	912.81
3-14-2		921.59		921.64
3-16-2		915.27		915.78
3-20-2		914.35		918.92
3-10-2	850	814.22	800	814.25
3-14-2		812.82		812.97
3-16-2		810.52		810.63
3-20-2		807.10		811.94
3-10-2	750	763.53	750	763.56
3-14-2		763.72		763.91
3-16-2		758.10		758.37
3-20-2		757.26		761.94

$$F(t) = \begin{cases} 10\,Sin(2\pi t), & 0 \leq t \leq 2 \\ 0, & t > 2 \end{cases}$$

The input and output in the neural net are the obtained responses and the corresponding physical parameters. So, the input layer will have the neurons in Figure 2a as the maximum responses for each story and -1 for the augmented neuron as required for the present rule of the ANN. The output layer contains the corresponding stiffness parameters of the system.

When the network is trained for a given accuracy, the corresponding (converged and final) weights of the connections of the neurons are saved. These trained converged weights are again tested for accuracy for the pattern that had been used in the training. Consequently some of the results of the testing for the given initial condition (case (1)(a) and (b)) have been incorporated in Figures 3 and 4 for initial stiffness parameters as $K_1 = 2500$, $K_2 = 2400$ and $K_1 = K_2 = 1000$ respectively, where parameters from neural and desired are shown for each case. It may be seen that the neural results are comparable with the desired (i.e. the used training pattern). Similarly testing for the problem with the considered forcing function (case (2)(a) and (b)) are given in Figures 5 and 6 for initial stiffness parameters $K_1 = 2200$, $K_2 = 2100$ and $K_1 = K_2 = 1000$. In order to have the accuracy of the results, the correlation coefficient between the desired and neural computations is found and those are 0.987 for all the training patterns giving acceptable weight matrices for the accuracy of the said problems.

Figure 5. Testing of the neural weights using the input pattern (Ground displacement, Initial parameter, K_1=2200, K_2=2100)

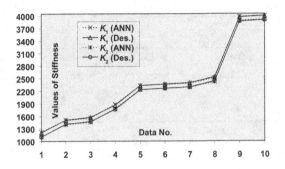

Figure 6. Testing of the neural weights using the input pattern (Ambient vibration, Initial parameter, K_1=1000, K_2=1000)

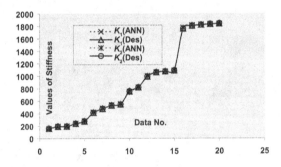

Figure 7. New pattern testing using the converged weights of the network (Ambient vibration, Initial parameter, K_1=2500, K_2=2400)

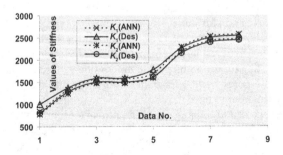

Figure 8. New pattern testing using the converged weights of the network (Ambient vibration, Initial parameter, K_1=1000, K_2=1000)

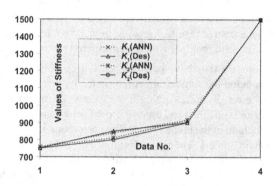

Figure 9. New pattern testing using converged weights of the network (Ground displacement,Initial parameter, K_1=2200, K_2=2100)

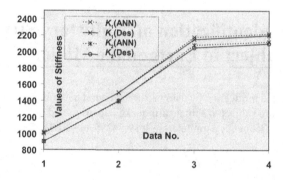

Figure 10. New pattern testing using the converged weights of the network (Ambient vibration, Initial parameter, K_1=1000, K_2=1000)

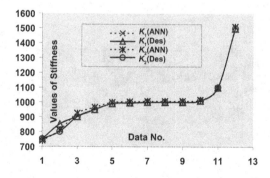

Next, new patterns are generated using equations (5) and (6) and these are fed into the network with the above generated converged weights. Figures 7 and 8 show the identified stiffness parameters from some new patterns for the first problem, that is, with only the initial condition (ambient vibration). It is clear from these Figures that the identified parameters are reliable as compared to the desired results. Again the new pattern testing has been shown in Figures 9 and 10 for the second problem with the considered forcing function.

Extensive numerical computations have been carried out by considering various neural architectures viz. by changing the number of neurons in the hidden layer, and so forth. This was done to see how the results are changing if we change the number of neurons. It is known generally that there is no direct and precise way of determining the most appropriate number of neurons to include in the hidden layer. The corresponding results for the case of Problem (1) (b) is given in Table 1. The first column of this table contains the architecture of the model as 3-N-2, where N is the number of neurons in the hidden layer. Here the value of N has been considered as 10, 14, 16 and 20. The identified parameters from each of the above architecture of the neural net is given in Table 1. It is to be noted here that the results are not good if the number of nodes are taken less than 16 in the hidden layer. Therefore, in all of the example problems the results are incorporated by taking N=16 to have the acceptable accuracy.

Response Identification of Multistory Structure Subject to Earthquake Motions

In this head, response identification by using artificial neural network has been investigated for a two-story structure subject to actual earthquake data triggered in India. The two-story structural system with frequency parameters $\omega_1 = 19.54395$, $\omega_2 = 51.16673$ and damping $= 0.1$

Figure 11. Uttarkashi Earthquake at Barkot in NE direction (Maximum response = 0.931m/s²)

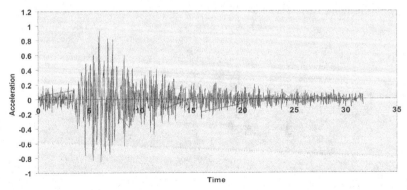

Figure 12. (a) Response comparison between neural and desired for first story, (b) response comparison between neural and desired for second story

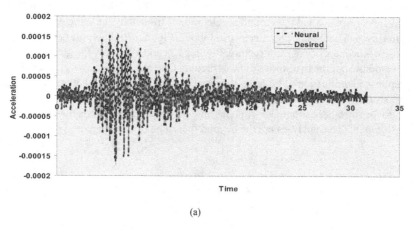

(a)

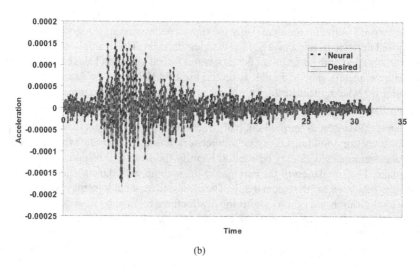

(b)

(10%) has been considered here for the investigation. Indian earthquake viz. the Uttarkashi earthquake occurred on October 20, 1991, (maximum ground acceleration = 0.931 m/sec/sec) at Barkot in NE (north–east) direction as given in Figure 11 have been considered for the training. First the ground acceleration of Uttarkashi earthquake was used to compute the response for the first and second-story structure using the usual procedure. After obtaining responses for first and second story, the first-story response and the ground acceleration is trained by the said ANN model for the mentioned structural system. This training was done for the total time range 0 to 31.72sec (1,587 points, earthquake period), taking the continuous activation function with accuracy 0.0005. When the training is completed, the weight matrices are stored. The ANN results have been obtained then by direct use of the trained weight matrix for the said earthquake ground acceleration. The neural and desired results obtained after training for first story of the structure are shown in Figure 12a. Similarly, the second-story response and the ground acceleration is trained by the same ANN model for a structural system with same parameters used above. Figure 12b shows the neural and desired response of second story of the structure. Figures 12a and 12b show the good identification of the responses for the two stories of the structure. One can use also a single model of ANN for response identification of both the stories at a time. The methodology may be extended for tall structures and, if trained, then it may also predict the dynamical behaviour of the structure for any other earthquake without going in to the detail analysis for each earthquake data. Although, these analyses are the beyond the scope of this chapter.

Conclusion

The present chapter demonstrates the application of soft computing technique such as ANN for the identification of structural parameters of multistory shear buildings by solving the forward vibration problem. Only the design parameters of the system is utilized in a proposed iterative training of ANN for the identification. The example problems namely ambient vibration with some initial condition and forced vibration have been analysed for different parameters. The main aim is on the use of ANN by a new *iterative scheme,* in particular for system identification problems and obtaining the weight matrices similar to memory matrices.

The chapter also includes the prediction of structural responses subject to actual earthquake data for multistory buildings using neural network model. The model when trained for a particular earthquake data may be used to identify the structural responses for any other earthquake. The simulation of the earthquake data, along with the structural responses of multistory buildings, are the recent trends. The discussed methodology may be used in future for the identification and health monitoring of structures for unknown earthquakes that may trigger, with good generalization of the present soft computing models.

Acknowledgments

The author would like to thank Director, C.B.R.I., for giving me permission to publish this chapter, and to Ms. Pallavi Gupta for drawing some of the Graphs. Thanks are also due to the Department of Earthquake Engineering, IIT, Roorkee for some of the data that are made available.

References

Bani-Han, K., Ghaboussi, J., & Schneider, S. P. (1999). Experimental study of identification and control of structures using neural network: Part 1 : Identification. *J. Earthquake Engng. and Struct. Dyn.*, *28*, 995-1018.

Biggs, J. M. (1964). *Introduction to structural dynamics*. McGraw-Hill.

Brownjohn, J. M. W. (2003). Ambient vibration studies for system identification of tall buildings. *J. Earthquake Engng. And Struct. Dyn.*, *32*, 71-96.

Chakraverty, S. (2004a). Modelling for identification of stiffness parameters of multistory structure from dynamic data. *J. Sci. Ind. Res.*, 63, 142-148.

Chakraverty, S. (2004b, September 7–9). Neural modelling based identification of structural parameters of multistory shear buildings. In *Proceedings of the Seventh International Conference on Computational Structures Technology (CST)*, Lisbon, Portugal.

Chakraverty, S. (2005a). Identification of structural parameters of multistory shear buildings from modal data. *J. Earthquake Engng and Struct. Dyn.*, *34*, 543-554.

Chakraverty, S. (2005b). Identification of structural parameters of multistory shear buildings by an iterative training of neural networks. Architectural Science Review (Submitted).

Chakraverty, S., Marwala, T., & Gupta, P. (2006). Response prediction of structural system subject to earthquake motions using artificial neural network. *Asian Journal of Civil Engineering*, *7*(3), 301-308.

Chakraverty, S., Sharma, R. K., & Singh, V. P. (2003). Soft-computing approach for identification of dynamic systems. *J. New Build. Mat. & Const. World, 9*(2), 50-56.

Chakraverty, S., Singh, V. P., & Sharma, R. K. (2005). Regression based weight generation algorithm in neural network for estimation of frequencies of vibrating plates. *Computer Methods in Applied Mechanics and Engineering*, *195*, (pp. 4194-4202).

Chassiakos, A. G., & Masri, S. F. (1991). Identification of the internal forces of structural systems using multiplier networks. *Comput. Systems Eng., 2*, 125 -134.

Chassiakos, A. G., & Masri, S. F. (1996). Modelling unknown structural systems through the use of neural networks. *J. Earthquake Engng. And Struct. Dyn., 25*, 117-128.

Chopra, A. K. (1981). *Dynamics of structures, a primer*. Berkeley, CA: Earthquake Engineering Res. Inst..

Datta, A. K., Shrikhande, M., & Paul, D. K. (1998). System identification of buildings—a review. In *Proceedings of the Eleventh Symposium on Earthquake Engineering*, Roorkee, India.

Huang, C. S. (2001). Structural identification from ambient vibration measurement using the multivariate AR model. *J. Sound and Vib., 241*(3), 337-359.

Huang, C. S., Hung, S. L., Wen, C. M., & Tu, T. T. (2003). A neural network approach for structural identification and diagnosis of a building from seismic response data. *J. Earthquake Engng. And Struct. Dyn., 32*, 187-206.

Ibanez, P. (1979). Review of analytical and experimental techniques for improving structural dynamic models. *Welding Research Council Bulletin, 249*.

Loh, C. H., & Tou, I. C. (1995). A system identification approach to the detection of changes in both linear and non-linear structural parameters. *J. Earthquake Engng. And Struct. Dyn., 24,* 85-97.

Lus, H., Betti, R., & Longman, R. W. (1999). Identification of linear structural systems using earthquake induced vibration data. *J. Earthquake Engng. And Struct. Dyn., 28,* 1449-1467.

Marwala, T., & Chakraverty, S. (2006). Fault classification in structures with incomplete measured data using autoassociative neural networks and genetic algorithm. *Current Science, 90*(4), 542-548.

Masri, S. F., Bekey, G. A., Sassi, H., & Caughey, T. K. (1982). Nonparametric identification of a class of nonlinear multidegree dynamic systems. *J. Earthquake Engng. And Struct. Dyn.,* 10, 1-30.

Masri, S. F., Chassiakos, A. G., & Caughey, T. K. (1992). Structure-unknown non-linear dynamic systems: Identification through neural networks. *Smart Materials Structures, 1,* 45-56.

Masri, S. F., Chassiakos, A. G., & Caughey, T. K. (1993). Identification of nonlinear dynamic systems using neural networks. *Journal of Applied Mechanics, 60,* 23-133.

Masri, S. F., Sassi, H., & Caughey, T. K. (1982). Identification and modelling of nonlinear systems. *Nuclear Engineering and Design, 72,* 235-270.

Masri, S. F., Smyth, A. W., Chassiakos, A. G., Caughey, T. K., & Hunter, N. F. (2000). Application of neural networks for detection of changes in nonlinear systems. *Journal of Engineering Mechanics, 126*(7), 666-676.

Mathur, V. K., Chakraverty, S., & Pallavi, G. (2004). Response prediction of typical rural house subject to earthquake motions using artificial neural network. *Journal of Indian Building Congress, 11*(2), 99-105.

Narendra, K. S., & Parthasarathy, K. (1990). Identification and control of dynamical systems using neural networks. *IEEE Trans. Neural Networks, 1,* 4-27.

Natke, H. G. (1982). *Identification of vibrating structures.* Berlin: Springer.

Qi, G. Z., Yang, J. C. S., & Amini, F. (1997). Neural network for identification and control of civil engineering structures. In N. Kartam, I. Flood, & J. J. Garrett, Jr. (Eds.), *Artificial neural networks for civil engineers: Fundamental and applications* (pp. 92-123). ASCE.

Quek, S. T. (1999). System identification of linear MDOF structures under ambient excitation. *J. Earthquake Engng. And Struct. Dyn.,* 28, 61-77.

Rumelhart, D. E., Hinton, G. E., & Williams, R. J. (1986). Learning representations by back-propagation errors. *Nature, 323,* 533-536.

Sanayei, M., McClain, J. A. S., Wadia-Fascetti, S., & Santini, E. M. (1999). Parameter estimation incorporating modal data and boundary conditions. *Journal of Structural Engineering, 125*(9), 1048-1055.

Schalkoff, R. J. (1994). *Pattern recognition: Statistical, structural and neural approaches.* New York: McGraw-Hill.

Schoukens, J., & Pintelon, R. (1991). *Identification of linear systems: A practical guideline to accurate modeling.* New York: Pergamon Press.

Wang, Y.-J., & Lin, C.-T. (1998). Runge-Kutta neural network for identification of dynamical systems in high accuracy. *IEEE Transactions on Neural Networks, 9*(2), 294-307.

Wrobleski, M., & Yang, H. T. Y. (2003). Identification of simplified models using adaptive control techniques. *Journal of Structural Engineering, 129*(7), 989-997.

Yang, J. N., Lei, Y., Pan, S., & Huang, N. (2003). System identification of linear structures based on Hilbert-Huang spectral analysis, Part 1: Normal modes. *J. Earthquake Engng. And Struct. Dyn., 32,* 1443-1467.

Yuan, P., Wu, Z., & Ma, X. (1998). Estimated mass and stiffness matrices of shear building from modal test data. *J. Earthquake Engng. And Struct. Dyn., 27,* 415-421.

Zurada, J. M. (1994). *Introduction to artificial neural systems.* West.

Chapter XV

Application of Neurocomputing to Parametric Identification Using Dynamic Responses

Leonard Ziemiański, Rzeszów University of Technology, Poland

Bartosz Miller, Rzeszów University of Technology, Poland

Grzegorz Piątkowski, Rzeszów University of Technology, Poland

Abstract

The chapter focuses on the applications of neurocomputing to the analysis of identification problems in structural dynamics, the main attention is paid to back-propagation neural networks. The analysed problems relate to (a) application of dynamic response to parameter identification of structural elements with defects modelled as a local change of stiffness or material loss; (b) updating of FEM models of beams, including the identification of material parameters and parameters describing possible defect; (c) identification of circular void or supplementary mass in vibrating plates; (d) identification of a damage in frame structures using both eigenfrequencies and elements of eigenvectors as input data. In the examples involving the experimental measurements the application of a random noise to increase the not sufficient number of data is proposed. The presented results have proved the proposed method capable of carrying out the appointed task and indicated good prospects of neuro-computing application to dynamics of structures.

Introduction

Artificial neural networks (ANNs) have features associated with their biological origin (Haykin, 1999). They are massively parallel and can process not only crisp but also noisy and incomplete data. Moreover, ANNs can be used for both approximation and classification purposes (Paez, 1993). Thanks to their generalization features they can be applied to the analysis of problems which obey certain rules learned by the networks during the training process. Such features enable us to apply ANNs in both direct and inverse analysis. ANNs advantages offer, in fact, complementary possibilities to standard computational methods, finite element method (FEM) in particular.

Computer simulations of artificial neural networks, called for short neurocomputing, can be used efficiently in the analysis of various problems of structural engineering. This concerns especially a standard feedforward neural network called in literature "multilayer perceptron" or "back-propagation neural network" (BPNN). BPNN is used for mapping of its inputs into outputs without an a priori assumed format of the approximated relations. This ability of BPNNs opens the door for implicit modelling of structural relations. Another possibility is associated with formulation of outputs as identified structural or material parameters, which corresponds to identification of structural systems (inverse analysis). For detailed explanation of neural networks basics, see Chapter XVI in this book. Following the topics of the chapter, the main attention will be focused on the applications of neurocomputing to the analysis of identification problems (inverse problems). For such purposes, NNs can be used either as an independent tool or they can interact with standard computational methods.

Different types of ANNs can be used in the inverse analysis of structural mechanics problems. The majority of engineering applications is related to taking advantage of the multilayer perceptron. The main part of the chapter is focused on ANNs applications to the identification problems of structural engineering.

The authors' attention is focused on parameter identification related to the so-called explicit modelling. It means that certain structural or material characteristics are adopted as functions of unknown *parameters,* which are to be *calibrated* during the identification process.

The problems considered above and corresponding tools are discussed on examples of engineering problems with particular attention to structural experimental mechanics.

In the Department of Structural Mechanics of the Rzeszów University of Technology, Poland, different nondestructive methods of materials and structural element testing have been developed. Special attention has been paid to measurements of dynamic responses to impact excitations. After transformation of records from time to spectral domain (fast Fourier transformation (FFT) was used) natural eigenfrequencies can be obtained. The values of eigenfrequencies are functions of unknown parameters corresponding to the analysed problem. If the eigenfrequencies are used as inputs of BPNN the unknown, sought parameters can be obtained as outputs from this NN.

The above sketched approach is discussed with respect to problems of damage parameter identification for simple structures, identification of various parameters (mass, material constant) and updating of dynamic models.

The damage parameter identification was also performed by wave propagation technique, using records from time domain without their transformation to spectral domain. Sets of patterns were generated by the FEM program or formulated on the base of results obtained from tests on laboratory specimens.

The basic approach presented in this chapter is to detect changes in dynamic behaviour of the structure that may be characterised by (a) the natural frequencies and mode shapes, (b) frequency response function, and (c) wave propagation. The reported problems deal with damage assessment of multistory frames and updating of dynamic model of beam. The achieved results indicated good prospects of neurocomputing application to dynamics of structures.

Measures of Error

In the chapter the most common measure of error is root mean square error (*RMSE*) defined as a root of MSE error:

$$MSEV = \frac{1}{V \cdot M} \sum_{p=1}^{V} \sum_{i=1}^{M} (t_i^{(p)} - y_i^{(p)})^2, \tag{1}$$

$$RMSEV = \sqrt{MSEV}, \tag{2}$$

where $t_i^{(p)}, y_i^{(p)}$ – *i-th* output from *p-th* pattern, either target value t or value obtained from neural networks y, M – the number of outputs. The letter V denotes the number of patterns taken into account during learning ($V=L$ and *MSEL, RMSEL*) or testing ($V=T$, *MSET, RMSET*).

There are also used statistical parameters which describe the accuracy for each output separately: standard error $St\varepsilon_i$ and linear regression coefficient R_i:

$$St\varepsilon_i = \sqrt{\frac{1}{V} \sum_{p=1}^{V} (t_i^{(p)} - y_i^{(p)})^2}, \quad R_i = \frac{\sum_{p=1}^{V} (t_i^{(p)} - \bar{t}_i)(y_i^{(p)} - \bar{y}_i)}{\sqrt{\sum_{p=1}^{V} (t_i^{(p)} - \bar{t}_i)^2 \sum_{p=1}^{V} (y_i^{(p)} - \bar{y}_i)^2}} \tag{3}$$

where $\bar{t}_i, \bar{y}_i$ – mean values of sets $\{t_i^{(p)}\}, \{y_i^{(p)}\}$ for fixed subscript i.

Additionally, standard deviation σ_i and relative errors ep_i are applied to assess network accuracy:

$$\sigma_i^2 = \frac{1}{V} \sum_{i=1}^{V} (ep_i - \bar{e}_i)^2 \text{ for } \bar{e}_i = \frac{1}{V} \sum_{i=1}^{V} ep_i \tag{4}$$

where:

$$ep_i = \left| 1 - \frac{y_i^{(p)}}{t_i^{(p)}} \right|.$$

Another measure of error was employed to assess to the accuracy of eigenfrequencies prediction by an updated model. The *RMSE* error of eigenfrequencies obtained from the updated model was applied, the error defined by the formula:

$$RMSE_{k-n} = \sqrt{\frac{\sum_{i=k}^{n}\left(\frac{f_{0i} - f_i}{f_{0i}}\right)^2}{n - k + 1}}, \tag{5}$$

where k – the first of the considered eigenfrequencies, n – the last of the considered eigenfrequencies, i – the number of eigenfrequency, $(i=k, k+1,...,n)$, f_{0i} – i-*th* eigenfrequency obtained from the measurements, f_i – i-*th* eigenfrequency obtained from the numerical simulations. For other measures of error, see Chapter XVI.

Data and their Preprocessing

The patterns for ANN learning and testing can be taken from:

- **Computer simulation of direct problems** (pseudo-experimental data)
- **Tests** on laboratory models or **measurements** on full-scale structures (e.g., real buildings)
- **Generation of noisy data**, that is, superposing artificial perturbation (e.g., adding random noise) on experimental or pseudo-experimental data

Original data are usually preprocessed in order to generate sets which can be efficiently explored during the ANN training or testing process. Data preprocessing can be related to:

- Standard **analytical transformation** (scaling, normalization, transition from time to spectral domains, calculating geometric characteristics of selected bands of investigated signal)
- Applications of special ANNs either to computing the neural outputs and using them as inputs in the **cascade package** or to **input data compression** in order to diminish the size of master ANNs (**replicators**)
- Introduction of **artificial noise** which enables increasing $K+1$ times the number of available patterns

For each of original patterns $p_j = \{x_j, y_j\}$, j being the number of investigated pattern, K new patterns are created according to the formula:

$$\tilde{p}_j^k = \{\tilde{x}_j^k, y_j\}, \quad \tilde{x}_j^k = (1 + \sigma\{n_k\}) \quad \text{for } k=1,...,K. \tag{7}$$

The noise set $\sigma\{n_k\}$ is K-times randomly selected from the range $[-3\sigma, 3\sigma]$ assuming normal probability density function with mean value $\mu = 0$ and variance σ. The new set of patterns consists of all p_j and $\tilde{p}_j^k$ patterns. By introducing an artificial noise to sets of patterns their number can be increased. This approach can be used efficiently in the formulation of networks with the parameter number corresponding to the number of learning patterns. The approach has two advantages: (a) a network formulated with a larger amount of patterns has, in general, better generalization properties than a network trained on a small number of patterns, and (b) a network trained on patterns with artificial noise can be more resistant (not sensitive) to perturbations of input variables.

Cascade BPNNs

Back-propagation neural network is a feed-forward, multilayer network of standard structure (i.e., neurons are not connected within the same layer, they are connected only with all the neurons of previous and subsequent layers, respectively). In a classical neural network the input layer serves only to introduce signals (values of input variables) while the hidden and output layers are composed of neurons (processing units). BPNNs discussed above can be called "one-level (standard) neural networks." In some cases of data sets it is possible to separate output variables and formulate a *cascade of BPNNs* (see Figure 1). In the cascade approach the first level network BPNN-I has only one scalar output, y_1, and after training this output is introduced as an additional input to the second-level network BPNN-II. Then both outputs y_1 and y_2 are used as additional inputs of the third level network BPNN-III.

Figure 1. The three-stage cascade network

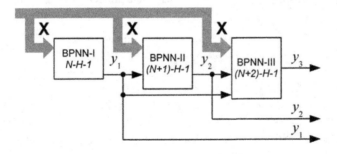

Replicator

The application of a time or frequency domain signal as a source of network input information requires prior preprocessing of the signal. One of the approaches presented in the chapter is the application of a specialized ANN called a signal replicator (see section "BPNN as a Replicator" in Chapter XVI). If the investigated signal is composed of N points, a network of N-h-N architecture is learned to replicate at the output the data given at the input. During the replication in a hidden layer, which contains fewer neurons than the number of inputs, the compression of the signal is performed. The network compresses information given at the input from N to h values, and then decompresses it back to N values. To obtain a network outputting a compressed signal the output layer of the learned N-h-N network is removed. The task of the output layer is taken over by so far hidden layer. Networks N-h give on the output signals condensed into h values.

Damage Detection Using Wave Propagation

Introduction

Nondestructive methods of detection of material properties change and damage in structural elements are an important and valuable tool. These methods allow estimating the state of a structure as well as predicting the time of safe usage. An ultrasonic method is one of the most often used nondestructive methods (Thompson, 1983). For structures such as rods, plates and shells another widely used method is based on structural waves propagation (Lee & Staszewski, 2003a, 2003b; Su & Ye, 2005; Su, Ye, Bu, Wang, & Mai, 2003). In this method (unlike in the classical method of ultrasonic testing), the wavelengths are large compared with the characteristic dimension of the structure, and a wave pulse propagates along the whole structure. Moreover, the excitation of structural waves and their velocity measurements are taken at different points (the velocity can be measured at a few different points) instead

Figure 2. The idea of structural wave propagation test

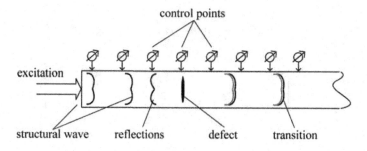

of surface scanning used in ultrasonic testing. It makes application of structural waves test easier in rods, beams, plates, and some other structures (Palacz & Krawczuk, 2005).

Figure 2 shows a structural wave propagation test performed to identify a defect inside a rod. In what follows a micro-impulse type excitation is considered, applied to a long rod in order to measure wave transition without reflections from rod clamped boundary. The changes of wave records, caused by disturbances and measured at control points, were taken as a base for the identification analysis.

ANNs were successfully applied to the identification of damage parameters, for both 1D and 2D structural models (Oishi, Yamada, Yoshimura, & Yagawa, 1995; Zang & Imregun, 2001; Zapico, González, & Worden, 2003; Ziemianski & Miller, 2000). Two analysed problems are discussed in short: (a) identification of local changes of stiffness in an elastic rod, and (b) identification of parameters of defect modelled by material loss in an elastic strip.

The FE code ADINA (ADINA, 2001) was applied to generate patterns as longitudinal wave records measured at control points, and then the MATLAB NN Toolbox (Demuth & Beale, 1998) and Levenberg-Marquardt learning algorithm were used. Neural networks with one or two hidden layers were tested. The input vectors consisted of preprocessed time signals, the outputs provided all parameters describing damage. The network identified some of these parameters with satisfactory precision, while others with a significant error. Cascade neural networks were built to improve the generalization abilities of networks (Ziemianski & Piatkowski, 2000).

Identification of Local Changes of Stiffness in an Elastic Rod

An elastic rod of constant cross-section and length of 8.0 m was investigated (see Figure 3). The rod was divided into 80 truss elements. The triangular micro-impulse of duration 200μs was applied to the rod and the wave velocity was calculated at control points. A group of finite elements with elasto-plastic material was inserted in the rod. The plastic-bilinear material model was used. The Young's modulus and strain hardening modulus were fixed. A failure was simulated by the change of yield stress $\sigma_0 \in [0.1, 0.9]$ varying with the increment 0.05. Width b and location l of failure were also varying (see Figure 3). Four different locations and two different widths were used.

Figure 3. Structural wave propagation test the rod with yielding zone

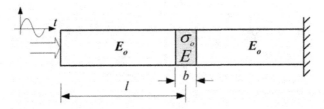

Figure 4. The results of identification of the yield stress parameter for standard (a) and cascade (b) networks

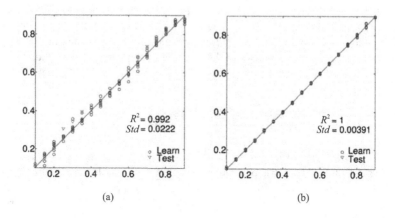

(a)   (b)

Table 1. Learning errors of yielding zone parameters identification

Stat. param.	Yield stress	Location	Width
	Standard net		
R^2	0.922	0.995	0.769
$St\ \varepsilon$	0.0222	0.0117	0.097
	Cascade net		
R^2	1.000	1.000	0.886
$St\ \varepsilon$	0.0039	0.0012	0.0680

The measurement points were located beyond the zone of yielding. For identification networks the input vectors consisted of compressed time signals. The compression was performed by specialized neural network called replicator (Haykin, 1999) of architecture 200-8-200. In an identification network, the output vectors consisted of three parameters of failure (σ_0, b, l). When standard nets were used, the yield strength parameter and the location of yield zone were identified correctly, while the width of defect was identified with much poorer accuracy. The application of cascade networks improved the identification of all parameters of defect (see Figures 4 and Table 1).

Another kind of stiffness change of a selected part of the rod was simulated by a change of the Young's modulus for one or several consecutive finite elements. The damage was placed in 1, 3, 5, or 11 neighbouring finite elements, so the width of defect was fixed as b = 0.1, 0.3, 0.5, 0.7 and 1.1 m. The location $l \in [2.1, 5.1]$ m was varying with the step Δl = 0.1 m. The ratio of moduli $ed = E /E_o \in [0.25, 1.75]$ was varied with the increment Δed = 0.1. A few thousand combinations of b, l, ed parameters were obtained in that way. For these combinations dynamic responses of the numerical model were calculated. A detailed analysis of the discussed problem was presented by Ziemianski and Piatkowski (2000).

Parameter Identification of Material Loss

A homogeneous steel strip of length 8.0 m and height 1.0 m was analysed by Ziemianski and Piatkowski (2000). Geometric parameters of a rectangular hole (see Figure 5), that is, depth $h \in [0.1, 0.8]$ m, width $b \in [0.1, 0.9]$ m, and location $l \in [4.1, 8.1]$ m, were identified. The triangular micro-impulse of duration 200μs was applied in the middle of the strip free edge and the velocity of the propagated wave was computed in time domain. The signals corresponding to the propagated wave velocities, computed by ADINA at four control points, were discretized for equal time increments Δt. Then N = 200 values of velocity v_{yi} were compressed into n = 12 values by means of the replicator. In this way $4 \times 12 = 48$ inputs were formulated for the network BPNN to identify three damage parameters h, l and b.

P = 1,640 patterns were computed using ADINA FE system, from among them 800 training and 840 testing patterns were randomly selected. Both standard and cascade nets were used at this stage. Better results were obtained from neural networks with two hidden layers. Stuttgart neural network simulator (SNNS; 1995) and resilient back-propagation (Rprop) learning algorithm (Haykin, 1999) were used for the intensive cross-validation which gave the following, two-hidden layer standard network: BPNN 48-12-8-3.

The width of defect has no influence on the quality of results. Damage identification is significantly better when cascade networks are used. In Table 2 the results of neural simulation are presented. The values of statistical parameters R^2 and $St\varepsilon = Std$ are shown in Table 2 for both the standard and cascade BPNNs. These results are also presented in Figure 6.

Figure 5. Structural wave propagation test the rod with notch

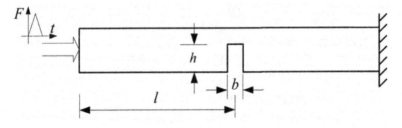

Figure 6. The results of identification of the depth (a, b), width (c, d) and location (e, f) of rectangular hole for standard (left column) and cascade (right column) networks

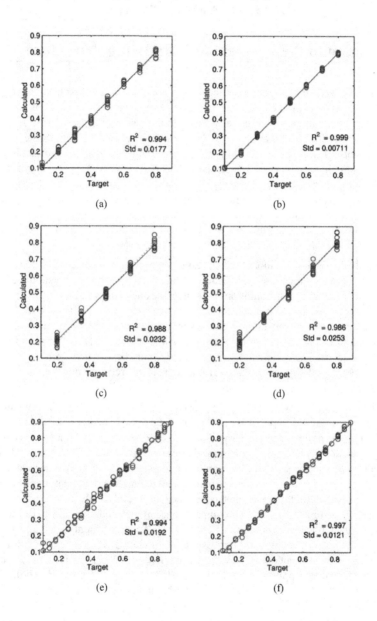

Updating of Finite Element Models of Beams and Frames

Introduction to the Problem and Updating Procedure

The current progress in both the development of new computational tools and in the construction of new, powerful computer hardware makes it possible to build very complex dynamic models. The time required to numerically simulate the response of a structure to static or dynamic excitations is shorter and shorter. Unfortunately, even very complicated models, with the number of degrees of freedom (DOF) exceeding hundreds of thousands, are in some cases not able to handle the present-day requirements. In some particular cases it is necessary to employ models which are capable of simulating or reproducing the behaviour of real structures with very high accuracy. In order to minimize the differences between the results of experimental measurements and numerical simulations some changes should be introduced to the model. The introduction of such changes is called *updating* (Friswell & Mottershead, 1996; Mottershead & Friswell, 1993; Natke, 1998).

In order to obtain an updated model which can be employed in a variety of different tasks (including, for example, damage identification), the model must be able to simulate the behaviour of the structure also in conditions other than the one considered during the updating. This may be achieved by careful selection of physical parameters of the model (e.g., the Young's modulus, mass density, joint rigidity, support stiffness) and the iterative "improvement" of these parameters with continuous physical interpretation of the changes introduced to model parameters. The parameters to be updated should be selected from among model parameters determined with the highest error and the measured structure responses should be sensitive to these parameters changes.

The crucial point of the whole updating procedure is an experiment. Unfortunately, in the majority of cases it is impossible to measure the response of the structure to external excitation at all the points necessary from the point of view of applied updating technique; especially the measurements of rotational degrees of freedom cause many difficulties (Ewins, 2000). Moreover, a significant number of degrees of freedom may be inaccessible. The amount of measurement data may be insufficient to perform the updating of the whole model, so the computational model must be reduced, or—rarely— the measurement model is expanded. Both approaches enable computational model updating and both introduce additional errors. However, in some cases the amount of measurement data is so small that the updating of reduced model is still impossible because the deterministic methods require an adequate amount of input data (Friswell & Mottershead, 1996). The presented updating problem may be solved applying neural networks (Chang, Chang, Xu, & To, 2002; Levin & Lieven, 1998; Lu & Tu, 2004). The proposed NN model updating procedure consists of the following steps:

- Generation of a set of training data vectors based on the dynamic model
- Training of the neural network with training data

Table 2. Comparison of statistical parameters of cross-section defect identification

Stat. param.	Height of defect		Location of defect		Width of defect	
	Learn	Test	Learn	Test	Learn	Test
	Standard net					
R^2	0.996	0.994	0.994	0.994	0.990	0.988
$St\ \varepsilon$	0.0149	0.0177	0.0193	0.0192	0.0215	0.0232
	Cascade net					
R^2	1.000	0.999	0.999	0.997	0.994	0.986
$St\ \varepsilon$	0.0043	0.0071	0.0094	0.0121	0.0165	0.0253

- Exposition of experimental data to obtain a set of changes
- Application of the changes to the original model in order to generate a new model
- Repetition of the previous steps if necessary

This procedure was initially tested on numerical data only (Ziemianski & Miller, 2000). In this chapter, the updating method involving experimental data is presented. The learning and testing patterns for ANNs were obtained from numerical simulations: the selected model parameters were varied in a reasonable range and some dynamic characteristics (e.g., eigen-frequencies) were calculated. Each group consisting of calculated dynamic characteristics and corresponding values of model parameters created one learning or testing pattern. The values of the parameters used to update the computational model of a structure were obtained from ANNs with the input vector consisting of dynamic characteristics (e.g., eigenfrequencies) acquired during laboratory experiments.

Updating of a Model of a Beam Hung on Two Links

The Structure and Initial Model

The first examined structure was a beam made of an aluminium alloy and hung on two elastic links (beam cross-section and one elastic link are shown in Figure 7a). The material properties of the aluminium alloy were as follows: volume mass density $\rho = 2743$ kg/m^3 and Poisson's ratio assumed as 0.33, while the Young's modulus was one of the parameters which were being updated.

The computational model built using FEM consisted of 24 beam elements and had 50 DOFs (see Figure 7b). Timoshenko model of a beam was used, thanks to type of the beam cross-section the influence of shearing forces was not negligible.

As a result of the numerical simulations 336 patterns were obtained by a change of the value of the Young's modulus E and a change of the value of shear correction factor k_s. The pat-

Figure 7. Beam hung on two links: (a) Laboratory and (b) computational models

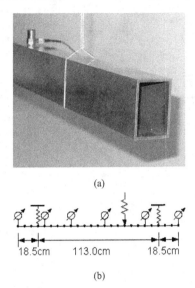

(a)

18.5cm   113.0cm   18.5cm

(b)

terns were divided into learning and testing sets. Randomly selected 10% of patterns were moved to the testing set, the remaining ones built up the learning set.

The vibrations of the laboratory model were excited by an impact, measured values of beam eigenfrequencies were as follows: $f_1 = 174.50$Hz, $f_2 = 467.45$Hz, $f_3 = 884.75$Hz, $f_4 = 1404.5$Hz, and $f_5 = 2009.0$Hz.

Updating of the Initial Model

ANNs trained using Levenberg-Marquardt algorithm or the radial basis function (RBF) networks were applied. Moreover, also ANNs trained using Rprop algorithm were tested, but the obtained results were of poorer accuracy. As the input information the eigenfrequencies or preprocessed frequency response functions (FRFs) were used. The networks updating the dynamic model had, according to the adopted approach, 3, 4, or 5 inputs (the first 3, 4, or 5 eigenfrequencies). The output vector consisted of the Young's modulus and shear correction factor, the obtained values were applied to the dynamic model in order to obtain the updated model. The results of this preliminary approach were satisfactory, the differences between eigenfrequencies obtained from the updated model and from the laboratory measurements were less than 1%.

Updating of a Beam with an Additional Mass

In the next example a computational model of the same beam with additional mass of 273.2g attached was updated. The additional mass was attached in a place corresponding to

Table 3. Results of linear mass density updating on the basis of eigenfrequencies (feed-forward ANNs)

Eigenfrequency	f_1 [Hz]	f_2 [Hz]	f_3 [Hz]	f_4 [Hz]	f_5 [Hz]
Measured value	167.5	439.4	881.3	1301.0	1923.4
Feed-forward 5-3-1 ANNs taught using LM algorithm					
Calculated value	166.7	438.8	877.2	1301.6	1943.8
Relative error	0.5%	0.1%	0.5%	0.0%	-1.0%
$RMSE_{1-5}$=5.6×10^{-3}					
RBF networks with iteratively determined number of hidden neurons					
Calculated value	166.6	438.7	877.2	1301.5	1943.7
Relative error	0.5%	0.2%	0.5%	0.0%	-1.0%
$RMSE_{1-5}$=5.7×10^{-3}					

the location of the 17th node of the initial computational model. Next, a finite element was introduced in this place, the length of which was equal to the width of the mass. The mass density of the new element was updated using the procedure described above. The input vector was composed of five eigenfrequencies, the output vector consisted of mass density of the new finite element. In order to simulate the measurement error in this example the numerical data were disturbed by an artificial, random noise.

The best results were obtained from the networks with 3 hidden neurons learned using patterns disturbed by a noise with the variance of 6×10^{-3}. The best results from among 50 independent runs of the learning of the ANN are shown in Table 3.

Both BPNN networks learned using Levenberg-Marquardt algorithm and RBF network gave very good results, the maximum error of prediction of the first five eigenfrequencies did not exceed 1%. It must be firmly stated that in the case of experimental verification the error of approximately 1% must be considered a very good result.

Updating of an Impaired Beam

In order to obtain more complete information about the investigated structure the measurements of the beam were repeated with six accelerometers attached to the beam. The total weight of all accelerometers was 61g, its influence on the eigenfrequencies was significant and it was taken into account during computational model preparation. However, during the updating the values of the Young's modulus and shear correction factor of a simple cross-section (with no additional mass) were considered to be equal to those obtained from the previous model.

The measurements were performed using a beam with or without defects. The defects were simulated by the removal of half of the thickness of both flanges on the length of one, two or three finite elements of the dynamic model or the complete removal of both flanges on the length of one, two or three finite elements.

Figure 8. Removed half of the thickness of the upper flange

Table 4. Results of impaired beam updating

Eigenfrequency	f_1 [Hz]	f_2 [Hz]	f_3 [Hz]	f_4 [Hz]	f_5 [Hz]
One-element defect – learning of 5-7-2 net, LM algorithm					
Measured value	166.9	442.5	835.6	1334.3	1891.9
Calculated value	165.7	442.6	840.5	1334.6	1891.5
Relative error	0.7%	0.0%	-0.6%	0.0%	0.0%
$RMSE_{1-5}$=4.2×10^{-3}					
Two-element defect – verification of 5-7-2 net					
Measured value	165.0	435.6	831.3	1329.5	1900.9
Calculated value	164.2	436.9	833.4	1319.8	1873.4
Relative error	0.5%	-0.3%	-0.3%	0.7%	1.5%
$RMSE_{1-5}$=7.9×10^{-3}					
Three-element defect – verification of 5-7-2 net					
Measured value	162.9	431.3	831.7	1328.3	1896.6
Calculated value	162.3	432.5	827.9	1307.6	1868.5
Relative error	0.4%	-0.3%	0.5%	1.6%	1.5%
$RMSE_{1-5}$=10.2×10^{-3}					

Figure 8 shows the beam with half of the thickness of both flanges removed on the length corresponding to the length of one finite element of the dynamic model. The defect was located in the place corresponding to the location of the 19th element of the model. The parameters to update were the volume mass density of the beam and the shear correction factor of the impaired cross-section. The depth of defect and its width were measured, so the geometrical characteristics of the defect were known. The values of the Young's modulus and shear correction factor were considered to be equal to the ones from updating of the beam with one accelerometer.

The input vector consisted of five eigenfrequencies disturbed by an artificial noise, the output vector consisted of the values of the volume mass density of the whole beam (including accelerometers) and shear correction factor of the impaired cross-section. The results obtained from the updated model are shown in Table 4. The updated value of volume mass density was $\rho = 2964$ kg/m^3, the updated value of shear correction factor of the impaired cross-section was $k_{s2} = 0.258$. The results of identification were verified by the comparison of eigenfrequencies measured on a laboratory model with defects introduced and eigenfrequencies obtained from a model with already updated parameters, two- or three-element defects were investigated. Some other examples of the defects investigated are presented by Miller and Ziemianski (2003).

Identification of Beam Defects Location

Using the same models another approach to model updating was performed. The assumption was made that material properties were known. The unknown quantities were the location of defect (the number of appropriate finite element) and the defect depth. Updating was performed using as testing data the data describing the beam with removed half of the thickness of both flanges on the length of one finite element.

The ANNs output vector consisted of the values of the remaining thickness of the flanges, shear correction factor of impaired cross-section and the distance from the defect to the symmetry axis. The differentiation between symmetrically located defects would be possible if also the eigenforms were included in the input vectors.

The distance from the symmetry axis was determined faultlessly in all 20 separate runs of learning. The value of the flanges thickness was properly identified by the ANNs as $g = 1.2$mm. The value of shear correction factor obtained from the ANNs was 0.4597. The eigenfrequencies obtained from the updated model are shown in Table 5. A similar example,

Table 5. Results of defect identification

Eigenfrequency	f_1 [Hz]	f_2 [Hz]	f_3 [Hz]	f_4 [Hz]	f_5 [Hz]
Measured value	166.9	442.5	835.6	1334.3	1891.9
Defect identification (depth and shear corr. factor) – 5-9-2 ANN					
Calculated value	165.3	441.5	840.6	1335.5	1891.5
Relative error	1.0%	0.2%	-0.6%	-0.1%	0.0%
RMSE$_{1-5}$=5.2×10^{-3}					
Defect identification (depth, shear corr. factor, location) – 5-7-3 ANN					
Calculated value	165.7	442.9	842.8	1348.7	1929.7
Relative error	0.7%	-0.1%	-0.9%	-1.1%	-2.0%
RMSE$_{1-5}$=11.2×10^{-3}					

presented by Miller and Ziemianski (2002), shows the identification of the location of the additional mass attached to the structure.

The presented updating algorithm was created using some numerical examples (Miller & Ziemianski, 2002). The verification of experimental data presented herein shows that the neural network procedure is able to deal with the data obtained from numerical simulations and from experimental measurements at the same time. Proper updating of the dynamic model of an aluminium cantilever beam shows that due to the constant physical interpretation of model improvements the proposed method can be applied in a variety of problems, including defects identification. In the presented examples both the location and size of the defect were identified. The presented procedure could also be very helpful in experiment planning, since it could point at optimal locations of accelerometers and external excitation point (Miller & Ziemianski, 2002).

Detection of Void and Masses in Vibrating Plates

The assessment of a structure's state is a task which requires continuous investigation and its development. There are a lot of methods based on non-destructive testing of structural elements, some of them use dynamics parameters of the monitored structure, such as modal characteristics (resonance frequencies), response of structure to impulse excitation and structural waves analysis (Yagawa & Okuda, 1996).

One of the most important objectives of these methods is the detection and localization of changes, failures and defects in the structural elements. This subchapter presents the possibility of application of neural networks to non-destructive detection of a circular void or additional mass in cantilever plates. Detection method is based on the analysis of eigenfrequencies, both standard and cascade networks (Piatkowski & Ziemianski, 2002; Waszczyszyn & Ziemianski, 2001) with the Levenberg-Marquardt learning algorithm were applied.

Identification of a Circular Hole Location

The geometry of a unit thickness rectangular plate in question is presented in Figure 9. For each location of the void a separate finite element model (2D solid plane stress elements, only in-plane vibrations of the plate) was prepared using FE code ADINA (ADINA, 2001). The numerical models were used to obtain the pseudo-experimental data necessary during the identification procedure.

The internal defect investigated herein was a circular hole (Piatkowski & Ziemianski, 2003) whose location was defined by two coordinates (OY and OZ) of the hole's centre. The coordinates and the hole's diameter were unknown parameters to be identified by the inverse analysis. Due to the symmetry of the eigenproblem solution the centres were located below or on the axis of symmetry of the plate. Figure 9 shows the investigated locations of the hole.

Figure 9. Numerical model of the plate, scheme, main dimensions

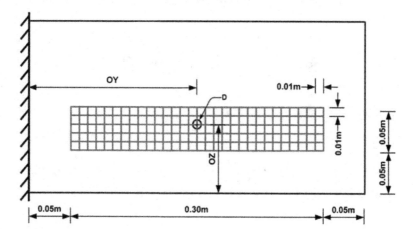

Figure 10. Results for OY > OZ cascade network of architecture 5-7-1 > 6-7-1

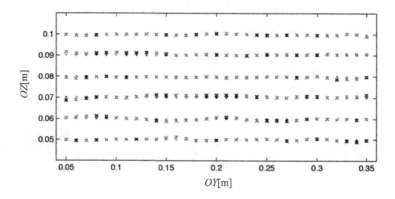

In the first neural experiment two coordinates of the centre of void were identified on the basis of eigenfrequencies, the diameter was considered known (D = 2mm). The location of the void was varied with a step of 10mm, 186 different positions of the hole were taken into account (see Figure 9).

The obtained results showed that acceptable accuracy could be achieved using cascade network (Piatkowski & Ziemianski, 2002) with an input vector composed of relative changes of the first five eigenfrequencies calculated as follows:

$$\Delta f_i = \frac{f_i - f_i^0}{f_i^0}, \quad i=1,...,5 \tag{8}$$

Figure 11. Results for D > OY > OZ cascade network of architecture 5-7-1 > 6-7-1 > 7-7-1

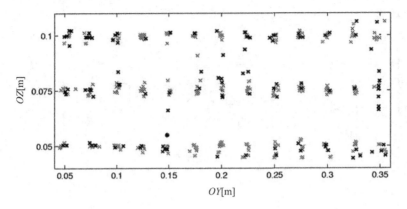

Figure 12. Histograms for (a) D diameter of hole, (b) OY coordinate, (c) OZ coordinate. Results for D > OY > OZ cascade network of architecture 5-7-1 > 6-7-1 > 7-7-1

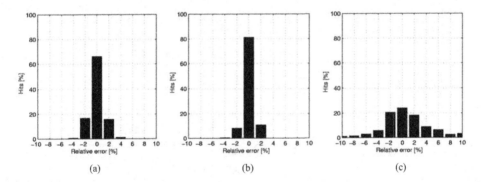

where f_i^0 is *i-th* eigenfrequency of a homogeneous plate (without a void) and f_i is *i-th* eigenfrequency of a plate with a void.

The maximum prediction errors were 1.4mm and 1.5mm for training and testing data sets, respectively. The identified locations are shown in Figure 10, voids centres are marked with **x** sign in grey or black colours for training or testing patterns, respectively.

In the next neural experiment the identification of all three parameters of void {*OY, OZ, D*} was performed, nine diameters in the range of {4mm, 6mm,..., 20mm} for each of 39 investigated locations of void were considered, the total number of patterns was thus 351. Relying on the previous experiment a cascade networks were applied, the parameters' prediction order was as follows: $D > OY > OZ$.

The results are presented in Figure 11—multiple **x** marks relate to different diameters of a void at a given location. One can notice that the vertical OZ coordinate of a void was predicted with a higher error than the horizontal one. The histograms presented in Figure 12 show that the maximum relative error of the OY coordinate identification was $\pm 2\%$, while the OZ coordinate was predicted with a relative error of $\pm 10\%$. The third parameter of the void, its diameter, was identified with a relative error of $\pm 4\%$.

Detection of Additional Mass

Two laboratory models—a steel one and an aluminium one—fixed to a massive stand by high-tensile bolts were built, as shown in Figure 13 (Waszczyszyn & Ziemianski, 2005). The out-plane vibrations of the plate were forced with a modal hammer, the response of a structure was measured by sensors which location is shown in Figures 13 through 15. The additional mass, whose position was to be identified, was equal to 1% or 3% of the overall mass of the plate, for steel or aluminium alloy models respectively (in what follows the values corresponding with the aluminium model will be presented in parenthesis). The mass was located in turn in each of the 27 (40) nodes of the measurement grid (see Figures 14 and 15). The acceleration signals measured in time domain were transformed by means of FFT into frequency domain and analysed in some selected bands—the harmonic with the highest amplitude inside each band was selected.

The total number of frequency characteristics for the steel plate was 216 (eight characteristics for each of 27 locations of an additional mass), five harmonics were selected inside each frequency characteristics so the overall number of considered values was 216×5. The error

Figure 13. Experimental model of the steel plate

Figure 14. Scheme of steel plate, experimental grid, main dimensions

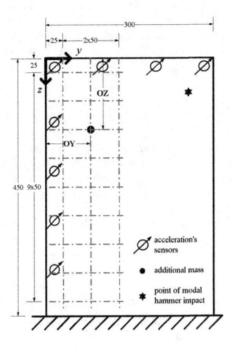

Figure 15. Scheme of aluminum alloy plate, experimental grid, main dimensions

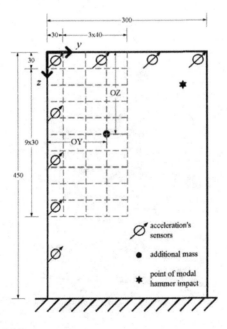

Figure 16. Results of identification of additional mass at steel plate, standard network 5-5-2

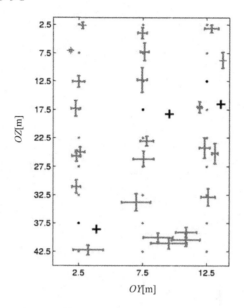

Table 6. Statistical parameters for networks 5-5-1 and 5-5-2

Stat. param.	5-5-1		5-5-2	
	OY	OZ	OY	OZ
R^2	0.992	0.991	0.939	0.980
$St\,\varepsilon$	2.97	2.60	8.21	3.75

biased harmonics were eliminated, then the average for each band was calculated and finally 27 input vectors consisting of 5 elements were prepared for training of neural networks.

In case of the aluminium alloy plate the number of frequency characteristics was 1,528 (40 locations of mass, eight sensors, at least three impacts). The characteristics were analysed in six selected bands, so 1,528×6 characteristics were obtained. The frequencies were averaged inside each band using six 2-10-1 networks, a separate one for each investigated band. Consequently, 40 input vectors, each consisting of six elements, were prepared. Due to the small number of patterns (27 or 40) some additional ones, up to the overall number of 400, were obtained by adding the artificial Gaussian noise with variance of 0.001 to the patterns obtained from the laboratory experiment.

Figure 17. Results of identification of additional mass at steel plate, cascade network 5-5-1 > 6-5-1

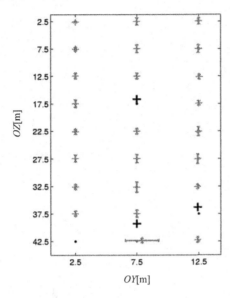

Two coordinates of location of additional mass on the steel plate were determined either by two separate 5-5-1 networks or by single 5-5-2 network. The statistical parameters describing the obtained results are collected in Table 6.

The results of identification of mass location are presented in Figures 16 and 17. The centre points of crosses show the average response of the network, the horizontal and vertical bars are proportional to standard deviation of determination of *OY* and *OX* coordinates, respectively. The best accuracy was obtained from *OY* > *OZ* cascade neural network of architecture 5-5-1 > 6-5-1 (the results of training - grey-coloured marks - and testing - black coloured marks) are shown in Figure 17.

The measurements were conducted also on another laboratory model of a steel plate. The dimensions of the plate and the location of sensors were the same, the difference lies in the grid describing the investigated locations of an additional mass (see Figure 18). The application of another grid gave 24 additional experimental patterns. Three definitions of input vector were employed:

1.   The input vector was composed of frequencies taken directly from both measurement sets.

2.   Similarly as in (1), but the input vector contained an additional element determining the measurement set.

Figure 18. Scheme of steel plate, new experimental grid, main dimensions

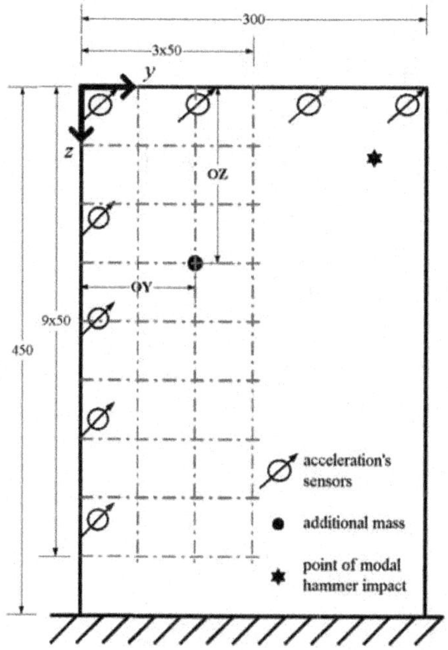

3. One of the measurement sets was transformed into the second one according to formula:

$$f = \left(f^2 - f_{min}^2\right)\frac{f_{max}^1 - f_{min}^1}{f_{max}^2 - f_{min}^2} + f_{min}^1,$$ (9)

where f_{max} – maximum of measured frequencies, f_{min} – minimum of measured frequencies, upper index (1 or 2) denotes measurement set (the first or the second, respectively).

The best accuracy was achieved with input vectors prepared in the third way. Selected results are presented in Figure 19.

For the aluminium alloy plate the results were analogous to those obtained for the steel model, the highest accuracy of additional mass location was obtained from the cascade network fed with input vector disturbed by a random noise with variance of 0.001. The results for training (grey marks) and testing patterns (black marks) are shown in Figure 20.

Figure 19. Identified locations of additional mass, cascade network 7-7-1 > 8-7-1

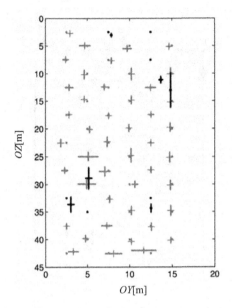

Figure 20. Results of identification of additional mass at aluminum alloy plate - cascade network 6-5-1 > 7-5-1

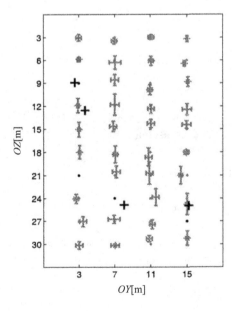

Modal Characteristics-Based Identification of Parameters in Steel Structures

Detection of Damage in Multistory Frames

The importance of fault detection in structures which are in danger of fatigue failure is evident. That is why the problem of efficient detection and localisation of damages within structures is getting more and more important. Most of the methods proposed for damage assessment are based on a mathematical model of the structure, established to give information about the correlation between damage and change in dynamic behaviour. Fang, Luo, and Tang (2005) as well as Kao and Hung (2003) showed that neural networks may be successfully applied to structural damage identification. This subsection presents a possibility of using neural network for damage detection in a multistory frame based on changes of modal properties (Ziemianski, 1999). The relative changes of natural frequencies—the most fundamental and simplest among the modal parameters—are adopted here. In the problem discussed below, an extension of the response parameters is shown (i.e., besides eigenfrequencies, also components of eigenforms are used as input variables).

Eigenfrequencies as Input Variables

A simple 8-story plane frame with rigid beams was investigated by Ziemianski and Harpula (1999), see Figure 21. Damage was defined as a decrease of elasticity modulus in both frame columns of each story. Thus, due to the frame symmetry, the damage was assumed to be on one side of the axis of symmetry.

Assuming six values of the damage parameters $ed = E_d/E$, eight locations of damaged column and one pattern corresponding to undamaged frame, altogether 49 patterns were obtained. A testing set was composed of 32 patterns assuming values of damage parameter ed different than for learning patterns. New training patterns were obtained as the sum of original patterns and the artificial random noise, the number of training patterns was increased to 539 after 10 selections of random noise per one "perfect" pattern. The testing set was formulated on the basis of the original 32 testing patterns; using an artificial noise the testing set composed of 352 patterns was obtained. One-hidden-layer networks were analysed by means of the SNNS simulator (SNNS, 1995) using the Rprop algorithm. The outputs defined the damage parameter and the story with damaged column.

On the basis of the intensive cross-validation procedure described by Ziemianski and Harpula (1999) the number of eigenfrequencies adopted on input was 7 and the number of hidden neurons was estimated as 27. Figure 22 shows that the errors of the network learning and testing were nearly the same $MSEL \approx MSET \approx 0.9 \times 10^{-3}$. Two conclusions can be drawn: (a) the neural approximation is strongly affected by the noise introduced to the sets of patterns used, and (b) thanks to the inserted artificial noise we can formulate more efficient networks.

Figure 21. Scheme of a plane frame

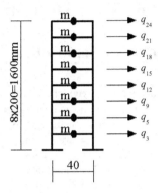

Figure 22. Influence of random artificial noise on network errors

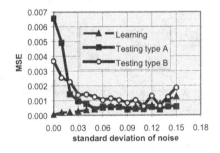

Eigenfrequencies and Components of Eigenforms as Input Variables

Besides eigenfrequencies, the vectors of eigenforms (or their selected components) can be used as network inputs. It was analysed by Ziemianski and Harpula (1999), where the number of eigenfrequencies and corresponding eigenforms was limited to two (i.e., to ω_1, ω_2, q^1, q^2).

From an extensive analysis only selected results are presented below. In Table 7 there are listed four types of networks called for short *U*1, *U*2, *U*3, and *U*4. In Table 8, there are shown the errors of neural approximation after a very extended cross-validation process on the basis of the training and testing patterns discussed in the previous point. In Table 8, there are shown also the standard deviations σ_i computed for the relative errors ep_i.

Table 7. Networks with eigenfrequency / eigenform inputs

Symbols of BPNNs	Structure of networks	Components of the input vector
U1	8-12-2	$\omega_1, \omega_2, q_6^1, q_{15}^1, q_{24}^1, q_6^2, q_{15}^2, q_{24}^2$
U2	7-13-2	$\omega_1, q_6^1, q_{12}^1, q_{15}^1, q_{18}^2, q_{21}^2, q_{24}^2$
U3	4-15-2	$\omega_1, q_6^1, q_{15}^1, q_{24}^1$
U4	4-15-2	$\omega_1, \omega_2, q_{24}^1, q_{24}^2$

Table 8. Errors related to networks with inputs listed in Table 7

Network	MSEL $\cdot 10^3$	MSET $\cdot 10^3$	Stiffness damage ed		Location of damage sn	
			max e_1	σ_1	max e_2	σ_2
U1	0.168	0.321	3.82%	0.0101	4.97%	0.0115
U2	0.413	0.249	5.81%	0.0148	4.01%	0.0116
U3	1.057	1.105	6.35%	0.0191	7.84%	0.0269
U4	0.739	1.102	3.82%	0.0127	10.44%	0.0268

Comparing MSE errors one can state that the application of components of eigenforms enables the use of smaller networks than in case of input vector consisting only of eigenfrequencies. Moreover, the networks U3 and U4 supported on one or two eigenfrequencies (but enriched by the components of eigenforms) can give better approximation than the network discussed in previous point for 7 eigenfrequency inputs. The results of numerical experiments discussed above confirm that the richer and more complete the dynamic responses are adopted as the network inputs the better neural approximation can be achieved. But on the other hand, we have to take into account measurement possibilities with respect to tests carried out on laboratory models or on real structures.

Conclusion

The chapter discusses an application of artificial neural networks to the field of identification problems. On the basis of the achieved results the following conclusions can be stated:

1. Back-propagation neural networks seem to be an efficient tool for implicit modelling and parameter identification of materials and structures. The accuracy of prediction of dynamic responses of investigated structures using material or structural parameters obtained from networks is very high, the comparison with measurements data shows that the errors from procedures involving neurocomputing are at the level of a few percentage points only.

2. Cascade networks enable us to extend the potential and significantly improve the accuracy of results in the identification analysis.

3. The proposed methods of data preprocessing (i.e., the use of data contaminated by an artificial random noise and data compression by means of specialized networks) improve the generalization abilities of applied networks and enable the neurocomputing application even to cases when the number of patterns is not sufficient to apply networks trained with unprocessed input data.

4. The discussed hybrid approach to the FE model updating, in which BPNNs are used for the inverse analysis, seems to be very promising also for identification of material and structural parameters. The presented examples verified by laboratory measurements show that this approach can be applied when the classical computational model updating is not possible due to insufficient amount of input data.

The obtained results showed that the methodology was able to identify the damage in simple structures, update the finite element model of beams and frames, identify the mass in vibrating plates.

References

ADINA Theory and modeling guide. (2001). Watertown: Author.

Chang, C. C., Chang, T. Y. P., Xu, Y. G., & To, W. M. (2002). Selection of training samples for model updating using neural networks. *Journal of Sound and Vibration, 249*(5), 867-883.

Demuth, H., & Beale, M. (1998). *Neural network toolbox user's guide, Version 3.0.* Natick, MA: MathWorks.

Ewins, D. J. (2000). *Modal testing: theory, practice and applications* (2nd ed.). Baldock: Research Studies Press.

Fang, X., Luo, H., & Tang, J. (2005). Structural damage detection using neural network with learning rate improvement. *Computers and Structures, 83*(25/26), 2150-2161.

Friswell, M. I., & Mottershead, J. E. (1996). *Finite element model updating in structural dynamics.* Dordrecht, Holland: Kluwer Academic.

Haykin, S. (1999). *Neural networks, a comprehensive foundation* (2nd ed.). Upper Saddle River, NJ: Prentice Hall.

Kao, C. Y., & Hung, S-L. (2003). Detection of structural damage via free vibration responses generated by approximating artificial neural networks. *Computers and Structures, 81*(28/29), 2631-2644.

Lee, B. C., & Staszewski, W. J. (2003a). Modelling of Lamb waves for damage detection in metallic structures: Part I. Wave propagation. *Smart Materials and Structures, 12*(1), 804-814.

Lee, B. C., & Staszewski, W. J. (2003b). Modelling of Lamb waves for damage detection in metallic structures: Part II. Wave interactions with damages. *Smart Materials and Structures, 12*(1), 815-824.

Levin, R. I., & Lieven, N. A. J. (1998). Dynamic finite element model updating using neural networks. *Journal of Sound and Vibration, 210*(5), 593-607.

Lu, Y., & Tu, Z. (2004). A two-level neural network approach for dynamic FE model updating including damping. *Journal of Sound and Vibration, 275*(3/5), 931-952.

Miller, B., & Ziemianski, L. (2002). Application of neural networks to the structural model updating. In H. A. Mang, F. G. Rammerstorfer, & J. Eberhardsteiner (Eds.), *Proceedings of the Fifth World Congress on Computational Mechanics* (Vol. 1, p. 315).

Miller, B., & Ziemianski, L. (2003). Neural networks in updating of dynamic models with experimental verification. In L. Rutkowski & J. Kacprzyk (Eds.), *Advances in soft computing* (pp. 766-771). Heidelberg, Germany: Physica-Verlag.

Mottershead, J. E., & Friswell, M. I. (1993). Model updating in structural dynamics: A survey. *Journal of Sound and Vibration, 167*(2), 347-375.

Natke, H. G. (1998). Problems of model updating procedures: A perspective resumption. *Mechanical Systems and Signals Processing, 12*(1), 65-74.

Oishi, A. K., Yamada, K., Yoshimura, S., & Yagawa, G. (1995). Quantitative nondestructive evaluation with ultrasonic method using neural networks and computational mechanics. *Computational Mechanics, 15*, 521-523.

Paez, T. L. (1993). Neural networks in mechanical system simulation, identification and assessment. *Shock and Vibration, 1*(2), 177-199.

Palacz, M., & Krawczuk, M. (2005). Analysis of longitudinal wave propagation in a cracked rod by the spectral element method, *Computers and Structures, 80*, 1809-1816.

Piatkowski, G., & Ziemianski, L. (2002). Identification of circular hole in rectangular plate using neural networks. In *Proceedings of AI-Meth Symposium on Methods of Artificial Intelligence*, Gliwice, Poland (pp. 329-332).

Piatkowski, G., & Ziemianski, L. (2003). Neural network identification of a circular hole in the rectangular plate. In L. Rutkowski, & J. Kacprzyk (Eds.), *Advances in soft computing* (pp. 778-783). Heidelberg, Germany: Physica-Verlag.

SNNS Stuttgart neural network simulator: User manual, version 4.1. (1995). Stuttgart, Germany: University of Stuttgart.

Su, Z., & Ye, L. (2005), Quantitative damage prediction for composite laminates based on wave propagation and artificial neural networks. *Structural Health Monitoring, 4*(1), 57-66.

Su, Z., Ye, L., Bu, X., Wang, X., & Mai, Y. W. (2003). Quantitative assessment of damage in a structural beam based on wave propagation by impact excitation. *Structural Health Monitoring, 2*(1), 27-40.

Thompson, R. B. (1983). Quantitative ultrasonic nondestructive evaluation methods. *Journal of Applied Mechanics, 50*(4B), 1191-1201.

Waszczyszyn, Z., & Ziemianski, L. (2001). Neural networks in mechanics of structures and materials: New results and prospects of applications. *Computers & Structures,* 79, 2261-2276.

Yagawa, G., & Okuda, H. (1996). Neural networks in computational mechanics. *Archives of Computational Methods in Engineering,* (3/4), 435-512.

Zang, C., & Imregun, M. (2001). Structural damage detection using artificial neural networks and measured FRF data reduced via principal component projection. *Journal of Sound and Vibration, 242*(5), 813-827.

Zapico, J.L., González, M.P., & Worden, K. (2003). Damage assessment using neural networks. *Mechanical Systems and Signal Processing, 17*(1), 119-125.

Ziemianski, L. (2005). Neural networks in the identification analysis of structural mechanics problems. In Z. Mroz, & G. E. Stavroulakis (Eds.), *Parameter identification of materials and structures* (CISM Lecture Notes Vol. 469, pp. 265-340). Wien: Springer.

Waszczyszyn, Z., & Ziemianski, L. (2005) Application of neural networks in dynamic of structures. In *Proceedings of the Fourth International Scientific Colloquium CAx Techniques,* Bielefeld, Germany (pp. 509-516).

Ziemianski, L., & Harpula, G. (1999). The use of neural networks for damage detection in eight story frame structure. In *Proceedings of the Fifth International Conference on Engineering Applications of Neural Networks,* Warsaw, Poland (Vol. 1, pp. 292-297).

Ziemianski, L., & Miller, B. (2000). Dynamic model updating using neural networks. *Computer Assisted Mechanics & Engineering Sciences, 4,* 68-86.

Ziemianski, L., & Piatkowski, G. (2000). Use of neural networks for damage detection in structural elements using wave propagation. In B. H. V. Topping (Ed.), *Computational engineering using metaphors from nature* (pp. 25-30). Edinburgh, Scotland: Civil-Comp.

Chapter XVI

Neural Networks for the Simulation and Identification Analysis of Buildings Subjected to Paraseismic Excitations

Krystyna Kuźniar, Pedagogical University of Cracow, Poland

Zenon Waszczyszyn, Rzeszów University of Technology, Poland

Abstract

The chapter deals with an application of neural networks to the analysis of vibrations of medium-height prefabricated buildings with load-bearing walls subjected to paraseismic excitations. Neural network technique was used for identification of dynamic properties of actual buildings, simulation of building responses to paraseismic excitations as well as for the analysis of response spectra. Mining tremors in strip mines and in the most seismically active mining regions in Poland with underground exploitation were the sources of these vibrations. On the basis of the experimental data obtained from the measurements of kinematic excitations and dynamic building responses of actual structures the training and testing patterns were formulated. It was stated that the application of neural networks enables us to predict the results with accuracy quite satisfactory for engineering practice. The results presented in this chapter lead to a conclusion that the neural technique gives new prospects of efficient analysis of structural dynamics problems related to paraseismic excitations.

Introduction

Structural vibrations induced by ground motion can be caused not only by earthquakes but also by human activity. Some of the sources of paraseismic excitations, as for instance traffic vibrations and industrial explosions, may be inspected and controlled. On the other hand, mining tremors resulting from underground raw mineral material exploitation are random events. Although these tremors are strictly connected with the activity of man, they differ considerably from other paraseismic vibrations. Neither the moment and place of their occurrence nor the magnitude can be foreseen, like in the case of earthquakes.

The estimation of building dynamic properties (first of all the periods of natural vibrations) and dynamic response of building subjected to kinematical excitations are very important problems of structural dynamics. The basic approach is related to full-scale measurements on actual buildings (Ciesielski, Kuźniar, Maciąg, & Tatara, 1992). In many cases such an experimentally supported analysis can be superior to the computational analysis. There are, obviously, many problems related to organization and costs of the tests on real buildings in the natural scale.

On the other hand, a structural dynamic analysis involves many tasks (e.g., formulation of a real model of building). Besides, the problem is related to very complex structures such as buildings, in particular—prefabricated buildings. There are a lot of difficulties with material, structural and load modelling. In spite of extensive development of computational methods (first of all the finite element method) and progress in computer software and hardware, the analysis of building vibration problems is far from satisfactory from the structural engineering point of view. That is why there are attempts to explore nonstandard, codisciplinary approaches in the analysis of the mentioned problems. From this point of view artificial neural networks (ANNs) seem to be a new tool, very prospective for solving the problems.

The chapter deals with dynamic analysis of medium height prefabricated buildings with load bearing walls subjected to paraseismic excitations. Mining tremors in strip mines and in the most seismically active mining regions in Poland with underground exploitation—Upper Silesian Coalfield (USC) and Legnica-Glogow Copperfield (LGC)—were the sources of these vibrations. The results of long-term experimental monitoring of actual structures were synthetically collected. The created database of the experimental data obtained from the measurements of kinematic excitations and response building vibrations makes it possible to use them to design the neural analysers for complete considerations of the building dynamic problems. The main problems discussed in the chapter deal with the application of neural networks to identification of dynamic properties of actual buildings, simulation of building responses to the paraseismic excitations, mapping of seismic parameters into response spectra on the ground level and simulation of response spectra on building basement from input spectra on the ground level.

At the beginning of the chapter, some basic information on paraseismic excitations is given .Then basics of artificial neural networks (ANNs) is briefly presented. This section is somewhat extended since it should be a base for other chapters of this book in which ANNs are used. Three following sections are devoted to applications of ANNs in the analysis of various problems related to buildings subjected to paraseismic excitations.

Paraseismic Excitations

Experimental Tests

Firings of explosive in nearby quarries (strip mines) and mining tremors caused by underground exploitation were the sources of the building vibrations.

In case of explosions in quarries the investigated buildings were located at a distance of 300m to 1200m from the site of the explosions. The blowing charges were fired in different places on the walls of excavations. Hence, various distances and angles of the waves reaching the analysed buildings were obtained. The explosions were carried out using the technology of long, almost vertical and horizontal holes. The quantity of the explosives varied from 300kg to 2100kg in case of vertical holes, and 45kg in the case of horizontal ones. In the technology of long vertical holes the holes' length of 5-24m and diameters of 90mm-180mm were applied. The diameters of horizontal holes were 42mm. Instantaneous or millisecond detonators were used.

Underground mining causes series of negative effects in the surrounding environment. Rock bursts are one of them. A large quantity of energy is emitted in the time of tremor which is the result of rocks bursting over mining excavations. The energy causes propagation of seismic waves that reach the surface of the earth. These induce the buildings vibrations subsequently. Although these tremors are strictly connected with human activity, they differ considerably from other paraseismic vibrations. Like in the case of earthquakes, the moment of their occurrence cannot be foreseen. Hence, mining tremors resemble tectonic earthquakes in their character, so it is also difficult to study them. But the analysis is a little easier because they occur in restricted areas, usually limited to mining regions.

There are several mining regions in Poland. Two of them (i.e., USC with underground coal exploitation and LGC with copper ore exploitation) are the most seismically active. In both fields the underground exploitation of raw mineral materials is, in many places, carried out under densely populated urban areas, also under cities. The strongest mining tremors induce the building damages. In recent years, the seismicity of LGC region has increased considerably as a result of deeper exploitation (1,000m under the earth surface) and hard blanket rocks. Mining tremors resulting from underground raw mineral material exploitation in LGC induce the surface horizontal vibrations reaching even 0.2 acceleration of gravity (g) and vertical components reaching 0.3g. The large scale of the effects might be shown by the fact that the intensity of surface vibrations caused by strong mining tremors in Poland (in LGC region with underground copper ore exploitation) is greater than the predicted (and taken into consideration in structural design) intensity of vibrations from earthquakes in

Table 1. Parameters of analysed mining tremors

Region	Energies E [J]	Epicentral distances r_e[m]
USC	$2 \times 10^4 - 4 \times 10^6$	123 – 1196
LGC	$1 \times 10^6 - 2 \times 10^9$	400 – 5089

the Slovak Republic, Czech Republic, and Germany. The seismicity in mines of USC and LGC regions has recently been monitored and the records of surface vibrations are taken from seismological stations. A comparison of a great amount of records from both regions leads to a conclusion that typical vibrations in the two mining regions differ significantly. So USC and LGC regions are considered separately.

The ranges of parameters of mining tremors which were the sources of measured vibrations are shown in Table 1.

Building vibrations were caused by the above mentioned seismic-type (paraseismic) vibrations. Typical, residential, prefabricated, with load bearing walls, five-story buildings were tested. Some of the buildings were erected in large panel technology, other buildings in large block technology. All the buildings have basements and continuous footings founded on different soils. In each of the buildings every story is 2.70m or 2.80m high. As an example the plan and sectional elevation of a selected building is shown in Figure 1.

Full-scale measurements have been made many times over a period of a several years (Ciesielski et al., 1992; Kuźniar, Maciąg, & Waszczyszyn, 2005). The performed experimental tests span two groups of measurements: dynamic measurements of horizontal vibration components in the directions parallel to transverse and longitudinal axes of the buildings. In the case of mining tremors, the vertical vibration components were also analysed because of their significant intensity. The seismographs and accelerometers were placed at the ground

Figure 1. Building WK-70 type: (a) Plan, (b) sectional elevation and points of vibration measurements

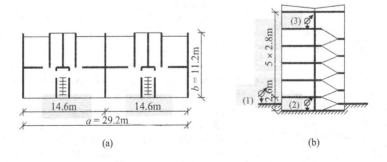

Figure 2. The displacement vibration records of BSK type building caused by an explosion in a nearby quarry (transverse direction): (a) Ground level, (b) basement level, (c) fourth floor level

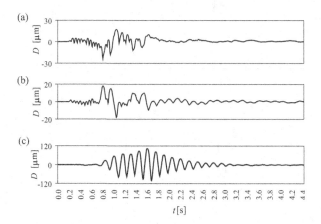

in front of the building (in a few meters distance) and on some levels of the building. The measurement points are shown in Figure 1b: (1) at the ground level outside the building, (2) at the basement level inside the building, (3) on the 4th floor level. In Figure 2 there are shown displacement records corresponding to vibrations in transverse direction measured at three points shown in Figure 1b.

Figure 3. Response spectra: (a) Displacement response spectrum from vibrations induced by firing of explosives in nearby quarry, (b) acceleration response spectrum from mining tremor in LGC region, (c) normalized acceleration response spectrum from mining tremor in LGC region

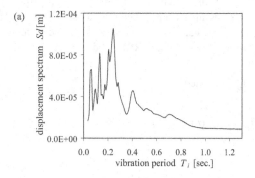

Figure 3. continued

(b)

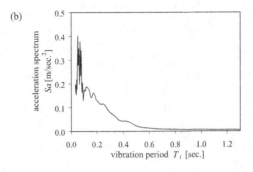

(c)

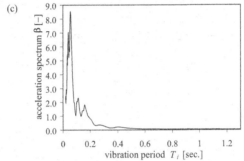

Response Spectra

In regions with strong seismicity response spectra are often applied in structural design as well as to determine dynamic resistance of existing buildings. The draft of Eurocode8 (1994) is also based on the response spectrum method. In a response spectrum the information about kinematic excitation is included. The main idea of a response spectrum is the assumption of one degree of freedom model (oscillator) for the structure. The vibrations of this structure are induced by surface vibrations. So the response spectrum is the function mapping of the natural periods and assumed damping of oscillators into the maximal values of their response (accelerations, velocities, displacements). There are acceleration response spectra (ARS), velocity response spectra (VRS) and displacement response spectra (DRS) in use. Let us focus on ARS.

ARS is a function $Sa\ (T_i,\ \xi)$ defined as:

$$Sa\ (T_i) = \max_j |\ a\ (t_j\ ;\ T_i,\ \xi\ ) + a_g(t_j)\ |\ , \tag{1}$$

where:

$a_j = \partial^2 x / \partial t^2 \mid_{t\,=\,t_j}$ – acceleration computed for the 1DOF oscillator of the equation of motion:

$$\ddot{x} + 2\xi\,\omega_i\,\dot{x} + \omega_i^2 x = -a_g(t), \tag{2}$$

where: $\omega_i = 2\pi f_i = 2\pi / T_i$ – angular frequency, $T_i = 1/f_i$ – period of vibrations, ξ – damping coefficient, $a_g(t)$ – excitation applied to structure, corresponding to ground acceleration. Having a measured accelerogram we can digitize it and apply to compute displacements $x_j = x(t_j)$ for fixed values of frequencies f_i or periods of vibration T_i and damping coefficient ξ. ARS is usually scaled to the dimensionless values $\beta(T_i)$, where $\beta = Sa/a_{gmax}$ for $a_{gmax} = \max_j |a_g(t_j)|$.

In engineering practice recording of real excitation time history for each building is not possible for economic and practical reasons. Therefore, the averaged response spectra are applied. The averaged response spectrum prepared on the basis of a huge number of recorded vibrations for each seismic region is obtained as a curve (ARS is drawn versus periods of natural vibrations) consisting of arithmetic means of series of normalized discrete response spectra $\beta = Sa\,(T_i)\,/\,a_{gmax}$ from particular vibrations.

Figure 3 illustrates the above mentioned response spectra obtained for vibrations at the ground level. Fraction of damping coefficient $\xi = 2\%$ was assumed. Such an intensity of damping may be accepted for the analysed buildings (Ciesielski, Kuźniar, Maciąg, & Tatara, 1995).

Some Basics on Neural Network

Artificial neural networks, called for short neural networks (NNs), belong to "biologically" inspired methods, and together with fuzzy systems and genetic algorithms (or, more generally, evolutionary algorithms, systems and strategies) they are new tools for information processing. Their computer simulations create the so-called soft computing (also called intelligent computing (cf. e.g., Jang, Sun, & Mizutani, 1997). Neurocomputing has been applied to the analysis of a great amount of problems in science and technology (cf. e.g., Haykin, 1999). This concerns also mechanics of structures and materials (cf. Waszczyszyn, 1999; Waszczyszyn & Ziemianski, 2003, 2005). Many papers deal with using the neural network technique for dynamic problems, among others also to buildings subjected to seismic excitations (e.g., Chen, Tsai, Qi, Yang, & Amini, 1995; Cheng & Popplewel, 1994; De Stefano, Sabia, & Sabia, 1999; Ghaboussi & Joghataie, 1995; Ghaboussi & Lin, 1998; Huang, Hung, Wen, & Tu, 2003; Hung & Kao, 2002; Lin & Ghaboussi, 2001). The literature provides rather a very small number of papers describing vibrations of buildings subjected to paraseismic excitations. For example, traffic-induced ground motions are analyzed in Hao,

Table 2. Mechanical system problems and relevant data

Problems	Inputs	Outputs
1. Mechanical system (MS) response simulation	Excitation variables and system parameters	Response variables
2. MS excitation simulation (identification)	Response variables and system parameters	Excitation variables
3. MS parameter identification	MS excitation and response variables	Parameters of MS
4. MS excitation and/or response assessment	All relevant measures of system conditions to be assessed	Assessment of system conditions

Chow, and Brownjohn (2004), surface-ground motions induced by blasts in jointed rock mass are discussed in Hao, Wu, Ma, and Zhou (2001), and a problem related to ground vibrations generated by a high-energy explosive loading is presented in Liu and Luke (2004).

NNs have features associated with their biological origin. They are simulated on standard computers as massively parallel and they can process not only crisp, but also noisy and incomplete data. Neurocomputing can be used for the approximation and classification purposes. NNs have generalization features (i.e., they can be used for the analysis of problems, which obey certain rules which are learned during the NN training process).

NNs can be applied to the analysis of basic types of problems which were classified by Paez (1993) as the simulation, identification and assessment problems (cf. Table 2).

Different types of NNs, applied also in the analysis of structural dynamics problems, are discussed in extended literature (cf. e.g., Jang et al., 1997; Haykin, 1999; Rojas, 1996; Waszczyszyn, 1999; Waszczyszyn & Ziemianski, 2003, 2005).

From among many structures (architectures) of NNs the most popular is the feed-forward multilayer NN also called a "multilayer perceptron" (Haykin, 1999) or "back-propagation NN" (BPNN) (cf. Waszczyszyn, 1999). The last name is related to the basic idea of this network, corresponding to the NN parameters computing (i.e., to the error back-propagation; see Figure 4a).

Back-Propagation Neural Network (BPNN)

BPNN is a network of standard structure (i.e., neurons are not connected in the layer but they join the layer neuron with all the neurons of previous and subsequent layers, respectively). In Figure 4a there is shown an example of a BPNN composed of an input layer, two hidden layers and an output layer. The input layer serves only to introduce signals (values of input

Figure 4. (a) Three-layer BPNN, (b) single neuron i in layer l, (c) identity function, (d) sigmoid function, (e) bipolar sigmoid activation function

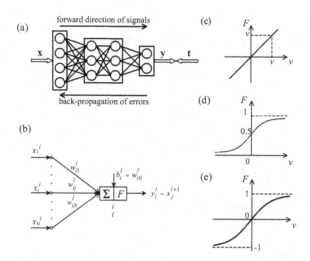

variables) and the hidden and output layers are composed of neurons (processing units) of the NN structure shown in Figure 4b.

The activation potential v_i^l of a single neuron i in the layer l is cumulated in the summing block Σ and transformed by the activation function F to have only one scalar output y_i^l, see Figure 4b:

$$y_i^l = F^l(v_i^l), \qquad v_i^l = \sum_{l=1}^{H_l} w_{ij}^l x_j + b_i^l = \sum_{l=0}^{H_l} w_{ij}^l x_j$$

$$y_i^l = F^l(v_i^l), v_i^l = \sum_{l=1}^{H_l} w_{ij}^l x_j + b_i^l = \sum_{l=0}^{H_l} w_{ij}^l x_j , \tag{3}$$

where: l – layer superscript $l = 1, \ldots , NL$ and NL – number of layers (hidden and output layers); w_{ij}^l – weights of connections, b_i^l – threshold parameter (bias) that can be treated as weight w_{ij}^0 corresponding to the unit signal $x_0 = 1$ (cf. Figure 4b), H_1 – number of neurons in layer l).

From among many of the used activation functions only three functions are shown in Figures 4c, 4d and 4e:

1. **Identity:**

$$F(v) = v ;$$ (4a)

2. **Sigmoid (logistic or binary sigmoid) function:**

$$F(v) = \frac{1}{1 + \exp(-\sigma v)} \in (0, 1) , \quad \frac{dF}{dv} \equiv F'(v) = \sigma F (1 - F) \text{ for } \sigma > 0;$$ (4b)

3. **Bipolar sigmoid:**

$$F(v) = \frac{1 - \exp(-\sigma v)}{1 + \exp(-\sigma v)} \in (-1, 1) , \quad \frac{dF}{dv} \equiv F'(v) = \frac{\sigma}{2}(1 + F)(1 - F) \text{ for } \sigma > 0.$$ (4c)

BPNN has a standard structure that can be written in short:

$$N - H_1 - H_2 - \dots - H_{NL-1} - M,$$ (5)

where: N – number of inputs, H_l – number of neurons in the *l-th* hidden layer for $l = 1,\dots,$ $NL-1$; M – number of outputs.

The values of weights w_{ij}^j and biases b_i^l are called network parameters. The number of BPNN parameters equals the total number of generalized weights (NN connections and neuron biases):

$$NNP = NNW + NNB = (N \times H_1 + \sum_{l=1}^{NL-1} (H_l \times H_{l+1})) + \sum_{l=1}^{NL} H_l,$$ (6)

where: $H_{NL} = M$ – number of output neurons. For instance, the network shown in Figure 4a has $NNW = 4 \times 3 + 3 \times 3 + 3 \times 2 = 27$, $NNB = \Sigma_h H_h = 3 + 3 + 2 = 8$, thus the total number of NN parameters is $NNP = NNW + NNB = 27 + 8 = 35$. The number NNP can be used as a measure of the NN size. For the NN design purposes it is of value to satisfy the inequality (well-based formulation):

$$NNP \leq L \times M ,$$ (7)

where: L – number of training patterns.

Training and Testing of BPNN

The values of weights w_{ij}^l and biases b_i^l can be put in order as components of the vector of generalized weights:

$$\mathbf{w} = \{w_j \mid j = 1,..., NNP\}. \tag{8}$$

The computation of network parameters is called training (learning) process. It is based on a training set of patterns L, composed of known pairs of inputs and outputs (targets) (i.e., input and output vectors $\mathbf{x}^{(p)}$, $\mathbf{t}^{(p)}$, respectively). After the training of the network it is tested on the testing set T. The training and testing pattern sets can be written in the following form:

$$L = \{(\mathbf{x}^{(p)}, \mathbf{t}^{(p)}) \mid p = 1,..., L\}, \; T = \{(\mathbf{x}^{(p)}, \mathbf{t}^{(p)}) \mid p = 1,..., T\}, \tag{9}$$

where: L, T – number of patterns in training and testing sets, respectively.

The components of weight vector w_j (network parameters) are iteratively computed by means of the following formula:

$$w_j(s+1) = w_j(s) + \Delta w_j(s), \tag{10}$$

where: s – the number of iteration step.

Gradient and Rprop Learning Methods

Let us write the learning formula for the weight increment in the following form:

$$\Delta w_j = -\eta g_j, \tag{11}$$

where: $g_j = \partial E / \partial w_j$ – gradient of network error function (called the "least-mean-square error"):

$$E = \frac{1}{2} \sum_{p=1}^{V} \sum_{i=1}^{M} (t_i^{(p)} - y_i^{(p)})^2 = \frac{1}{2} \sum_p \sum_i (\delta_i^{(p)})^2, \tag{12}$$

This formula is related to the output vectors $\mathbf{y}^{(p)}$ and the target vectors $\mathbf{t}^{(p)}$. The error $\delta_i^{(p)}$ corresponds to the *i-th* output for *p-th* pattern and $V = L$, T.

Formula (12) corresponds to the gradient method of steepest descent. This method is sensitive to values of the learning rate η. For small values of η the iteration process can be slowly convergent or even divergent below certain values of η_{crit}. That is why many other learning methods were formulated (cf. Rojas, 1996).

From among them the resilient-propagation (Rprop) method is recommended (cf. neural network computer simulators; e.g., MATLAB NN Toolbox [Neural Network Toolbox, 2001]; SNNS [Stuttgart Neural Network Simulator]; Rojas, 1996; Zell, 1998). The method is associated with the application of local learning rates η_j in the learning formula:

$$\Delta w_j(s) = -\eta_j(s) \, \text{sgn} \, g(g_j(s)), \tag{13}$$

where:

$$\eta_j(s) = \begin{cases} \min(\eta^+ \eta_j(s-1), \, \eta_{max}) & \text{for} & g_j(s) \, g_j(s-1) > 0, \\ \max(\eta^- \eta_j(s-1), \, \eta_{min}) & \text{for} & g_j(s) \, g_j(s-1) < 0, \\ \eta_j(s-1) & \text{otherwise}. \end{cases} \tag{14}$$

The fixed parameters used in (13) are: $\eta^+ = 1.2$, $\eta^- = 0.5$, $\eta_{max} = 50$, $\eta_{min} = 10^{-6}$ (cf. Neural Network Toolbox, 2001). The Rprop formula is of heuristic type and it is frequently used for the NN training in the case of a large number of training patterns.

During the iteration process all the patterns $p = 1, \ldots, L$ are presented. One, forward transmission of signals for all the patterns and back propagation of errors is called an epoch. The iteration is ended according to stopping criteria.

There are different stopping criteria (SC), corresponding to different error measures. One of the frequently applied SC is associated with a fixed number of epochs S that corresponds to the stabilization of error value during the training process. Such a stopping value S is evaluated on the base of initial computation during the NN design process. Another SC corresponds to reaching of comparable training and testing errors.

Application of Kalman Filtering for the Network Learning

In this chapter Kalman filters (KFs) are applied as a new refined method of NN learning. KFs are algorithms well known in the automatics and signal processing for the analysis of discrete dynamic processes (cf. Haykin, 1999, 2001). The algorithms are related to time series and their stochastic analysis. Unlike to the standard temporal neural networks KF is based only on one time delay (i.e., besides the current discrete pseudo-time i the history of process is explored with respect to the previous time instant $i - 1$). Moreover, all values of the used variables $z_k(i)$ can be analytically calculated in a recurrent form as functions of variables $z_k(i-1)$. Original KF were derived only for the linear dynamic processes (cf. Kalman, 1960), and due to the application of NNs nonlinear KFs can be also considered. This possibility was explored for formulating new algorithms for the NN learning.

In this chapter, the decoupled extended Kalman filter algorithm (DEKF) is used (Haykin, 2001). The algorithm is based on two equations: (1) process equation and (2) measurement equation:

$$\mathbf{w}_j(i+1) = \mathbf{w}_j(i) + \boldsymbol{\omega}(i) , \tag{15a}$$

$$\mathbf{y}(i) = \mathbf{h}(\mathbf{w}(i), \mathbf{x}(i)) + v(i) , \tag{15b}$$

where: $i = 1,2,\ldots, I$ – discrete pseudo-time parameter, j – number of neuron in NN; $\mathbf{w}(i)$ – state vector (one-column matrix) corresponding to the set of vectors $\mathbf{w}$ of synaptic weights and biases for n neurons of NN; $\mathbf{h}$ – nonlinear input-output relation; $\mathbf{x}, \mathbf{y}$ – input/output vectors; $\boldsymbol{\omega}(i)$, $v(i)$ – Gaussian process and measurement noise with mean and covariance matrices defined by:

$$E[\boldsymbol{\omega}(i)] = E[v(i)] = 0 , \; E[\boldsymbol{\omega}(i)\,\boldsymbol{\omega}^{\mathrm{T}}(l)] = \mathbf{Q}(i)\,\delta_{il} , \; E[v(i)\,v^{\mathrm{T}}(l)] = \mathbf{R}(i)\,\delta_{il} . \tag{16}$$

In the analysis decoupling of the state vector to j groups is performed, where: $- j\text{-}th$ group composed of synaptic weights of $j\text{-}th$ node of the network was follows. Then a model described by equation (15) is formulated in recurrent form for each of the neuron groups as the following form of DEKF algorithm:

$$\mathbf{A}(i) = [\mathbf{R}(i) + \sum_{j=1}^{g} \mathbf{H}_j^{\mathrm{T}}(i)\, \mathbf{P}_j(i)]^{-1} ,$$
$$\mathbf{K}_j(i) = \mathbf{P}_j(i)\, \mathbf{H}_j(i)\, \mathbf{A}(i) ,$$
$$\boldsymbol{\varepsilon}(i) = \mathbf{y}(i) - \hat{\mathbf{y}}(i) , \tag{17}$$
$$\hat{\mathbf{w}}_j(i+1) = \hat{\mathbf{w}}_j(i) + \mathbf{K}_i(i)\,\boldsymbol{\varepsilon}(i) ,$$
$$\mathbf{P}_j(i+1) = \mathbf{P}_j(i) - \mathbf{K}(i)\, \mathbf{H}_j^{\mathrm{T}}(i)\, \mathbf{P}_j(i) + \mathbf{Q}_j(i) ,$$

where: $\mathbf{K}_j(i)$ – Kalman gain matrix; $\mathbf{P}_j(i)$ – approximate error covariance matrix; $\boldsymbol{\varepsilon}(k) = \mathbf{y}(k) - \hat{\mathbf{y}}(i)$ – error vector, $\mathbf{y}(i)$ – target vector for the $i\text{-}th$ presentation of a training pattern; $\hat{\mathbf{w}}_j(i), \hat{\mathbf{y}}_j(i)$ – $i\text{-}th$ estimate of weight vector and output vector, $\mathbf{H}_j$ – matrix of current linearization of Equation (15b) which takes the form:

$$\mathbf{H}_j(i) = - \,\partial\, \mathbf{h}_j(i,\mathbf{w}) / \partial\, \mathbf{w}_j \mid \mathbf{w}_j = \hat{\mathbf{w}}_j , \tag{18}$$

and equation (16) can be written as:

$$E[\boldsymbol{\omega}(i)\,\boldsymbol{\omega}^{\mathrm{T}}(l)] = \mathbf{Q}(i)\,\delta_{il} , \; E[v(i)\,v^{\mathrm{T}}(l)] = \mathbf{R}(i)\,\delta_{il} . \tag{19}$$

The term 'extended' KF (EKF) is used because in the measurement equation (15b) a nonlinear output-input relation is due to the introduction of the vector-function **h**.

Measures of Errors

Besides the least-square-error E defined in (12) there are applied other error measures for evaluation of the accuracy of neural approximation. The most popular are the mean-square-error (MSE) and root-mean-square-error ($RMSE$):

$$MSE(V) = \frac{1}{VM} \sum_{p=1}^{V} \sum_{i=1}^{M} (\overline{t}_i^{(p)} - \overline{y}_i^{(p)})^2, \quad RMSE = \sqrt{MSE}, \tag{20}$$

where: $t_i^{(p)}$, $y_i^{(p)}$ – target and neurally computed i-th outputs for p-th pattern. The values of $\overline{t}_i^{(p)}$ and $\overline{y}_i^{(p)}$ are usually scaled. In case if the sigmoid activation function (4b) is used in the output layer then the range [0.1, 0.9] is recommended for the output variables.

For estimation of neural prediction the following relative errors are used:

$$\text{avr } eV = \frac{1}{V} \sum_{p=1}^{V} ep, \quad \max eV = \max_{p} ep, \quad \text{where: } ep = \left| 1 - y^{(p)}/ t^{(p)} \right| \times 100 \,\%. \tag{21}$$

The numerical efficiency of the trained network can be evaluated by the success ratio (SR) using the following formula:

$$SR = \frac{NBep}{V} \times 100\%, \tag{22}$$

where: $NBep$ – number of patterns corresponding to modules of relatitive errors (i.e., $|ep| \le Bep$ for the assumed restrained error Bep; $V = L, T, P$ – number of patterns of the considered sets). SR corresponds to the cumulative curve which is used in statistics.

From among the statistical parameters the standard error $st\,\varepsilon$ and linear regression coefficient r are frequently used with respect to the set of pairs $\{(t_i, y_i)^{(p)} \mid p = 1,..., V\}$:

$$r(V) = \frac{\displaystyle\sum_{p=1}^{V} \sum_{i=1}^{M} (t_i^{(p)} - \hat{t}_i)(y_i^{(p)} - \hat{y}_i)}{\sqrt{\displaystyle\sum_{p=1}^{V} \sum_{i=1}^{M} (t_i^{(p)} - \hat{t}_i)^2 \sum_{i=1}^{M} (y_i^{(p)} - \hat{y}_i)^2}} \tag{23}$$

where: $\hat{t}_i, \hat{y}_i$ – mean values of sets $\{t_i^{(p)}, y_i^{(p)}\}$ for the fixed output i.

Some Remarks on Design of BPNNs

There are many problems concerning design of NNs and methods of their analysis (cf. e.g., Rojas, 1996; Jang et al., 1997; Haykin, 1999; Waszczyszyn, 1999). During the design process we should take into account the over-fitting of the neural approximation. This numerical effect occurs if training results fit very well the target outputs data but testing errors are high. Such an effect take place if the values of the number of the network parameters NNP are close to the number of training patterns L. That is why it is recommended (cf. Haykin, 1999), to have in formula (7) much bigger L (let say two or three times) than NNP is.

The analysis of data and selection of training and testing patterns are of primary importance. The training and testing sets should be statistically representative so usually they are randomly selected from a set $\mathcal{P} = \mathcal{L} \cup \mathcal{T}$ of input/output pairs of known patterns. In case of a small number of P patterns we can apply so-called multifold cross-validation method (cf. Haykin, 1999). The method depends on random selection of small testing sets $\mathcal{T}$, which is repeated several times and then an average testing error is computed. This approach is similar to the booth-strap validation method (e.g., Rojas, 1996).

To end this point, it is worth emphasizing that what is the most important is evaluation of the testing errors since the main goal is the neural prediction, that is, application of the trained network for predicting outputs for inputs which were not used in the training process (generalization properties of the trained network).

Reduction of the Input Space Dimensionality

One of possible methods to avoid overfitting of the NN approximation is the reduction of the number of inputs. This approach enables us to design smaller networks of the number of BPNN parameters NNP lower than for networks without reduction of inputs. From among several methods which reduce the input space dimensionality only two, frequently applied methods are presented as follows: (a) application of BPNN as the replicator, (b) principle components analysis of the input space (cf. Haykin, 1999).

BPNN as a Replicator

BPNN discussed in the previous section can be classified as a hetero-associative memory network. This means that the trained BPNN can perform the mapping of input vectors $\mathbf{x}^{(p)} \in \mathcal{R}^N$ into the output vectors $\mathbf{y}^{(p)} \in \mathcal{R}^M$: $\mathbf{x}_{N \times 1}^{(p)} \rightarrow \mathbf{y}_{M \times 1}^{(p)}$, where: $p = 1, \ldots, P$ – number of patterns. It was proved that BPNN is a general approximator which can be used also for the auto-associative mapping:

$$\mathbf{x}_{N \times 1}^{(p)} \xrightarrow{\quad BPNN \quad} \tilde{\mathbf{x}}_{N \times 1}^{(p)}, \tag{24}$$

where: $\tilde{\mathbf{x}}_{N \times 1}^{(p)} = \mathbf{x}_{N \times 1}^{(p)} + \boldsymbol{\varepsilon}_{N \times 1}^{(p)}$ output vector with the error of neural approximation.

Figure 5. (a) BPNN as a replicator, (b)(c) splitting of replicator into compressor and decompressor, respectively

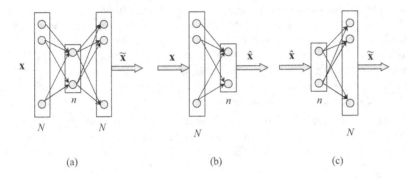

(a)   (b)   (c)

BPNN corresponding to mapping (24) is called a replicator (Haykin, 1999). Let us consider a simple replicator with only one hidden layer, Figure 5a. In case the number of hidden neurons $H = n$ is much smaller than the number of inputs/outputs, that is, $n \ll N$, the replicator can be used for data compression with the compression ratio $CR = N / n$.

After the training the replicator can be split into the compressor and decompressor of data, as shown in Figure 5b, c. The compressing of data into the vector $\hat{x}$ is performed by the first part of replicator (compressor) and its decoding to the output vector $\tilde{x}$ can be made by means of the second part of the replicator (decompressor).

Principal Component Analysis

Principal component analysis (PCA) is related to a linear transformation of the input vector components to the principal directions of the input space. The transformation can lead to decreasing of the input space dimensionality. It is possible by means of eigenanalysis to select only those principal components which preserve the most important features of the original input space (Haykin, 1999).

PCA starts from computation of the autocorrelation matrix **R:**

$$\mathbf{R} = \frac{1}{L}\sum_{p=1}^{L}\mathbf{x}^{(p)\mathrm{T}}\mathbf{x}^{(p)}, \tag{25}$$

where: $\mathbf{x}^{(p)} \in \mathcal{R}^{N}$ input vector of the *p-th* training pattern. Then the linear eigenvalue problem is analysed:

Figure 6. Illustration of principal components for data group from two-dimensional space

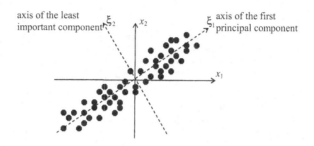

$$\mathbf{R}\,\mathbf{v}_j = \lambda_j\,\mathbf{v}_j \text{ for } j = 1,\ldots, N. \tag{26}$$

Eigenvalues are put in order $\lambda_1 > \lambda_2 > \ldots > \lambda_N$ and then relative eigenvalues m_j are computed:

$$m_j = \lambda_j \Big/ \sum_{k=1}^{N} \lambda_k \tag{27}$$

as measures for evaluating the significance of each eigenvalue λ_j.

In case there are great differences of eigenvalues we can preserve only the highest ones and formulate the vector of a reduced number of eigenvalues:

$$\boldsymbol{\lambda}_{(K\times 1)} = \{ \lambda_k \mid k = 1, \ldots, K\} \quad \text{for } K < N. \tag{28}$$

The corresponding eigenvectors $\mathbf{v}_k$ are used for computing the principal components:

$$\xi_k = \mathbf{v}_k^{\mathrm{T}}\mathbf{x} \;\; \text{or} \;\; \xi_k^{(p)} = \mathbf{v}_k^{\mathrm{T}}\mathbf{x}_k^{(p)} \;\; \text{for } k = 1,\ldots, K \text{ and } p = 1,\ldots, L. \tag{29}$$

The principal components are the projections of original input vectors $\mathbf{x}$ on principal directions related to unit eigenvectors $\mathbf{v}_k$ (versors of the turned axes k). This is shown in Figure 6 corresponding to 2D output space. The principal directions 1 and 2 are related to eigenvectors of maximal and minimal variances of data, respectively.

Table 3. Experimentally measured natural periods of buildings and relative errors of their neural prediction

Building	Direction	Pattern number[*]	Measured natural period T_1 [sec]	Relative errors $ep_1 = (1 - y^{(p)}/t^{(p)}) \times 100$ [%] for networks				
				4-4-1 $\mathbf{x}_{(4\times1)} = \{C_z,b,s,r\}$	2-3-1 $\mathbf{x}_{(2\times1)} = \{C_z,b\}$	4-3-1 $\mathbf{x}_{(1\times1)} = \{\mu_s,\mu_M,\mu_L,b\}$	1-2-1 $\mathbf{x}_{(1\times1)} = \{\xi_1\}$	2-3-1 $\mathbf{x}_{(2\times1)} = \{\xi_1,\xi_2\}$
1		2	3	4	5	6	7	8	9
DOMINO -68(I)	transv.	1	0.256	-5.7	-4.8	-4,1	1.7	-2.3
	longitud.	2	0.230	1.2	5.1	4.0	-7.6	-0.4
DOMINO -68 (II)	transv.	3	0.256	-5.7	-4.8	-4.1	1.7	-2.3
	longitud.	4	0.230	1.2	5.1	4.0	-7.6	-0.4
WUF-T -67-SA/V	transv.	5[T]	0.253	-7.6	-6.1	-5.0	0.5	-2.7
	longitud.	6	0.204	-4.8	-6.5	-7.8	-22.2	-4.9
WUF-GT 84(I)	seg.I transv.	7	0.175	-2.3	-2.8	-4.9	-7.0	-7.5
	seg.I longitud.	8	0.185	-1.0	-6.5	4.0	-1.1	-4.4
	seg.II transv.	9	0.180	-8.6	0.1	-2.0	-4.0	-4.0
	seg.II longitud.	10[T]	0.169	-4.2	-30.9	1.3	-9.2	-13.4
WUF-GT 84(II)	seg.I transv.	11	0.157	3.5	-6.3	-5.2	-5.9	-3.1
	seg.II transv.	12	0.180	2.7	7.3	8.3	7.6	9.2
	seg.II longitud.	13	0.177	-1.3	-24.8	-7.2	7.7	-1.3
C/MBY/V(I)	transv.	14	0.172	2.2	6.7	6.7	5.2	4.8
	longitud.	15	0.192	2.2	9.5	9.5	13.4	-4.4
C/MBY/V(II)	transv.	16[T]	0.185	-0.7	-4.9	-19.3	-9.1	-9.6
	longitud.	17	0.213	-2.3	4.5	0.9	5.3	7.0
C/MBY/V(III)	transv.	18[T]	0.227	5.6	2.3	-6.4	0.4	-1.1
	longitud.	19	0.233	-1.5	2.6	0.3	3.0	4.6
BSK(I)	seg.I transv.	20	0.155	1.7	-9.8	-9.7	-12.5	-3.9
	seg.I longitud.	21	0.233	-0.9	6.8	1.4	-2.4	-2.6
	seg.II transv.	22	0.155	1.7	-9.8	9.7	-10.4	-0.7
	seg.II longitud.	23	0.233	2.4	7.8	3.0	3.0	2.6
BSK(II)	seg.II transv.	24	0.156	10.5	-9.1	-9.0	-9.7	-0.1
	seg.II longitud.	25	0.233	-0.2	7.8	3.0	3.0	2.6
WWP	transv.	26	0.270	-1.1	2.3	2.2	7.5	1.6
	longitud.	27[T]	0.294	10.8	18.3	19.4	7.5	-5.5
WBL	transv.	28	0.294	-5.8	10.3	10.1	15.1	4.9
	longitud.	29	0.263	5.1	0.2	0.0	5.1	-1.0
WK-70	transv.	30	0.256	0.0	-3.1	-3.2	2.5	-4.3
	longitud.	31	0.227	-2.7	-5.7	-4.3	-9.1	6.1

Note: [*] *T – testing pattern*

Application of Neural Networks to the Dynamic Analysis of Actual Buildings

Prediction of Dynamic Properties of Medium Height Prefabricated Buildings

Natural periods, vibration damping and mode shapes of natural vibrations describe dynamic properties of structures. In the dynamic analysis of buildings subjected to paraseismic excitations it is necessary very often to estimate their fundamental periods of natural vibrations. A corresponding empirical formulae or computer programs are used in evaluation of the technical state of buildings subjected to paraseismic excitations (cf. Ciesielski et al., 1992).

From the dynamic full-scale tests performed on the analysed prefabricated five-story buildings the data of natural periods were obtained among other results. The tests included measurements of horizontal vibration components in two mutually perpendicular directions parallel to the longitudinal and transverse axes of the buildings. Because of very small damping in the investigated buildings (Ciesielski et al., 1995), the differences between the free vibrations and the eigenvibrations were considered negligible.

The dominant periods of recorded vibrations were accepted to be the natural ones. The results of experimental tests show that the buildings vibrate practically only with the fundamental natural periods. Sometimes, but only exceptionally, the second frequency is observed in the phase of forced vibrations. The duration and amplitudes of these vibrations are very small, so higher frequencies are not considered. In order to determine the fundamental periods of vibrations, various methods of records processing were used—fast fourier transformation (FFT) and spectral analysis in particular. The experimentally (i.e., computed on the basis of measured records) evaluated fundamental vibration periods of the buildings are collected in column 4 of Table 3.

Application of NNs for building natural periods identification was the next step in the dynamic analysis of buildings considered and the main problem associated with NN identification was a proper selection of input variables.

The identification problem is formulated as a relation between structural and soil parameters, and the fundamental period of building. In the light of full-scale tests of the analysed buildings it can be stated (Maciag, 1986) that the soil-structure interaction plays an important role in vibrations of medium height buildings. The foundation flexibility is expressed by the coefficient of an elastic uniform vertical deflection of the subgrade C_z. The next representative parameter is the building dimension in the direction of vibrations b. Other parameters correspond to the equivalent bending stiffness $s = \sum_i EI_i/a$ and equivalent shear stiffness $r = \sum_i GA_i/a$, where: E, G – elastic and shear moduli, respectively; I_i, A_i – moment of inertia and cross-sectional area of the i-th wall in the building plan, a – length of building. For instance in case of transverse direction of the building DOMINO-68(I) the above mentioned parameters make respectively: $C_z = 50\text{MPa}$, $b = 10.6\text{m}$, $s = 44100\text{MNm}$, $r = 3786\text{MN/m}$ with the corresponding value of $T_1 = 0.256[\text{sec.}]$ and in case of longitudinal direction of the building C/MBY/V(I): $C_z = 267\text{MPa}$, $b = 13.8\text{m}$, $s = 24515\text{MNm}$, $r = 1497\text{MN/m}$, $T_1 = 0.192[\text{sec.}]$.

Application of Standard BPNNs

The above considered parameters were adopted as the input vector $\mathbf{x}$ of BPNN (Kuźniar, Maciąg, & Waszczyszyn, 2000):

$$\mathbf{x}_{(4\times1)} = \{C_2, b, s, r\}, \tag{30}$$

and the building fundamental natural period was a scalar output y:

$$y = T_1. \tag{31}$$

$P = 31$ patterns from the experimental data were obtained for the input vector (30) (cf. Table 3). With a very small set of patterns a network was designed introducing a modification of the multifold cross-validation procedure (Haykin, 1999). From the total number of $P = 31$ patterns, $T = 5$ patterns of a testing set were randomly selected, whereas the remaining $L = 26$ patterns were assigned as a learning set. This procedure was repeated one hundred times. The network of structure BPNN: 4-4-1, Rprop learning method and sigmoid activation in all layers were used. The number of network parameters $NNP = 25$ was slightly smaller than the number of training patters $L = 26$. The computations were performed by means of SNNS computer simulator (Zell, 1998).

The accuracy of neural approximation was evaluated by $MSE(V)$ error (20). Following the multifold cross-validation method computations were repeated a hundred times using different randomly selected training and testing sets of pattern numbers $L=26$, $T=5$, respectively. The corresponding mean training and testing errors were $MSE(L)=0.000075$ and $MSE(T)=0.000327$.

In Table 4 there are put together errors $MSE(V)$ computed for the testing patterns marked in column 3 of Table 3 by superscripts T since they gave error $MSE(T) = 0.000320$, very

Table 4. Errors of natural building periods identification using different types of neural networks

No.	Input parameters	Network (BPNN)	$MSE(V) \times 10^4$		avr eV [%]			$st\varepsilon(P)$	$r(P)$
			L	T	L	T	P		
1	b, C_2, s, r	$4 - 4 - 1$	0.8	3.2	3.0	5.8	3.5	0.011	0.964
2	b, C_2	$2 - 3 - 1$	2.5	12.0	6.5	12.5	7.5	0.020	0.873
3	μ_S, μ_M, μ_L, b	$4 - 3 - 1$	1.4	9.8	4.9	10.3	5.8	0.017	0.916
4	$\xi_1(b, s, r, C_2)$	$1 - 2 - 1$	3.2	1.9	7.0	5.3	6.7	0.017	0.901
5	$\xi_1(b, s, r, C_2), \xi_2(b, s, r, C_2)$	$2 - 3 - 1$	0.7	2.3	3.5	6.5	4.0	0.010	0.972

Figure 7. Success Ratio SR vs. relative restraint error Bep for back-propagation neural prediction of fundamental natural periods of buildings

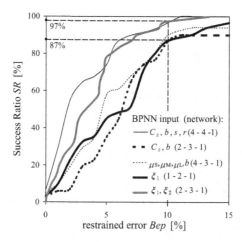

close to the mean error 0.000327, computed in the multifold cross-validation procedure. The training and testing sets completed in such a way were fixed for later computations which gave errors listed in Tables 3 and 4. Because of small number of testing patterns the statistical parameters were computed for the total number of patterns $P = 31$.

Following the remarks expressed in Ciesielski et al. (1992) a simpler BPNN was also examined using the following two inputs:

$$\mathbf{x}_{(2\times1)} = \{C_z, b\} \tag{32}$$

and output (31).

The errors obtained by BPNN: 2–3–1 are listed in Tables 3 and 4. In Figure 7 there are shown graphics $SR(ep)$, that is, functions of the success ratio defined by formula (22) for the prediction of fundamental natural periods of prefabricated medium height buildings obtained using various BPNNs. The graphics shown in Figure 7 enable an evaluation of what percentage of patterns $SR[\%]$ gives the neural prediction with the error not greater than $Bep[\%]$. For instance, the BPNN: 4–4–1 gives good prediction for about 97% of neurally computed fundamental natural periods with the accuracy of the relative restrain error $ep \leq Bep = 10\%$. Corresponding figures for BPNN: 2–3–1 are $SR(10\%) = 87\%$. These results can be explained by the fact that the two-component input vector (32) includes less information on buildings than the four-component input vector (30).

Figure 8. Triangular membership functions for values C_z

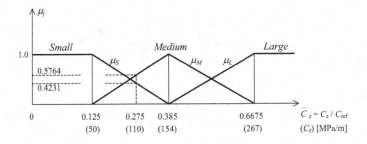

BPNNs with Fuzzy Inputs

The input parameter C_z reflects the nature of soil basement. The values of C_z are estimated as mean values from the ranges found experimentally. Those ranges are given in the Polish code (PN-80/B-03040, 1980) for different soils of small, medium and large stiffnesses. This is, in fact, the introduction of the linguistic variable associated with the fuzzy character of the coefficient C_1 instead of crisp values C_z:

$$C_1 = \{\mu_S, \mu_M, \mu_L\} , \qquad\qquad (33)$$

where: μ_S, μ_M, μ_L – values of membership functions for small, medium and large rigidity of soil, respectively corresponding to triangular membership functions shown in Figure 8 (Kuźniar & Waszczyszyn, 2002).

Figure 9. Training errors MSE(L) for networks with input vectors $\mathbf{x}_{(2\times1)} = \{C_z, b\}$ and $\mathbf{x}_{(4\times1)} = \{\mu_S, \mu_M, \mu_L, b\}$

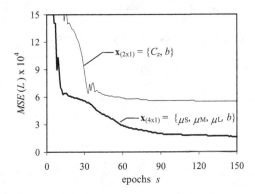

In what follows the linguistic variable C_1 is used instead of the crisp variable C_z (i.e., the input vector **x** has four components):

$$\mathbf{x}_{(4\times1)} = \{\mu_S, \mu_M, \mu_L, b\}. \tag{34}$$

The back-propagation neural network BPNN: 4–3–1 corresponding to input vector (34) was trained by the same set of $L = 26$ patterns as the previous network BPNN: 2–3–1 with crisp variable C_z and input vector given in form (32). In Tables 3 and 4 there are collected errors computed for this network. In Figure 7 the corresponding graphic $SR(ep)$ is also shown.

On the basis of results obtained it can be concluded that the introduction of linguistic variable (33) can lead to a better accuracy of neural identification than that by means of crisp values of the soil-structure interaction parameters C_z. These results can also be explained by the increase of the number of network parameters but with no increase of physically interpreted input variables.

In Figure 9, the results of training process are shown for BPNN with different input vectors corresponding to two physical parameters C_z and b. As can be seen, the iteration process associated with the training of network with linguistic variable (33) is much more quickly convergent than for network with crisp values of C_z. Thus the introduction of a linguistic variable C_1 instead of crisp variable C_z causes a decrease of the number of training epochs and increase of success ratio SR (see Figure 7).

Computation of Fundamental Natural Periods Using PCA

In the case of BPNN: 4–4–1 and input vector (30) the number of network parameters is $NNP = 25$, versus the number of training patterns $L = 26$. The corresponding figures for the network with linguistic variable BPNN: 4–3–1 are $NNP = 19 \leq L = 26$. This means that these networks, according to the evaluation (7), are close to the limit of generalization capacity. That is why we are interested in design of smaller networks corresponding to decreasing of the input number. In what follows the application of the principal component analysis (PCA) is discussed as a one of the possibilities to perform data compression of the input data. This leads to the reduction of number of network parameters and improvement of the network generalization properties (Kuźniar & Waszczyszyn, 2006).

For $P = 31$ patterns $\mathbf{x}^{(p)} = [C_z^{(p)}, b^{(p)}, s^{(p)}, r^{(p)}]^T$ ($p = 1, ..., P$), the correlation matrix $\mathbf{R} \in \mathcal{R}^4 \times \mathcal{R}^4$ was set up. Four eigenvalues and eigenvectors corresponding to them were computed:

$\lambda_1 = 32029, \lambda_2 = 1548, \lambda_3 = 81, \lambda_4 = 0.1,$

$\mathbf{v}_1 = [0.0852, 0.0589, 0.0019, 0.9946]^T, \mathbf{v}_2 = [-0.3472, -0.9339, -0.0072, 0.0850]^T,$

$\mathbf{v}_3 = [\ 0.9339, -0.3526, 0.0070, -0.0592]^T, \mathbf{v}_4 = [0.0092, 0.0044, -0.9999, 0.0009]^T.$

$$\tag{35}$$

In the case considered, according to formula (27), the relative contributions of the principal components to the total variance of data are:

$$m_1 = 0.9516 \ (95.2\%), \ m_2 = 0.04599 \ (4.6\%), \ m_3 = 2.4 \cdot 10^{-3} \ (0.24\%), \ m_4 = 3 \cdot 10^{-6} (0.0003\%).$$

$$(36)$$

Looking at (36) it is clear that the first principal component reaches more than 95% part of the total variance of data. Thus the first principal component is predominant. Therefore, considering the building parameters, it is sufficient to take only the greatest principal component $\xi_1(C_z, b, s, r)$ using formula (29). The other principal components can be neglected because they do not affect the information substantially.

Thus the neural network BPNN: 1–2–1 was trained for the natural periods identification using only one input $x_{(1\times1)} = x = \xi_1(C_z, b, s, r)$. The training and testing was performed using the same patterns selection as in previous Point, that is, with $T = 5$ testing patterns marked in Table 3. The results of application of this network were compared in. Table 3, Table 4 and Figure 7 with the previously computed results for a case when the input vectors contained building parameters with no compression: $x_{(4\times1)} = \{C_z, b, s, r\}$, $x_{(2\times1)} = \{C_z, b\}$.

Looking at the curves of success ratio, see Figure 7, it is clear that in every analysed version of BPNNs, at least 90% of patterns are identified with the error not greater than 15%. The comparison of network errors is presented in Table 4. The best results are confirmed for BPNN: 4–4–1 with four-elements input vector $x_{(4\times1)} = \{C_z, b, s, r\}$. However, the results for a very small network 1–2–1 (network number 4 in Table 4) with the first principal component produced quite satisfactory accuracy. This is visible in Figure 7 where success ratio for this network is $SR(10\%) = 87\%$. Attention should be paid to the fact that in this network the input information is compressed four times. The application of such a very small network 1–2–1 leads to even better prediction of fundamental periods of building natural vibrations than for the network BPNN: 2–3–1 with the input vector $x_{(2\times1)} = \{C_z, b\}$. Hence, due to application of PCA, the design of considerably smaller neural networks than those without data compression with no greater increase of the neural approximation errors is possible due to application of PCA.

In Table 4 and Figure 7 results of computations for two principal components $\xi_1 (C_z, b, s, r)$ and $\xi_2 (C_z, b, s, r)$ are also shown (Kuźniar & Waszczyszyn, 2006). The addition of the second principal component improves dramatically the results which are comparable with those obtained with no data compression. The use of network BPNN: 2–3–1 with inputs ξ_1 and ξ_2 is much better than in the case of inputs b and C_z. This can be explained by a more extended information which is incorporated in the principal components which are functions of four original inputs (30). This is also a reason why the approximation errors for the principal components ξ_1 and ξ_2 and network BPNN: 2–3–1 are nearly the same as for the mentioned original inputs used in the nearly twice bigger network BPNN: 4–4–1.

Empirical Formula

There are some proposals for an empirical formula for the calculation of the fundamental natural periods of the medium height building considered in this Point. The following formula was suggested in Ciesielski et al. (1992):

Table 5. Empirical formulae approximation errors

No.	Formula	avr eP [%]	max eP [%]	st$\varepsilon(P)$	$r(P)$
(33)	$T_1 = 0.98 / \sqrt[3]{C_z}$	12.8	53.5	0.045	0.768
(34)	$T_1 = 1.2 / \sqrt[3]{C_z + 0.003 \cdot (s+r)/b}$	11.9	20.5	0.029	0.793
(35)	$T_1 = 0.238 + 0.08\bar{\xi}_1 - 0.1165\bar{\xi}_1^2 + 0.03\bar{\xi}_1^3$	8.9	23.7	0.024	0.811

$$T_1 = \frac{0.98}{\sqrt[3]{C_z}}, \tag{37}$$

and in Kuźniar et al. (2000) the following formula was derived:

$$T_1 = \frac{1.2}{\sqrt[3]{C_z + 0.0003(s+r)/b}}, \tag{38}$$

where the variables correspond to building and soil parameters introduced as the inputs (30). In Table 5 approximation errors corresponding to formulas (37) and (38) are shown.

PCA gives another opportunity of formulating a neural based empirical formula. Due to one principal component we have an implicit function $T_1 = y(\xi_1)$. This function can be approximated by any suitable function of the independent variable ξ_1. This is shown in Figure 10 where the 3rd order polynomial was used:

Figure 10. Fundamental period T_1 vs. the first principal component ξ_1

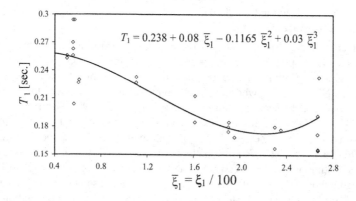

$$T_1 = 0.238 + 0.08\overline{\xi}_1 - 0.1165\overline{\xi}_1^2 + 0.03\overline{\xi}_1^3 \,, \tag{39}$$

where: $\overline{\xi} = (0.0852\ C_z + 0.0589\ b + 0.0019\ s + 0.9946\ r)/100$. The coefficients in (39) were computed by the LMS (least mean square) method using experimental points shown in Figure 10.

Looking at approximation errors listed in Table 5 it is clear that formula (39) gives accuracy comparable to that by formula (38) and quite satisfactory for engineering evaluation. It is worth emphasizing that the formulation of function (39) in 1D space is directly related to the results of measurements, better than in case of formula (38).

Simulation of Building Response to Seismic-Type Excitation

The neural network technique was also used for the simulation of building response to paraseismic excitations. In order to analyze this relation the description of excitation vibrations and information about the dynamic properties of a building are included into the input vector. As outputs different building responses are used. The maximal displacement or the displacement record in time domain for a selected floor were considered in Kuźniar and Waszczyszyn (2002, 2003).

Simulation of Building Floor Displacement

The sets of training and testing patterns were formulated on the basis of full-scale tests performed on actual structures. The measurements were carried out on the above mentioned five-story prefabricated buildings. The buildings were subjected to paraseismic excitations caused by explosions in nearby quarries. $P = 112$ patterns were associated with 112 selected measurements in time domain, corresponding to displacement records at the basement or ground levels as well as on the fourth floor, cf. Figure 2. Then all the recorded experimental data in form of vibrations in the time domain at the basement and ground level were pre-processed (Kuźniar & Waszczyszyn, 2002).

The displacement response spectra $Sd(T_i)$ were computed for the basement and ground levels, assuming the damping coefficient $\xi=2\%$ (Ciesielski et al., 1995). These data were compressed in order to diminish the neural network size. The replicator, was formulated as the network BPNN: 99–7–99, where the input/output layers corresponded to 99 discrete values of the displacement response spectrum $d_i = Sd(T_i)$ for $i = 1, ..., 99$ taken from the range $T \in [0.025, 1.005]$sec. These discrete values were computed for the constant period increment $\Delta T_i = 0.01$sec. The number of neurons in the hidden layer $ne = 7$ gives the compression ratio $N/ne=99/7=14.1$. The set of $P = 112$ patterns was randomly split into $L = 90$ and $T = 22$ training and testing patterns, respectively. The SNNS computer simulator (Zell, 1998), Rprop learning method and sigmoidal activation functions were used in this replicator. The training error $MSE(L) = 0.01403$, the testing error $MSE(T) = 0.00349$, the correlation coef-

Figure 11. Success ratio SR vs. relative error ep and errors computed for the neural prediction of maximal displacement D4 of the buildings' 4th floor

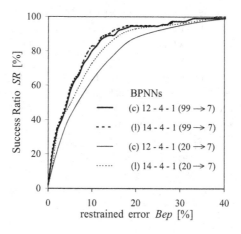

ficient $r(P)$=0.991 and the standard error st$\varepsilon(P)$= 0.109 were obtained according to formulae (23) where all the input and output variables were scaled into the range [0.1, 0.9].

The simulation problem was formulated as mapping of compressed values of excitation response spectrum $Sd(T_i)$ and structural parameters into the maximal displacement D_{max} = D4 of the fourth floor of the building in question. The excitation vibrations were recorded on the ground in front of the building and the other ones at the basement level. But in the contact ground - building foundation the reduction of vibrations was observed as a result of soil-structure interaction. Therefore, not only the excitation, e.g. response spectrum, but also the place of its registration had to be described. Thus the following input vector and scalar output were adopted:

Table 6. Errors for BPNN simulation of maximal displacement D4 using compression of input data $\{d_i\}$ and crisp (c) or linguistic variable (l) for soil parameter

Compression $N \rightarrow n$	BPNN (c) or (l)	Stopping number of epochs s	$MSE(V) \times 10^3$		avr eV [%]		$r(P)$
			L	T	L	T	
$20 \rightarrow 7$	(c) 12 - 4 - 1	22000	1.37	0.49	9.1	12.2	0.964
	(l) 14 - 4 - 1	13000	1.11	0.21	7.2	7.8	0.972
$99 \rightarrow 7$	(c) 12 - 4 - 1	15000	1.08	0.15	6.6	6.7	0.973
	(l) 14 - 4 - 1	12000	1.02	0.13	6.4	7.3	0.974

$$\mathbf{x}_{(12\times1)} = \{d_i \mid i = 1,..., 7; C_z, b, s, r, p\}, y = D4 \tag{40}$$

where: d_i – compressed values of excitation response spectrum; C_z, b, s, r – variables as in (30); p – parameter related to the place of recorded excitation (it was evaluated $p = 0.4$ for vibrations recorded at the ground and $p = 0.7$ for vibrations at the basement level); $D4$ – maximal displacement of the building 4th floor.

The same $L = 90$ and $T = 22$ patterns as those used for the replicator training were applied to the training and testing of a master network (BPNN-M). The BPNN-M: 12–4–1 had sigmoid activation functions and was trained by Rprop method. The success ratio and the errors for the considered network are shown in Figure 11.

A more extended analysis was performed in Kuźniar and Waszczyszyn (2002). The training and testing errors for different BPNNs are listed in Table 6. In order to shorten the networks description the master networks are marked as (c) BPNN or (l) BPNN where (c) corresponds to the standard networks with crisp inputs and (l) was added for inputs with linguistic variable C_1. The compression procedure $N \rightarrow n$ was performed for different number of discrete

Figure 12. (a) Scheme of data pre-processing and training phase of neural networks, (b) operational phase of the trained neural networks

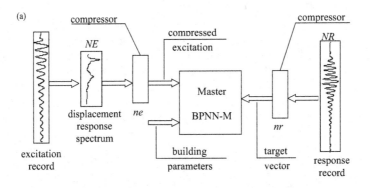

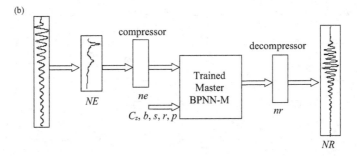

points $N = 20$, 99 of the displacement response spectra. On the basis of results shown in Table 6 we can conclude that from the engineering point of view the discussed network (c)12–4–1 with compression 99 $\to$ 7 gives quite satisfactory results. The network with linguistic variable (l) BPNN: 14–4–1 gives a little smaller errors.

Prediction of Displacement Records of Building Floor

The knowledge of the maximal displacement of the highest floor of the building is very useful in the approximate and quick estimation of dynamic responses of buildings. The full information of the response is given by records of vibrations in time domain. This problem was analyzed in Kuźniar and Waszczyszyn (2003) where besides the compression of the input data corresponding to paraseismic excitations, also the output displacement records (responses) were compressed in order to train the master network BPNN-M. Thereby the input and output vectors were composed of a smaller number of components (Figure 12a corresponds to the training phase).

Following the approach, the building was characterized by additional variables which complete the input vector of BPNN-M corresponding to (40). The same approach was used for formulating the output vector $y_{(nr \times 1)} \in \mathcal{R}^{nr}$, where nr is the number of compressed outputs. After the master network is trained and tested (training phase), it can be used for prediction of new records of displacements on the 4-th floor (i.e., dynamic excitations are mapped into diagrams of displacement response records on the selected floor of buildings [operational phase], Figure 12b).

The building response corresponds to the displacement record in time domain $D(t)$ related to vibrations of the fourth floor. Similarly as for excitations, the displacement records associated with the responses are compressed using the neural replicator. Each of the vibrations in time domain was described by $NR = 620$ discrete displacements D_i; $i = 1, ..., 620$. The replicator $620 - nr - 620$ was trained and tested on the same $L = 90$ and $T = 22$ patterns which were applied to data compression. The computer simulator SNNS and Rprop learning method were used (Zell, 1998). After an extensive validation, $nr = 9$ sigmoidal neurons were accepted in the hidden layer so the compression ratio of building response was $620/9 = 68.9$. The errors for the replicator BPNN: 620–9–620 were: $MSE(L) = 0.00018$, $MSE(T) = 0.000056$, avr $eP = 1.6\%$. It was stated that 96% of neurons had relative maximal errors less than 5%.

The compressed displacements cr_i ($i = 1, 2,...,9$) of the 4th building floor were used as components of the output vector of BPNN-M:

$$y_{(9 \times 1)} = \{cr_i \mid i = 1, 2,..., 9\} \tag{41}$$

where: cr_i – compressed response of building.

The same sets of patterns as those described in the previous Point were used for the training and testing of BPNN-M: 12–11–9 composed of sigmoid neurons. The Rprop learning

Figure 13. Comparison of measured and neurally simulated displacement records on the fourth floor of buildings: (a) With larger amplitudes, (b) with smaller amplitudes

(a)

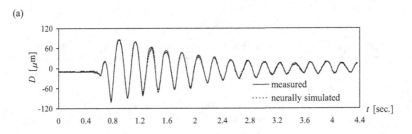

(b)

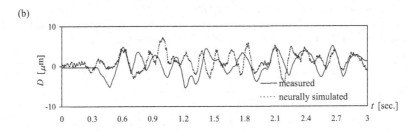

method and SNNS simulator were explored. The obtained errors were: $MSE(L) = 0.00075$, $MSE(T) = 0.00026$, avr $eL = 4.5\%$, avr $eT = 2.7\%$.

After the master BPNN was trained, its output values had to be decompressed in order to simulate the displacement records in time domain. The decompression was performed by a part of replicator called decompressor BPNN: 9–620, see Figure 5. More details about the decompressor formulation can be found in Kuźniar and Waszczyszyn (2003).

The testing process led to the conclusion that the neural simulation was very close to measured records for the majority, i.e., to about 70%, of the considered $P = 112$ patterns. The agreement was satisfactory for records with larger amplitudes of measured responses, Figure 13a and not satisfactory for smaller amplitudes, Figure 13b.

As a final conclusion of the discussed approach we can state that the neural networks can be efficiently used for prediction of not only selected displacements of a monitored building but they can also be used for predicting also records of vibrations in time domain without application of the dynamic analysis of the whole structure.

Neural Analysis of Response Spectra

Mapping of Seismic Parameters into Response Spectra from Mining Tremors

The dynamic analysis of the building based on normalized acceleration spectra is obviously approximate. On the other hand, the prediction of real vibration effects of expected mining tremors is very difficult. It seems that artificial neural networks technique can be efficiently applied to an analysis of this problem. This approach is shown with respect to the application of neural networks for evaluation of a relation between mining tremor energies, epicentral distances and acceleration response spectra (Kuźniar et al., 2005).

Experimental data in form of acceleration records are related to two seismic regions in Poland, i.e. 1) Upper Silesian Coalfield (USC) and 2) Legnica-Glogow Copperfield (LGC). The records in time domain corresponding to both USC and LGC were first pre-processed. From all the accelerations, the normalized acceleration response spectra $\beta(T_i; \xi) = Sa(T_i; \xi) / a_{gmax}$ were computed according assuming the damping coefficient $\xi = 2\%$. The vibration periods $T_i \in [0.02, 1.3]$sec. were computed using relation $T_i = 1/f_i$, where frequencies f_i were taken with the step $\Delta f = 0.25$Hz. Thus 198 discrete values of $\beta(T_i)$ were obtained from each of the analysed record.

The following input vector was adopted:

$$\mathbf{x}_{(3\times1)} = \{E, r_e, T_i\} \tag{42}$$

where: E – mining tremor energy, r_e – epicentral distance, T_i – vibration period for $i = 1,...,$ 198.

The corresponding value of normalized acceleration response spectrum β computed for the T_i vibration period was expected as the output of neural network:

$$\mathbf{y}_{(1\times1)} = \beta(T_i). \tag{43}$$

Table 7. Errors of neural networks for USC and LGC regions

Region	Network	$MSE(V) \times 10^3$		avr eV [%]		$r(P)$	$st\varepsilon(P)$
		L	T	L	T		
USC	3-7-1	3.80	4.00	14.1	13.0	0.737	0.062
	3-10-5-1	3.25	4.50	12.9	14.2	0.764	0.059
	3-6-12-5-1	3.10	4.10	12.8	12.7	0.781	0.057
LGC	3-12-6-1	6.00	7.80	21.3	25.6	0.562	0.080
	3-15-25-12-1	5.40	8.30	19.7	24.1	0.617	0.077

Figure 14. Success ratio SR vs. relative restraint error Bep for neurally predicted response spectra

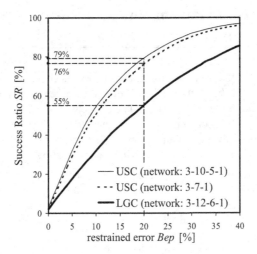

BPNNs designed by means of SNNS simulator, Rprop learning method and sigmoid activation functions were formulated separately for USC and LGC regions. Making use of the recorded experimental data in form of ground accelerations in time domain, 146 normalized acceleration response spectra (ARS) $\beta = \beta\,(T_i)$ were computed for USC region. The set of ARS was randomly split into about 80% and 20% training and testing patterns (i.e., 116 spectra were used for formulating the training patterns and the remaining 30 spectra were explored for formulation of the testing patterns). This means that $L = 116 \times 198 = 22968$ patterns and $T = 30 \times 198 = 5940$ patterns were obtained for the training and testing of BPNNs. For LGC region the corresponding figures were: $L = 102 \times 198 = 20196$ and $T = 26 \times 198 = 5148$.

In Table 7 final results of extensive numerical design of BPNNs of different architectures are listed. Smaller neural approximation errors were obtained for the USC region. The measurements collected for LGC enabled us to design BPNNs of lower accuracy. As can be seen, $MSE(V)$, avr eV and max eV errors are not very sensitive to increase of the network size (measured by the number of network parameters NNP calculated according to formula (6)). The two hidden layer networks BPNN: 3–10–5–1 and BPNN: 3-12–6–1 could be recommended. In Figure 14 the success ratios $SR[\%]$ are depicted for these networks. If we assume the restrain error $Bep = 20\%$ then $SR(20\%) \approx 79\%$ for USC but for LGC the neural prediction corresponds only to $SR(20\%) \approx 55\%$. Additionally for USC the SR curve related to BPNN: 3–7–1 is shown with $SR(20\%) \approx 76\%$.

Figure 15 illustrates normalized ARS corresponding to selected records taken from USC and LGC regions. It is visible that the normalized response acceleration spectra, neurally predicted on the basis of energies and epicentral distances only, are very close to the response spectra computed conventionally (target normalized ARS).

Figure 15. Comparison of computed on the basis of experimental data and neurally predicted normalized acceleration response spectra: (a) USC region, E=2×10⁴J, r_e=294m; (b) LGC region, E=1.4×10⁷J, r_e=1345m

(a)

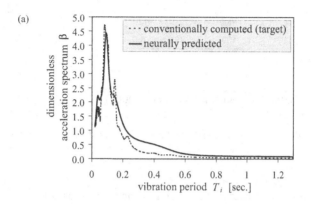

(b)

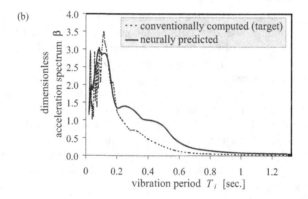

It is worth mentioning that response spectra have been recently computed by BPNNs trained by means of Kalman filtering (Krok & Waszczyszyn, 2005). Because of the scope of this chapter this problem is shortly discussed only in the next point.

Simulation of the Response Spectra in the Soil-Structure Interaction Problem

Using the data from measurements performed on actual structures, the neural technique was applied for evaluation of a relation between the response spectra from the ground vibrations and corresponding response spectra from the vibrations recorded at the same time on the

Table 8. Errors of training and testing processes for neural analysis of soil-structure interaction using displacement response spectra

BPNN	Learning algorithm	Number of epochs	$MSE(V) \times 10^4$		avr eV [%]		$r(T)$
			L	T	L	T	
6-5-1	Rprop	7000	13.7	12.5	13.7	13.3	0.879
2-5-1	DEKF	500	3.9	6.5	10.8	10.4	0.958

basement level of the building (Kuźniar & Maciąg, 2004). This way can be treated as the soil-structure interaction NN analysis.

The temporary window approach is adopted taking symmetric back and forward time-delays for the input variables. This approach is also called the moving average method (MAM) (cf. Haykin, 1999), or the subpicture idea (cf. Kuźniar & Maciag, 2004), known in literature on the analysis of control systems and picture transmission. The method is related to the application of static approach corresponding to the back-propagation learning method without time-delay outputs.

The problem considered is related to the transmission of vibration from the ground level to vibration of the building basement. DRS on the ground and basement levels are computed from monitored records of vibration. The soil-structure interaction is considered as mapping $Sdg(T_i) \rightarrow Sdb(T_i)$, where digitized values of DRS corresponds to the period of natural vibrations T_i. Displacement records were obtained from the measurements related to medium height (five-story), prefabricated buildings, subjected to paraseismic excitations caused by explosions in nearby quarry. Two kinds of DRS were computed (Kuźniar & Maciag, 2004),that is, DRSg related to the ground level at monitored buildings and corresponding DRSb associated with vibrations of the building basement. DRSg were adopted as inputs of the neural networks and DRSb were corresponding outputs. The corresponding discrete values $Sdg(T_i)$ and $Sdb(T_i)$ were computed for 198 periods of natural vibrations $T_i \in$ [0.02,1.3]sec, quite similarly as perviously.

The following input vector and scalar output were proposed:

$$\mathbf{x}_{(6\times1)} = \{ Sdg(i–2), S_{dg}(i–1), S_{dg}(i), S_{dg}(i+1), S_{dg}(i+2), T_i \} , y = Sdb(T_i) \tag{44}$$

where: $i = 3,...,$ 196 – successive number of vibration period.

The set of 10 pairs of DRS values $\{\{Sdg(i) \mid i = 1,..., 198\}, \{Sdb(i) \mid i = 3,..., 196\}\}$ was randomly split into an equal number of 5 training and 5 testing sets, respectively. The corresponding numbers of training and testing patterns were $L = T = P/2 = 5\times194 = 970$. These patterns were used for the training of BPNNs composed of sigmoid values. The Rprop learning method was used in the applied SNNS simulator. After an extensive cross-validation procedure the network BPNN: 6–5–1 was designed. The training and testing errors are shown in Table 8. All errors and statistical parameters were computed for the outputs scaled to the range [0.1, 0.9].

Figure 16. Displacement response spectra for selected learning and testing spectra DRS l#1 and DRS t#3, corresponding to the measured spectra on ground level (input DRSg) and basement level (target DRSb),and computed spectra by means of Kalman filtering (DEFF DRSb) and Rprop learning method (BPNN DRSb)

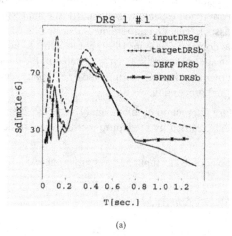

(a)

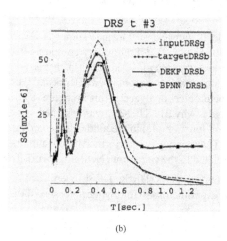

(b)

The accuracy of NN prediction of DRS on the building basement seems to be satisfactory from the engineering point of view if the average relative errors avr eL = avr $eT \approx 13\%$ and correlation coefficient $r(T)$ = 0.879 are taken into account. The graphics of the selected testing DRS are shown in Figure 16. As can be seen the satisfactory accuracy was achieved for the fundamental periods of vibrations from the interval [~0.03, ~0. 3]sec. In case of the 5-story building discussed, (cf. Table 3), the fundamental periods are within the interval [0.155, 0.294]sec. which is well placed in the range of accuracy which was obtained by means of the trained network BPNN: 6–5–1.

Neural Prediction of DRS by Means of Kalman Filtering

In paper by Krok and Waszczyszyn (2005), the neural Kalman filtering was applied to the BPNN learning. This approach is based on the theory of stochastic processes and is used in the control theory as an approach especially suitable for the analysis of discrete dynamic systems. This approach associated with temporal neural networks was used for the approximation of records in time domains (Haykin, 1999) but it seems to be also very prospective for NN approximations in spectral domain (as it was stated in the quoted paper by Krok & Waszczyszyn, 2005). Kalman filtering (KF), originated by Kalman (1960), is widely explored in the automatics and signal processing.

Below, in the soil-structure interaction problem, analyzed above by the standard approach, that is, by means of BPNN learned by the Rprop method, the application of DEFF algorithms is discussed.

DRS discrete values were adopted as input and output variables:

$$\mathbf{x} = \{Sdg(i-1), Sdb\ (i-1)\}\ , y = Sdb\ (i)\ , \tag{45}$$

where: $Sdg\ (i-1)$ – value of DRSg at the ground level for discrete time $i-1$; $Sdb\ (i-1)$, Sdb (i) – values of DRSb at the basement for $i-1$ and i discrete times for $i = 2,3,...,198$.

The autoregressive time-delay input $Sdb(i-1)$ was assumed as a variable well fitting to the character of the Kalman filtering method. Preliminary computations were performed using two types of neural networks suggested in Haykin (2001), that is, BPNN and recurrent layer neural network (RLNN). Contrary to results obtained in Krok and Waszczyszyn (2005) it was stated that in the case of the problem analyzed in this Point the superiority of a more refined RLNN was not proved. That is why BPNN: 2-5-1 was designed using sigmoid neurons with bipolar sigmoidal activation function (4c) in the hidden layer and linear activation function (4a) in the output neuron. The training was performed by the authors' procedures written in the MATLAB language related to the simulator (Neural Network Toolbox, 2001).

The same randomly selected five pairs (DRSg, DRSb) were used for the network training. On the basis of numerical experiments the following functions of the Gaussian noises were then found:

$$\mathbf{Q}(i) = 0.01\ \exp(-(s-1)/50)\mathbf{I}, \qquad R(i) = 7\ \exp(-(s-1)/50)\ , \tag{46}$$

where : $\mathbf{I}$ – unit matrices of dimension (3×3) for the $j = 1, 2,..., 5$ neurons of the hidden layer and (6×6) matrix for the output; s – number of the epoch in training process. The stopping criterion was established with respect to the fixed number of epochs S corresponding to the testing error $MSE(T) < \varepsilon_{adm}$ for output values scaled to the [0.1, 0.9]. After introductory computation the stopping criterion was related to $S = 300$ assuming $\varepsilon_{adm} = 1\times10^{-4}$.

After extensive preliminary training the network BPNN: 2-5-1 was designed using the algorithm DEKF as a learning method. After the network was trained the mean value $Sdb(1) = 0.2577$ was established in order to start with the testing process. The training and testing errors are listed in Table 8 and corresponding graphics of selected DRS are shown in Figure 16.

The most striking results concern the increased accuracy of neural simulation obtained by means on the Kalman algorithm DEKF versus results of computation in which standard Rprop learning method was used; see Table 8 and Figure 16. Another advantage of the Kalman filtering application corresponds to a lower number of training epochs. A great efficiency of the Kalman filtering learning method was resulted also from the introduction of the autoregressive input variable $Sdb(i-1)$. This effect was proved in Krok and Waszczyszyn (2005).

Summing up the obtained results we can draw a conclusion that the application of the theoretically well based Kalman filtering as a learning method seems to be a prospective approach to increase the accuracy of neural approximation in the analysis of response spectra.

Final Remarks

1. The chapter is devoted to the application of neural networks to the analysis of five-story prefabricated building vibrations caused by paraseismic excitations. The excitations were related to mining tremors and explosions in nearby quarries. The acceleration and displacement records measured on the ground level outside and inside the building on the levels of basement or fourth floor were the background for formulation of input and output variables. Two problems were discussed: (a) identification of fundamental natural periods and simulation of displacements on the fourth story; (b) simulation of response spectra on the ground level related to parameters of mining tremors, and RS corresponding to the soil-structure interaction. At the beginning of the chapter, selected basics of neural networks were discussed, related only to the standard multilayer back-propagation neural networks BPNNs. The use of BPNN as a replicator for data compression/decompression was shortly described. The basics on the application of the principal component analysis (PCA) and Kalman filtering in BPNNs were discussed in sections close to points in which these approaches were used.

2. The identification of natural periods was a difficult problem because of a small number of patterns from full-scale measurements on actual buildings. That is why the main attention was focused on diminishing or better specifying the number of inputs when BPNNs were used. It was shown that two basic input variables, that is C_z (coefficient of an elastic uniform vertical deflection of the subgrade) and b (building dimension in the direction of vibrations) gave satisfactory accuracy. Then the change of C_z from the crisp value to the linguistic variable C_1 was discussed. Introduction of PCA proved to be a very efficient approach. Even one PC gave neural approximation satisfactory from the engineering point of view. This was also a way to formulate an empirical formula for predicting the building natural periods.

3. The application of replicators as data compressors for the reduction of input number was shown on the example of neural simulation of floor displacements. The use of a part of the replicator for data decompression made it possible to predict also floor

displacement records as a function of building parameters and compressed input displacement response spectra.

4.   The last part of the chapter is devoted to the neural analysis of response spectra. It was stated that neural networks can be used for mapping of data of mining tremor energies and epicentral distances into response spectra from mining tremors. Then, which is the substantial advantage of the neural approach, the prediction of acceleration response spectra can be performed without recording of surface vibrations. The presented way of computation of acceleration response spectra can be mainly applied to the prognosis of mining tremors influences on structures. The use of BPNNs for the mapping of RS related to the soil-stucture interaction, i.e. the neural prediction concerns mapping DRSg → DRSb, where: DRSg is the displacemet RS on the ground level of the considered building and DRSb corresponds to the basement level inside the building, was also discussed. It was shown that the application of Kalman filtering to the training of BPNNs can give much more accurate results than those from BPNNs trained by the Rprop method standard learning method.

References

Chen, H. M., Tsai, K. H., Qi, G. Z., Yang, J. C. S., & Amini, F. (1995). Neural network for structural control. *Journal of Computing in Civil Engineering, 9*(2), 168-176.

Cheng, M., & Popplewell, N. (1994). Neural network for earthquake selection in structural time history analysis. *Earthquake Engineering and Structural Dynamics, 23,* 303-319.

Ciesielski, R., Kuźniar, K., Maciąg, E., & Tatara, T. (1992). Empirical formulae for fundamental natural periods of buildings with load bearing walls. *Archives of Civil Engineering, 38*(4), 291-299.

Ciesielski, R., Kuźniar, K., Maciąg, E., & Tatara, T. (1995). Damping of vibration in precast buildings with bearing concrete walls. *Archives of Civil Engineering, 40*(3), 329-341.

De Stefano, A., Sabia, D., & Sabia, L. (1999). Probabilistic neural networks for seismic damage mechanisms prediction. *Earthquake Engineering and Structural Dynamics, 28*(8), 807-821.

Eurocode 8 (1994). *Design provisions for earthquake resistance of structures. European Prestandard.* Brussels, Belgium: European Committee for Standarization.

Ghaboussi, J., & Joghataie, A. (1995). Active control of structures using neural networks. *Journal of Engineering Mechanics, 121*(4), 555-567.

Ghaboussi, J., & Lin, Ch.-Ch. J. (1998). New method of generating spectrum compatible accelerograms using neural networks. *Earthquake Engineering and Structural Dynamics, 27*(4), 377-396.

Hao, H., Chow, N., & Brownjohn, J. (2004). Field measurement and analysis of road filtering on traffic-induced ground motions. In D. Doolin, A. Kammerer, T. Nogami, R. B. Seed, & I. Towhata (Eds.), *11th International Conference on Soil Dynamics and Earthquake Engineering and the Third International Conference on Earthquake Geotechnical Engineering*, Berkeley, CA (Vol. 1, pp. 196-203).

Hao, H., Wu, Y., Ma, G., & Zhou, Y. (2001). Characteristics of surface ground motions induced by blasts in jointed rock mass. *Soil Dynamics and Earthquake Engineering, 21*(2), 85-98.

Haykin, S. (1999). *Neural networks: A comprehensive foundation* (2nd ed.). Upper Saddle River, NJ: Prentice Hall.

Haykin, S. (2001). *Kalman filtering and neural networks*. New York: Wiley.

Huang, C. S., Hung, S. L., Wen, C. M., & Tu, T. T. (2003). A neural network approach for structural identification and diagnosis of a building from seismic response data. *Earthquake Engineering and Structural Dynamics, 32*(2), 187-206.

Hung, S.-L., & Kao, C. Y. (2002). Structural damage detection using the optimal weights of the approximating artificial neural networks. *Earthquake Engineering and Structural Dynamics, 31*(2), 217-234.

Jang, J.-S., Sun, Ch.-T., & Mizutani, E. (1997). *Neuro-fuzzy and soft computing. A computational approach to learning and machine intelligence*. Upper Saddle River, NJ: Prentice Hall.

Kalman R. E. (1960). A new approach to linear filtering and prediction problems. *Trans. ASME, Ser.D, Journal of Basic Engineering, 82*, 34-45.

Krok, A., & Waszczyszyn, Z. (2005). Neural prediction of response spectra from mining tremors using recurrent layered networks and Kalman filtering. In K. J. Bathe (Ed.), *Third MIT Conference on Computational Fluid and Solid Mechanics* (pp. 302-305).

Kuźniar, K., & Maciąg, E. (2004). Neural network analysis of soil-structure interaction in case of mining tremors. In D. Doolin, A.Kammerer, T. Nogami, R. B. Seed, & I. Towhata (Eds.), *11th International Conference on Soil Dynamics and Earthquake Engineering and the 3rd International Conference on Earthquake Geotechnical Engineering*, Berkeley, CA (Vol. 2, pp. 829-836).

Kuźniar, K., Maciąg, E., & Waszczyszyn, Z. (2000). Computation of natural fundamental periods of vibrations of medium-height prefabricated buildings by neural networks. *Archives of Civil Engineering, 46*(4), 515-523.

Kuźniar, K., Maciąg, E., & Waszczyszyn, Z. (2005). Computation of response spectra from mining tremors using neural networks. *Soil Dynamics and Earthquake Engineering, 25*(4), 331-339.

Kuźniar, K., & Waszczyszyn, Z. (2002). Neural analysis of vibration problems of real flat buildings and data pre-processing. *Engineering Structures, 24*(10), 1327-1335.

Kuźniar, K., & Waszczyszyn, Z. (2003). Neural simulation of dynamic response of prefabricated buildings subjected to paraseismic excitations. *Computers & Structures, 81*(24/25), 2353-2360.

Kuźniar, K., & Waszczyszyn, Z. (2006). Neural networks and principal component analysis for identification of building natural periods. *Journal of Computing in Civil Engineering,20*(6), 431-436.

Lin, Ch-Ch. J., & Ghaboussi, J. (2001). Generating multiple spectrum compatible accelerograms using stochastic neural networks. *Earthquake Engineering and Structural Dynamics, 30*(7), 1021-1042.

Liu, Y., & Luke, B. A. (2004). Role of shallow soils in defining seismic response of a deep basin site subjected to high-energy explosive loading. In D. Doolin, A. Kammerer, T. Nogami, R. B. Seed, & I. Towhata (Eds.), *11th International Conference on Soil Dynamics and Earthquake Engineering and the 3rd International Conference on Earthquake Geotechnical Engineering*, Berkeley, CA (Vol. 2, pp. 17-24).

Maciąg, E. (1986). Experimental evaluation of changes of dynamic properties of buildings on different grounds. *Earthquake Engineering & Structural Dynamics, 14*(6), 925-932.

Neural Network Toolbox for Use with MATLAB User's Guide, Version 2. (2001). Math-Works.

Paez, T. L. (1993). Neural networks in mechanical system simulation, identification and assessment. *Shock and Vibration, 1*(2), 177-199.

PN-80/B-03040. (1980). *Foundation and machine support structures. Analysis and design* (in Polish). Polish Code Committee.

Rojas, R. (1996). *Neural networks: A systematic introduction.* New York: Springer-Verlag.

Waszczyszyn, Z. (1999). Fundamentals of artificial neural networks. In Z. Waszczyszyn (Ed.), *Neural networks in the analysis and design of structures* (pp. 1-51). New York: Springer.

Waszczyszyn, Z., & Ziemianski, L. (2003). Neural networks in mechanics of structures and materials—new results and prospects of applications. *Computers & Structures, 79*, 2261-2276.

Waszczyszyn, Z., & Ziemianski, L. (2005). Neural networks in the identification analysis of structural mechanics problems. In Z. Mróz & G. Stavroulakis (Eds.), *Parameter identification of materials and structures* (CISM Courses and Lectures No. 469, pp. 265-340). New York: Springer.

Zell, A. (Ed.). (1998). *SNNS: Stuttgart neural network simulator, user's manual, version 4.2.* Stuttgart, Germany: University of Stuttgart.

About the Authors

Nikos D. Lagaros is an assistant professor at the Faculty of Civil Engineering of the University of Thessaly and a research associate of the National Technical University of Athens. Dr. Lagaros is an active member of the structural-engineering research community, focusing on (a) nonlinear dynamic analysis of concrete and steel structures under seismic loading, (b) performance-based earthquake engineering, (c) structural design optimization of real-world structures, (d) seismic risk and reliability analysis, (e) neural network in structural engineering, (f) fragility evaluation of reinforced concrete structures, (g) inverse problems in structural dynamics, (h) parallel and distributed computing—grid computing technologies, (i) evolutionary computations and (j) geotechnical earthquake engineering.He has more than 130 publications, including 40 refereed international journal papers. He is a member of the editorial board of five international scientific journals and reviewer in nine scientific journals and 15 international conferences.

Yiannis Tsompanakis has received his MSc and PhD in civil engineering from the Department of Civil Engineering, National Technical University of Athens, Greece. He is currently an assistant professor of structural earthquake engineering at the Department of Applied Sciences, Technical University of Crete, Greece. He teaches undergraduate and postgraduate courses in structural mechanics and earthquake engineering, and he is a supervisor of diploma, master's, and doctoral theses. He is a reviewer for archival journals, and he has participated in the organization of several international congresses. He has published over 80 research papers in international journals, book chapters, and conference proceedings. He has been involved in many research and practical projects in the field of earthquake engineering and computational mechanics. His main research interests include: structural and geotechnical earthquake engineering, structural optimization, probabilistic mechanics, structural assessment, applications of artificial intelligence methods in engineering.

* * * * *

Giuseppe Acciani is an associate professor of electric circuits and electro-technology at Politecnico di Bari, Italy. He received the electrical engineering degree summa cum laude from the University of Bari. From 1982 to 1984 he worked for a Computer Research Centre (CSATA, Italy). In 1985 he joined the Electrical Engineering Department of the Technical University of Bari as an assistant professor, where he still works as associate professor. Currently he teaches the course Electric Circuits and Intelligent Systems for Industrial Diagnostics. His present main research interests include neural networks, in particular unsupervised networks for clustering, and soft computing for nondestructive diagnostics.

Sk. Faruque Ali received a bachelor's degree in civil engineering from Jadavpur University, Kolkata, India (2003). He is currently a doctoral research student at Department of Civil Engineering, Indian Institute of Science, Bangalore, India. His current research interests include intelligent structural control, multiobjective optimization, control of distributed parameter systems, bridge engineering, and nonlinear dynamics.

Arzhang Alimoradi, PhD, EIT, is a senior research engineer with John A. Martin and Associates, one of the largest structural and earthquake engineering firms in the world. A graduate of The University of Memphis in 2004 and a Herff College of Engineering fellow, he has collaborated with researchers from the MAE center (University of Illinois), John A. Blume Earthquake Engineering Center (Stanford), EERL (Galtech), and CERI (The University of Memphis). He is an active member of several ASCE and EERI technical committees. Dr. Alimoradi's specialty is in the areas of inelastic nonlinear dynamic response, design optimization, strong ground-motion selection and scaling, soft computing, and seismic isolation and energy dissipation devices.

Dominic Assimaki was born in Athens, Greece, in 1975. She received her BS in civil engineering from the National Technical University of Athens (Athens, Greece) in 1998, and her MS and ScD from the Department of Civil and Environmental Engineering at MIT (Cambridge, MA) in 2000 and 2004, respectively. During her doctoral studies, she also participated in the European Research Training Network SAFERR as a young researcher in GDS (Paris, January 2001-September 2002) and received a graduate research fellowship from the National Technical University of Athens (Athens, Greece, September 2002-August

2002). After graduating from MIT, she worked as a postdoctoral researcher at the Institute for crustal studies at the University of California, Santa Barbara (February 2004-June 2005). In July 2005, she joined the School of Civil and Environmental Engineering at the Georgia Institute of Technology as an assistant professor, where she conducts research in numerical methods in earthquake engineering and geophysics that include forward simulations of dynamic nonlinear soil response, soil-structure interaction, and scattering phenomena in heterogeneous media, as well as inverse problems. She is a member of the American Society of Civil Engineers, the Earthquake Engineering Research Institute, the American Geophysical Union, the Seismological Society of America, the International Association for Computer Methods and Advances in Geomechanics, and the Southern California Earthquake Center.

Alex H. Barbat is a professor of structural mechanics at the Technical University of Catalonia, Barcelona, Spain. Research activity has been developed mostly in the International Center for Numerical Methods in Engineering (CIMNE), Barcelona, Spain. Current research fields include seismic damage evaluation of structures, vulnerability and risk evaluation, active and passive structural control, and evaluation of the seismic behaviour of historical structures. He has published more than 60 articles in journals and various books. He is the president of the Spanish Association of Earthquake Engineering (AEIS). He collaborated in numerous Spanish and international research projects related to the mentioned research fields, among them, many of the European Commission.

Omar D. Cardona is a civil engineer of the National University of Colombia (UNC), Manizales, and doctor of earthquake engineering and structural dynamics of the Technical University of Catalonia. He is professor of the Institute of Environmental Studies of UNC, Manizales, and the University of Los Andes, Bogota. He is the former president of the Colombian Association for Earthquake Engineering and general director of the National Directorate of Risk Mitigation and Disaster Preparedness of Colombia. He is member of ACI committees 118, and 314 on computers and simplified design of RC buildings. In 2004 he was the winner of the U.N. Sasakawa Prize for Disaster Reduction.

Martha L. Carreño is a civil engineer and received an MSc in structural engineering of University of Los Andes, Bogotá, Colombia. She is a PhD student and research assistant of the Technical University of Catalonia at the program of earthquake engineering and structural dynamics. Her doctoral thesis is titled *Innovative Techniques for the Evaluation of Seismic Risk and Its Management in Urban Centres: Ex Ante and Ex Post Actions*. In the last 5 years, she has participated in several projects related to earthquake damage evaluation and disaster risk management assessment using computational intelligence techniques. She is also consultant of Ingeniar Cad/Cae Ltda. and of the Colombian Association for Earthquake Engineering.

Giuseppe Leonardo Cascella received an MSc degree with honours and a PhD in electrical engineering from the Technical University of Bari, Italy (2001 and 2005, respectively). He worked with the Getrag GmbH Systemtechnik, St. Georgen, Germany, on the automatic transmission, and is currently assistant researcher with Technical University of Bari. His research interests include artificial intelligence and advanced optimization techniques for computer vision and electric drives.

Snehashish Chakraverty is working in Central Building Research Institute, Roorkee, India. His research area includes computational and mathematical modelling related to vibration and building sciences. He received a PhD from IIT, Roorkee, in vibration of plates (1992). Then he went to ISVR, University of Southampton, UK (1996) and to Concordia University, Canada (1997-1999) for PDF. He was a visiting faculty at McGill and Concordia Universities in Canada. Dr. Chakraverty has undertaken a number of national and international collaborative research projects funded by different organisations. He published 62 research papers in journals, conferences, and one textbook. Various awards and fellowships, namely, the University Gold Medals, CSIR Young Scientist Award, BOYSCAST Fellowship, and Golden Jubilee Directors Award, have been awarded to him. Dr. Chakraverty is included in Who's Who in Computational Science and Engineering, U.K. Millennium Edition of WHO's WHO IN THE WORLD of Marquis Publications, USA. He is a supervisor of number of PhD and MSc students and is also a reviewer of various journals.

Christopher M. Foley, PhD, PE, is an associate professor of civil engineering at Marquette University in Milwaukee, Wisconsin. He currently teaches courses in seismic analysis of structural systems, linear and nonlinear structural analysis and bridge design/analysis. He is presently a member of the American Society of Civil Engineers Structural Engineering Institute's technical committees on compression and flexure members, optimal structural design, and emerging computing technology. He also currently sits on the American Institute of Steel Constructions Committee on Research and Connections Specification Task Committee. His primary areas of research are evolutionary computation and advanced-analysis-based design methodologies as applied in structural steel building systems.

Girolamo Fornarelli received his master's degree in electronic engineer and the PhD degree in electrical engineering from the Politecnico di Bari, where he is an assistant professor. His most research interests include analysis aspects of switching circuits, practical aspects, and development of neural networks and artificial intelligence, particularly in the field of nondestructive evaluation.

Ricardo O. Foschi received a degree in civil engineering from the University of Rosario, Argentina, 1962; a master's degree and PhD in applied mechanics, Stanford University, California, 1964 and 1966; and is, since 2003, Emeritus Professor of Civil Engineering at the University of British Columbia, Vancouver, Canada. He moved to this university in 1982, after working for the Canadian Government in wood mechanics research since 1967. In 1982 he was the recipient of the Marcus Wallenberg International Prize for wood products research and applications. He has maintained a strong interest in structural reliability and probabilistic methods applied to different branches of civil engineering, in particular earthquake engineering.

Michalis Fragiadakis graduated from the School of Civil Engineering of the National Technical University of Athens (NTUA). He holds two postgraduate degrees in the field of earthquake engineering and structural dynamics (Imperial College, London) and in the structural analysis and design of structures (NTUA), and recently received his PhD from NTUA. His research activity is focused primarily on earthquake engineering and on the nonlinear analysis of structures under both dynamic and static loading conditions. Furthermore, his interests extend to structural optimization (evolutionary algorithms) and structural reliability. He has authored and co-authored in total nine journal papers and has presented a number of papers in international conference proceedings.

Hitoshi Furuta is a professor in the Department of Informatics at Kansai University, Osaka, Japan. In 1980 he received his doctorate in engineering from Kyoto University, Japan. Before joining Kansai University in 1993, he worked for 18 years in the Department of Civil Engineering at Kyoto University. He was a visiting assistant professor at Purdue University, a visiting scholar at Princeton University, and a visiting professor at the University of Colorado at Boulder. His main areas of expertise are structural reliability, structural optimization, life-cycle cost analysis and design of bridge structures, and applications of soft computing including artificial intelligence, fuzzy logic, neural network, chaos theory, and genetic algorithm.

Rita Greco has been an assistant professor at Politecnico di Bari, Italy, since 2002. She received an MSc with honours in civil engineering in 1994 from Politecnico of Bari and a PhD in structural mechanics in 1999 from University of Naples, Italy. Currently, she teaches Structural Rehabilitation at the Faculty of Architecture of Bari. Her main research interests include stochastic dynamics, earthquake engineering and seismic protection techniques, vulnerability and fragility analyses, and nonlinear structural analysis.

Ali Heidari received his PhD from Kerman University in Iran in 1984, and he is currently assistant professor and head of constructional office in Shahrekord University, Iran. He has published over 35 papers in international journals and conferences on structural optimization under dynamic loads. He was chosen as the distinguished graduate of Kerman University in 1985.

Miguel R. Hernandez-Garcia was born in Tunja, Colombia, in 1976. He received a summa cum laude BS in civil engineering from the Universidad Industrial de Santander, Colombia, in 1998, an MS in structural engineering from Universidad de Los Andes, Colombia, in 2001, and an MS in structural mechanics from University of Southern California in 2004. He is pursuing the PhD degree in civil engineering at University of Southern California. Since 2003, he has been with the Structural Dynamics Group at University of Southern California, working on structural health monitoring, nonlinear system identification, statistical process monitoring, and machine learning.

Jorge E. Hurtado received a civil engineering degree by the National University of Colombia and a master's degree in earthquake engineering and doctoral degree in civil engineering by the Technical University of Catalonia (Spain). His theoretical research areas include nonlinear systems, stochastic systems, Monte Carlo simulation, stochastic mechanics, structural optimization, statistical learning, and artificial intelligence applications to nonlinear systems. His applied research fields are material modelling, earthquake engineering, and natural disasters. The results of these research activities have been published in leading journals like *Structural Safety, Probabilistic Engineering Mechanics, Computer Methods in Applied Mechanics and Engineering, Journal of Structural Engineering, Archives of Computational Methods in Engineering*, among others, as well as in numerous conference proceedings. He is also the author of the book *Structural Reliability—Statistical Learning Perspectives* (Springer, 2004).

Chan Ghee Koh is a professor at the Department of Civil Engineering, National University of Singapore. He received his PhD from the University of California, Berkeley, in 1986. His main research areas are structural dynamics, structural health monitoring, and system identification. He has more than 100 publications, including 55 refereed international journal papers. He was a recipient of the prestigious Marie Curie Fellowship (1994) awarded by the Commission of the European Communities and the IES Prestigious Publication Award (Best Paper in Theory, 1996) by the Institution of Engineers, Singapore. He was invited to deliver seven keynotes in the last 3 years including in UK and Greece. He is currently an associate editor of U.S.-based International *Journal on Structural Health Monitoring*, and an editorial board member of a new international journal called *Journal of Smart Structures and Systems*.

Kazuhiro Koyama graduated from the Department of Civil Engineering at Kansai University, Osaka. Japan, in 2003, and received a master's degree of informatics from Kansai University in 2005. Now he is working for NTT Comware Ltd. in Japan as a system engineer.

Krystyna Kuźniar is an associate professor of structural dynamics in Pedagogical University of Cracow, Poland; in 1980 received an MSc in civil engineering from Cracow University of Technology; in 1991 received a PhD in technical sciences from Cracow University of Technology; in 2005 received a DrSc in technical sciences, discipline in civil engineering, specialization in structural dynamics, from Cracow University of Technology. Dr. Kuźniar's main field of research is theoretical and experimental (full-scale tests) analysis of soil-structure interaction in case of mining tremors, dynamic properties and dynamic response of buildings subjected to kinematic excitations, and application of neurocomputing in structural dynamics. Dr. Kuźniar has over 20 publications in peer reviewed journals and over 30 publications in conference proceedings.

Mauro Mezzina is full professor in structural design since 1991 at Politecnico di Bari, Italy. Formerly the dean of the Faculty of Architecture, he is the director of department ICAR of Politecnico of Bari. He has been a member of the directional board of ANIDIS (Italian National Association for Earthquake Engineering), CTA (Board of Technicians of Steel

Construction), and a member of organizing and scientific committee of several national and international conferences on earthquake engineering, diagnostics, and vulnerability. He is member of the scientific board of the research centers Tecnopolis and CIRP. Currently, he teaches structural design in undergraduate courses and in the PhD program in structural engineering of Politecnico of Bari. His main research fields include earthquake engineering and seismic risk reduction; stochastic dynamics; diagnostics; vulnerability assessment; analysis of historical and modern masonry structures; nonlinear analysis of structures; and damage mechanics applied to the modelling of masonry structures.

Bartosz Miller is an assistant professor at Department of Structural Mechanics, Rzeszów University of Technology, Rzeszów, Poland. He received an MSc (1994) and PhD (2002) degrees from Rzeszów University of Technology, both in civil engineering. His research interests include the application of soft-computing methods, mainly neural networks, in the structural mechanics. His publications concern the updating of computational models and the detection and identification of failures and damages in engineering structures. He is the author of about 30 publications and conference papers concerning neural network applications in structural mechanics.

Manolis Papadrakakis is a professor of structural engineering at the School of Civil Engineering, National Technical University of Athens (NTUA). His research activity is focused on the development and the application of computer methods and technology to structural engineering analysis and design. He is involved in the following scientific activities: Editor of the International Journal Computer Methods in Applied Mechanics and Engineering (CMAME); honorary editor of the International Journal of Computational Methods (IJCM); president of the Greek Association for Computational Mechanics (GRACM); member of the editorial board of eight international scientific journals and reviewer in 40 scientific journals; fellow, corresponding member of the Executive Council and member of the General Council of the International Association for Computational Mechanics (IACM); and chairman of the European Committee on Computational Solid and Structural Mechanics (ECCSM).

Michael John Perry gradated from the National University of Singapore with first-class honours in civil engineering in 2003, under the Asia–New Zealand–Singapore Scholarship program. After receiving the award for the top civil engineering student, he has continued his studies under an NUS research scholarship. His area of PhD research is in developing genetic algorithm identification strategies for structural and offshore applications.

Shahram Pezeshk, PhD, PE, is the Emison distinguished professor of civil engineering at The University of Memphis, Tennessee. A graduate of the University of Illinois and University of California at Berkeley; he currently teaches graduate courses in geotechnical earthquake engineering and structural design. Pezeshk is the chairman of the Optimal Structural Design Committee of the American Society of Civil Engineers. He is the recipient of numerous awards of excellence in teaching and research and contributor to many national and international journals. His primary areas of research are engineering seismology; structural optimization, nonlinear structural response, and mid-America earthquake hazard analysis.

Grzegorz Piątkowski was awarded his MSc in 1994 and PhD in dynamic of structures in 2003 by Rzeszów University of Technology. Now he is assistant professor at Department of Structural Mechanics, Rzeszów University of Technology, Rzeszów, Poland. His research is oriented at the application of computational technology to structural mechanics—especially the damage detection in structures, the use of soft-computing methods in mechanics, and the modelling by finite element method and some aspects of data processing. He is the author of series of publications and conference papers concerning neural network applications in structural mechanics.

Ananth Ramaswamy received his Bachelor of Technology in civil engineering from Indian Institute of Technology, Madras, India, in 1985. He completed his MS and PhD degrees from University of California at Davis, and Louisiana State University (1986 and 1992, respectively). He is currently associate professor in the Department of Civil Engineering, Indian Institute of Science, Bangalore, India. His research interests include FRP composites; smart materials; structural vibration control; structural and shape optimization; reinforced, prestressed, and fiber reinforced concrete; and bridge engineering.

Eysa Salajegheh received his PhD in the area of structural engineering from the University of Surrey, England, in 1981. He is currently professor of civil engineering at the University of Kerman, Iran. He held the position of visiting scholar at the University of California, Santa Barbara, in 1985 and 1991. He has published over 150 research papers in international journals and conferences. His current research interests include structural optimization, earthquake engineering, and space structures. He has received numerous honours and awards and has been chosen as distinguished professor of Kerman University and Ministry of Science, Research and Technology of Iran.

Mauricio Sanchez-Silva was born in Bogotá, Colombia, in 1966. He received his BS (1989) and MS (1992) from Los Andes University (Colombia) and his PhD from the University of Bristol, U.K. (1996). Since 1996, he has been working as professor in the Department of Civil and Environmental Engineering at Los Andes University (Bogotá, Colombia). In addition, he has been visiting scholar at the Technical University Munich (Germany) and Texas A&M University (USA). His main areas of research are reliability and risk analysis with applications to structural safety and structural deterioration processes. Other areas of interest include statistical learning processes, structural optimization, life-cycle analysis, and disaster-engineering related issues.

Giuseppina Uva is associate professor of structural design at Politecnico di Bari, Italy. She received an MSc degree with honours in civil engineering in 1994 from the Politecnico of Bari and a PhD in computational mechanics in 1997, from University of Calabria, Italy. In 1999, she joined the department ICAR of Politecnico of Bari as assistant professor, where she presently works as associate professor. Currently, she teaches structural design and structural rehabilitation at the faculty of architecture of Bari. She is member of the teachers' board of the PhD Program in Computational Mechanics of the University of Calabria.

Her main research interests include earthquake engineering and seismic risk reduction; diagnostics; vulnerability assessment; analysis of historical and modern masonry structures; nonlinear analysis of structures; and damage mechanics applied to the modelling of masonry structures.

Zenon Waszczyszyn is a professor emeritus and full professor of Cracow University of Technology, Poland, and full professor of Rzeszów University of Technology, Poland; member of Polish Academy of Sciences, Warsaw, and Polish Academy of Science and Arts, Cracow. DHC of Budapest University of Technology and Economics, Hungary. In 1957–2005 he was associated with Cracow University of Technology, where he was awarded an MSc in civil engineering and a PhD in structural mechanics. His research interest has included nonlinear structural mechanics; stability of structures; theory of plasticity; shells and plates; and, especially, various computational methods. Since 1995, he has continued research and educational activity in the field of soft computing and applications of neural networks in civil and structural engineering. He is author and co-author as well as editor of eight books and about 250 papers written in English and Polish.

Leonard Ziemiański is professor and head of the department of structural mechanics at the Rzeszów University of Technology, Rzeszów, Poland. His research is concerned with the application of computational technology—especially artificial intelligence techniques—to structural mechanics. Leonard Ziemiański was awarded his MSc (1977) by Cracow University of Technology, Cracow, and PhD (1985) by AGH University of Sciences and Technology, Cracow. Current application area include the use of soft-computing methods in engineering, the artificial methods in mechanics, the structural health monitoring, and the damage detection in structures. He is the author of over 100 publications concerned with various aspects of engineering computation and experimental research.

Index